COMPLEXITY THEORY

COMPLEXITY THEORY
Current Research

Edited by

Klaus Ambos-Spies Steven Homer Uwe Schöning
Universität Heidelberg *Boston University* *Universität Ulm*

Published by the Press Syndicate of the University of Cambridge
The Pitt Building, Trumpington Street, Cambridge CB2 1RP
40 West 20th Street, New York, NY 10011-4211, USA
10 Stamford Road, Oakleigh, Melbourne 3166, Australia

© Cambridge University Press 1988

First published 1993

Printed in the United States of America

Library of Congress Cataloging-in-Publication Data available.

A catalogue record for this book is available from the British Library.

ISBN 0-521-44220-6 hardback

CONTENTS

Preface	vii
Reductions to Sets of Low Information Content V. Arvind, Y. Han, L. Hemachandra, J. Kobler, A. Lozano, M. Mundhenk, M. Ogiwara, U. Schoening, R. Silvestri, T. Thierauf	1
On Average P vs. Average NP J. Belanger and J. Wang	47
Additional Queries and Algorithmically Random Languages R. Book	69
Bounded Reductions H. Buhrman, E. Spaan and L. Torenvliet	83
Promise Problems and Guarded Access to Unambiguous Computation J-Y. Cai, L. Hemachandra and J. Vyskoc	101
The Complexity of Space Bounded Interactive Proof Systems A. Condon	147
Fixed Parameter Tractability and Completeness R. Downey and M. Fellows	191
Degrees of Unsolvability in Abstract Complexity Theory M. Kummer	227
On the Non-Uniform Complexity of the Graph Isomorphism Problem A. Lozano and J. Toran	245
Upper and Lower Bounds for Certain Graph Accessibility Problems on Bounded Alternating omega-Branching Programs C. Meinel and S. Waack	273
Associative Storage Modification Machines J. Tromp and P. van Emde Boas	291

PREFACE

On February 2 - 8, 1992 a workshop on Structure and Complexity was held at the Dagstuhl International Conference and Research Center outside the town of Wadern, Germany. The conference brought together forty leading researchers in complexity theory and some closely related areas of theoretical computer science and mathematics. It was a successful, productive conference and resulted in a great deal of useful interaction and collaboration.

The present volume consist of research papers solicited from those invited to this workshop. They comprise a cross-section of current research in complexity theory as done by some of the central active workers in this field. The eleven papers in this volume are representative of many of the important recent developments in complexity theory. They include papers on efficient reductions, interactive proofs, average case complexity, models of computations, unambiguous and alternating computation, and non-uniform complexity.

One of the central defining concepts of complexity theory is that of reductions. Reductions provide the main tool with which the complexity of combinatorial problems is compared. Every paper in this volume touches on reductions, at least in some part. They play a central role in several of the papers including that by Kummer which uses the most general reductions between sets, Turing reductions, to determine the degrees of unsolvability of fundamental sets arising in abstract complexity theory. The paper of Book studies reductions to algorithmically random languages and proves that allowing more queries to such languages increases the power of the reduction. Buhrman, Spaan and Torenvliet work with resource bounded reductions and show that small differences in the reductions result in differences in the associated complete sets. Finally the research of Downey and Fellows explores reductions which preserve the fixed parameter complexity of problems.

Two of the papers concern reductions to sparse languages, those containing (relatively) few elements. The paper of Arvind, et al, studies the existence of different types of efficient reductions to sparse sets and the implications for the problems being reduced. The work of Lozano and Torán considers whether a particular problem, the graph isomorphism problem, can be reduced to a sparse set.

Two of the papers concern quite new areas of complexity theory which have seen great excitement and progress in recent years. The paper by Condon surveys interactive proof systems which use space-bounded computation. The work of Belanger and Wang explores the structural complexity of average case complete problems, considering combinatorial problems which are hard on average rather than only hard in the worst case.

Another thread of complexity theory studies different models of computation and tries to delineate the limits of doing computation using differing models. In the present volume three papers consider particular computational models and study their capabilities. Cai, Hemachandra and Vyskoc consider computations which have access to unambiguous or to probabilistically defined oracles. And they compare the computational power of computations using such resources. The research of Meinel and Waack studies certain bounded branching programs, proving interesting lower and upper bounds on their computational strength. Finally the paper by Tromp and van Emde Boas concerns parallel complexity and defines and studies a new parallel model of computation based on a parallel version of the storage modification machine.

In sum, these papers give a state of the art view of several areas of complexity theory at this point in time. They point the way toward current trends and ask important question for future work in this field. Some of this research will be the concern of a second Dagstuhl meeting on Structure and Complexity to be held in February of 1994.

The Editors
June 21, 1993

Reductions to Sets of Low Information Content*

V. Arvind[†]   Y. Han[‡]   L. Hemachandra[§]

J. Köbler[¶]   A. Lozano[//]   M. Mundhenk[¶]   M. Ogiwara[**]

U. Schöning[¶]   R. Silvestri[††]   T. Thierauf[‡‡]

---

*Some of these results were presented at the 19th Intern. Colloq. on Automata, Languages, and Programming [AHH+92a] and the 12th FST& TCS Conf. [AKM92].

[†]Department of Computer Science and Engineering, Indian Institute of Technology–Delhi, New Delhi 110016, India. Work done while visiting Universität Ulm. Supported in part by an Alexander von Humboldt Research Fellowship.

[‡]Department of Computer Science, University of Rochester, Rochester, NY 14627, USA. Supported in part by the National Science Foundation under grant CCR-8957604.

[§]Department of Computer Science, University of Rochester, Rochester, NY 14627, USA. Supported in part by the National Science Foundation under research grants CCR-8957604 and NSF-INT-9116781/JSPS-ENG-207.

[¶]Abteilung für Theoretische Informatik, Universität Ulm, Oberer Eselsberg, D-W-7900 Ulm, Germany. Supported in part by the DAAD through Acciones Integradas 1991, 313-AI-e-es/zk.

[//] Department of Software (L.S.I.), Universitat Politècnica de Catalunya, Pau Gargallo 5, E-08028 Barcelona, Spain. Work supported in part by ESPRIT-II Basic Research Actions Program of the EC under Contract No. 3075 (project ALCOM) and by the DAAD through Acciones Integradas 1991, 313-AI-e-es/zk.

[**]Department of Computer Science and Information Mathematics, University of Electro-Communications, 1-5-1, Chofugaoka, Chofu-si, Tokyo 182, Japan. Work done in part while visiting SUNY–Buffalo. Supported in part by the National Science Foundation under research grant CCR-9002292 and the JSPS under research grant NSF-INT-9116781/JSPS-ENG-207.

[††]Dipartimento di Scienze dell'Informazione, Università degli Studi di Roma "La Sapienza," 00198 Rome, Italy. Work done in part while visiting the University of Rochester. Supported in part by Ministero della Pubblica Istruzione through "Progetto 40%: Algoritmi, Modelli di Calcolo e Strutture Informative."

[‡‡]Abteilung für Theoretische Informatik, Universität Ulm, Oberer Eselsberg, D-W-7900 Ulm, Germany. Work done in part while visiting the University of Rochester. Supported in part by a DFG Postdoctoral Stipend, by the National Science Foundation under grant CCR-8957604, and by the DAAD through Acciones Integradas 1991, 313-AI-e-es/zk.

1 Overview

This paper is concerned with two basic questions about sparse sets, and a related question about sets of low instance complexity:

Question 1 With respect to what types of reductions might NP have hard or complete sparse sets?[1]

Question 2 If a set A reduces to a sparse set, does it follow that A is reducible to some sparse set that is "simple" relative to A?

Question 3 With respect to what types of reductions might NP have hard or complete sets of low instance complexity, and, relatedly, what is the structure of the class of sets with low instance complexity?

With respect to the first and third questions, intuitively one would expect that even with respect to flexible reductions NP is unlikely to have complete sets whose information content is low. With respect to the second question, one might intuitively feel that the structure imposed on a set by the fact that it reduces to a sparse set makes it plausible that we can indeed find a simple sparse set that can masquerade as the original sparse set. These two intuitions are in many ways certified by the current literature, and by the results of this paper.

The rest of this section summarizes the results of this paper and compares them with previous work.

With regard to Question 1:

- We show that if any NP-complete set bounded truth-table reduces to a set that conjunctively reduces to a sparse set, then P = NP. Our result extends the strongest previously known result, which is due to Ogiwara and Watanabe: if any NP-complete set bounded truth-table reduces to a sparse set, then P = NP [OW91]. As a consequence of our result, if any NP-complete set conjunctively reduces to a sparse set,[2] then P = NP. The latter result has been obtained independently by Ranjan and Rohatgi [RR92].

---

[1] For the reductions we will discuss, the question of sparse hard sets is equivalent to asking what type of reductions might reduce many-one complete sets for NP to some sparse set; we will often use this formulation.

[2] A conjunctive reduction from A to B means that there is a Turing machine with oracle B that accepts A, and the Turing machine's acceptance mechanism is

We also show similar results for the classes UP, PP, $C_=P$, Mod_kP, and the class of nearly near-testable sets.

- One might ask whether in the above-mentioned result of Ogiwara and Watanabe the bounded truth-table case is optimal, or whether it can be extended by making the bound on the number of queries bigger than constant, for example, by making it some function that is $\omega(\log n)$. We show that there are relativized worlds in which the boolean hierarchy does not collapse and yet there are tally NP-complete sets with respect to such reductions. This provides relativized upper bounds to the work of Ko, Orponen, Schöning, and Watanabe [KOSW86,Orp90,KOSW90], of Ogiwara and Watanabe [OW91], and of the current paper. Our result strengthens a result of Homer and Longpré [HL91], who independently of the work of this paper showed that there are relativized worlds in which there are sparse sets that are NP-complete with respect to such bounds and yet (relativized) $P \neq NP$.

With regard to Question 2:

- In the context of recent comparisons between equivalence and reducibility to sparse sets [AHOW92,GW], it is interesting to know, for various classes of sets that reduce to sparse sets, the complexity of the easiest sparse sets to which such sets reduce. We show that any set A that disjunctively reduces (respectively, disjunctive bounded truth-table reduces, 2-truth-table reduces) to a sparse set in fact disjunctively reduces (respectively, disjunctive bounded truth-table reduces, 2-truth-table reduces) to a sparse set that is in P^{NP^A} (respectively, $P^{NP^A[\log]}$, $P^{NP^A[\log]}$).[3] Thus, for such sets, reducing to some sparse set implies reducing to some relatively simple sparse set. The nearest previous result is one of Allender, Hemachandra, Ogiwara, and Watanabe [AHOW92]: If $P = NP$ and set A 2-truth-table reduces to a sparse set, then A truth-table reduces to some sparse set that itself truth-table reduces to A. However, A does not *two*-truth-table reduce to

---

that it accepts if and only if every string it queries is a member of B [LLS75].

[3]The [log] means that there is an $\mathcal{O}(\log n)$ bound on the number of calls made to the NP^A oracle (see [Wag90]).

the particular sparse set constructed in [AHOW92]. Via censusfunctions, graph-coloring, and the Erdős-Rado sunflower lemma, our techniques avoid the level of explicit coding (and thus the complexity of reduction) required by previous methods.

With regard to Question 3:

- We completely characterize the sets of low instance complexity (that is, the class IC[log, poly] [KOSW86,Orp90,KOSW90]) in terms of reductions to tally sets: IC[log, poly] is exactly the class of sets that both disjunctively and conjunctively reduce to tally sets.

- We show that Orponen's result [Orp90] on 1-truth-table-complete sets for NP generalizes to disjunctive and conjunctive reductions: If P $\neq$ NP and A is $\leq_c^p$-hard or $\leq_d^p$-hard for NP, then $A \notin$ IC[log, poly].

Section 3 discusses the results associated with Question 1. Section 4 discusses the results associated with Question 2. Section 5 discusses the results associated with Question 3.

2 Notation

A set T is said to be a *tally* set if $T \subseteq 0^*$. $S^{=n}$ will denote the length n strings in S, and $S^{\leq n}$ (respectively, $S^{<n}$) will denote the strings in S of length at most (respectively, strictly less than) n. A set S is said to be *sparse* if it has at most polynomially many elements at each length: S is sparse if and only if for some polynomial p it holds that $(\forall n)[||S^{=n}|| \leq p(n)]$. We use TALLY and SPARSE to represent, respectively, the classes of tally and sparse sets. Tally and sparse sets have come to play a large role in modern complexity theory (see, e.g., the surveys [Mah86,Mah89,HOW92]).

Let $\langle \cdot, \cdot \rangle_2$ denote a (non-onto) pairing function over finite strings with the standard nice computability, and invertibility properties, and such that $(\forall x, y)[|\langle x, y \rangle_2| = 2|x| + |y|]$. For every $k \geq 2$, let $\langle y_1, y_2, \ldots, y_k \rangle$ denote $\langle k, \langle y_1, \langle y_2, \langle \ldots, \langle y_{k-1}, y_k \rangle_2 \cdots \rangle_2 \rangle_2 \rangle_2$.

The reductions discussed in this paper are polynomial-bounded reductions defined by Ladner, Lynch, and Selman [LLS75]. Table 1 lists the abbreviations we will use for various types of reductions.

Name	Notation
many-one	$\leq_m^p$
one truth-table	$\leq_{1\text{-}tt}^p$
k truth-table	$\leq_{k\text{-}tt}^p$
k disjunctive (truth-table or Turing)	$\leq_{k\text{-}d}^p$
bounded (truth-table or Turing)	$\leq_b^p$
$f(n)$ truth-table	$\leq_{f(n)\text{-}tt}^p$
conjunctive (truth-table or Turing)	$\leq_c^p$
disjunctive (truth-table or Turing)	$\leq_d^p$
bounded conjunctive (truth-table or Turing)	$\leq_{bc}^p$
bounded disjunctive (truth-table or Turing)	$\leq_{bd}^p$
Turing	$\leq_T^p$
nondeterministic conjunctive (truth-table or Turing)	$\leq_c^{NP}$

Table 1: Polynomial-time Reductions

We will use the following notation to describe downward closures of classes under various reductions.

Notation 2.1 [BK88,AHOW92] For any reducibility $\leq_r^p$ and any class of sets $\mathcal{C}$, let $R_r^p(\mathcal{C}) = \{A \mid (\exists B \in \mathcal{C})[A \leq_r^p B]\}$.

The interrelations among $R_r^p(\text{SPARSE})$ classes have been studied by Book and Ko [BK88], Ko [Ko89], Allender, Hemachandra, Ogiwara, and Watanabe [AHOW92], and Gavaldà and Watanabe [GW]. Figure 1 shows some of the inclusion structure among $R_r^p(\text{SPARSE})$ classes.

3 Sets Reducing to Sparse Sets

The study of sparse complete sets was sparked by the conjecture of L. Berman and J. Hartmanis [BH77] that there are no sparse NP-complete sets; they were motivated to make this conjecture since if it fails then there are NP-complete sets that are not polynomial-time isomorphic (and at that time they conjectured that all NP-complete

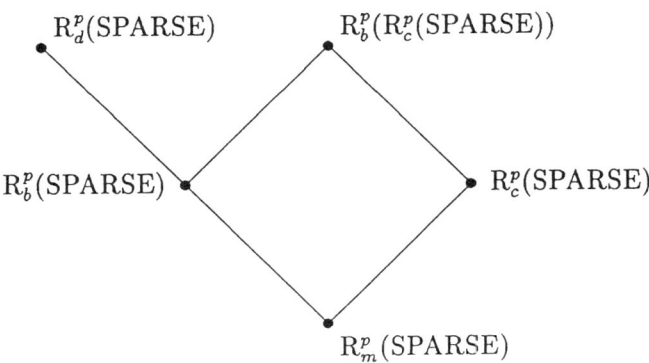

Figure 1: Inclusion structure of some reduction classes to sparse sets; all inclusions indicated are proper.

$R_m^p(\text{SPARSE}) \subsetneq R_b^p(\text{SPARSE})$ and $R_m^p(\text{SPARSE}) \subsetneq R_c^p(\text{SPARSE})$ are respectively from [BK88] and [Ko89]. From the result of [AHOW92] that $R_b^p(\text{SPARSE}) \subseteq R_d^p(\text{SPARSE})$ and the result of [GW] that $R_c^p(\text{SPARSE}) \not\subseteq R_d^p(\text{SPARSE})$, it follows immediately that $R_b^p(\text{SPARSE}) \subsetneq R_b^p(R_c^p(\text{SPARSE}))$. From the result of [Ko89] that $R_b^p(\text{SPARSE}) \not\subseteq R_c^p(\text{SPARSE})$, it follows immediately that $R_c^p(\text{SPARSE}) \subsetneq R_b^p(R_c^p(\text{SPARSE}))$.

It remains only to show that $R_b^p(\text{SPARSE}) \subsetneq R_d^p(\text{SPARSE})$. [AHOW92] shows that $R_b^p(\text{SPARSE}) \subseteq R_d^p(\text{SPARSE})$, and from this it follows that $\text{coSPARSE} \subseteq R_d^p(\text{SPARSE})$, which in turn implies $\text{SPARSE} \subseteq R_c^p(\text{coSPARSE})$, and hence $R_c^p(\text{SPARSE}) \subseteq R_c^p(R_c^p(\text{coSPARSE})) = R_c^p(\text{coSPARSE})$. If $R_b^p(\text{SPARSE}) = R_d^p(\text{SPARSE})$, then, since $R_b^p(\cdot)$ classes are closed under complement, $R_b^p(\text{SPARSE}) = R_b^p(\text{coSPARSE})$. The previous two sentences imply $R_c^p(\text{SPARSE}) \subseteq R_b^p(\text{SPARSE})$, and thus $R_c^p(\text{SPARSE}) \subseteq R_d^p(\text{SPARSE})$, contradicting the result of [GW].

sets were polynomial-time isomorphic, though recent work has dimmed hopes on that issue [JY85,KMR89]).

The first result along the lines of their sparseness conjecture was P. Berman's proof that P = NP if some tally set is NP-complete [Ber78]. This result was quickly followed by Fortune's proof that if there is a sparse coNP-complete set, then P = NP [For79]. Finally, Mahaney obtained the striking result that P = NP if any NP-complete set many-one reduces to a sparse set [Mah82].

Although Mahaney obtained the complete collapse of the polynomial hierarchy in the case of many-one reducibility, possible collapses in the case of more flexible reducibilities have remained an active research area. For the case of Turing reductions, it is known that the existence of sparse Turing-complete sets for NP would collapse the polynomial hierarchy to $P^{NP[\log]}$ [Kad89], and the existence of sparse Turing-hard sets for NP would collapse the polynomial hierarchy to $\Sigma_2^p \cap \Pi_2^p$ [KL80]; both these results are known to be essentially optimal with respect to relativizable proof techniques [Kad89,Hel86].

As just noted, for the cases of many-one and Turing reductions the consequences of sparse NP-complete sets are well-understood. However, with respect to reductions whose strength lies between Turing and many-one reductions, the question of extending Mahaney's many-one result has proved considerably more challenging. For the case of bounded truth-table reductions, Ukkonen [Ukk83] generalized Berman's result [Ber78] by showing that if there is a tally bounded truth-table hard set for NP, then P = NP. Yesha [Yes83] generalized Fortune's result [For79] by showing that if there is a sparse bounded positive truth-table hard set for coNP, then P = NP. Yesha also (partially) generalized Mahaney's Theorem [Mah82] by showing that if there is a sparse bounded positive truth-table complete set for NP, then P = NP. These results regarding bounded truth-table reductions have been recently subsumed by Ogiwara and Watanabe ([OW91], see also [Wat88] and [JY90, footnote 9]), who successfully extended Mahaney's result and showed that if there is a sparse bounded truth-table hard set for NP, then P = NP.

For the case of conjunctive reductions, Ukkonen [Ukk83] and Yap [Yap83] generalized Fortune's result [For79] by showing that if there is a sparse conjunctive hard set for coNP, then P = NP. Yap [Yap83] also (partially) generalized Mahaney's Theorem [Mah82] by showing that if there is a sparse set that is both conjunctive and disjunctive complete for NP, then P = NP. However, in the decade since Mahaney's Theorem, it has remained an open question whether his result can be extended to the case of conjunctive reductions. Section 3.1 resolves this question.

3.1 NP, PP, and $C_=P$

We show that if there is a sparse set that is conjunctive hard for NP, then P = NP (Corollary 3.4). In fact, in Corollary 3.5 we establish that if NP $\subseteq R_b^p(R_c^p(\text{SPARSE}))$ then P = NP, thus extending the result of Ogiwara and Watanabe [OW91] (see Figure 1).

Definition 3.1 [OW91] Let A be in NP, let W be a set in P, and let q be a polynomial such that $A = \{ x \mid (\exists w \in \Sigma^{q(|x|)}) [\langle x, w \rangle \in W] \}$. For $x \in A$ let $w_{max}(x) = \max\{ w \in \Sigma^{q(|x|)} \mid \langle x, w \rangle \in W \}$. We will say that $Left(A) = \{ \langle x, w \rangle \mid x \in A, w \in \Sigma^{q(|x|)}$ and $w \leq w_{max}(x)\}$ is the *left set of A*.[4]

Theorem 3.2 If $A \in$ NP and $Left(A) \in R_b^p(R_c^p(\text{SPARSE}))$, then A is in P.

Although the next result is a direct consequence of the above theorem, a simple direct proof for it can be found in the appendix of [AHH+92].

Theorem 3.3 If $A \in$ NP and $Left(A) \in R_c^p(\text{SPARSE})$, then A is in P.

From Theorem 3.3, it immediately follows that Mahaney's Theorem generalizes to the case of conjunctive reductions. Corollary 3.4 has been obtained independently by Ranjan and Rohatgi [RR92].

Corollary 3.4 If any NP-complete set conjunctively reduces to a sparse set, then P = NP.

Indeed, $Left(A) \leq_m^p A$ for any NP-complete set A, and thus the above theorems apply to the case in which some NP-complete (or NP-hard) set reduces (via the reductions named above) to some sparse set.

---

[4]The left set tacitly depends on the particular witness relation chosen.

Corollary 3.5 If any NP-complete set is in $R_b^p(R_c^p(\text{SPARSE}))$, then P = NP.

In the remainder of this section we prove Theorem 3.2, give applications of the obtained results to other complexity classes, and show relativized upper bounds to the above result.

The following characterization, due to Hausdorff, of the boolean closure of certain classes of sets plays a central role in our proof of Theorem 3.2.

Theorem 3.6 [Hau14,Wec85] Let K be any class of sets closed under finite unions and intersections that includes $\emptyset$ and Σ^*. Let $BC(K)$ be the closure of K under finite union, finite intersection, and complement. Every $A \in BC(K)$ can be represented as $A = \bigcup_{i=1}^{k}(A_{2i-1} \cap \overline{A_{2i}})$, where $A_j \in K$ (for $1 \leq j \leq 2k$) and $A_1 \supseteq A_2 \supseteq \cdots \supseteq A_{2k}$.

In order to use this characterization for sets in $R_b^p(R_c^p(\text{SPARSE}))$, we need to show that $R_b^p(R_c^p(\text{SPARSE})) = BC(R_c^p(\text{SPARSE}))$, and that $R_c^p(\text{SPARSE})$ is closed under finite unions and intersections.

Theorem 3.7 [KSW87] Let $\mathcal{K}$ be a class that contains P and is closed under many-one reductions. Then $BC(\mathcal{K}) = R_b^p(\mathcal{K})$.

By applying the recent result of Buhrman, Longpré, and Spaan (stated as Theorem 3.8 below), it remains only to prove the closure of $R_c^p(\text{TALLY})$ under union.

Theorem 3.8 [BLS92] $R_c^p(\text{SPARSE}) = R_c^p(\text{TALLY})$.

The following lemma is straightforward and is stated without proof.

Lemma 3.9 $R_c^p(\text{TALLY})$ is closed under finite unions and intersections.

Now we are ready to prove Theorem 3.2.
Proof of Theorem 3.2:
Let q be a polynomial and let P_A be a polynomial-time set such that $A = \{x \mid (\exists w \in \Sigma^{q(|x|)})[\langle x, w \rangle \in P_A]\}$. Recall that $Left(A) = \{\langle x, w \rangle \mid x \in A \wedge w \in \Sigma^{q(|x|)} \wedge w \leq w_{max}\}$, where $w_{max} = \max\{w \in \Sigma^{q(|x|)} \mid \langle x, w \rangle \in P_A\}$. In the following we describe an

algorithm for testing membership in A, that computes w_{max} (the lexicographically largest witness, if it exists) by a breadth-first search of the tree of prefixes of all potential witnesses. In order to do this we use the set $prefix(Left(A)) = \{\langle x,y \rangle \mid (\exists z)[\langle x,yz \rangle \in Left(A)]\}$. Each prefix y actually represents the interval of all possible extensions of y to length $q(|x|)$. It is not hard to see that $prefix(Left(A))$ is many-one equivalent to $Left(A)$ and thus $prefix(Left(A)) \in R_b^p(R_c^p(\text{SPARSE})) = R_b^p(R_c^p(\text{TALLY}))$.

By Theorems 3.6, 3.7, 3.8, and by Lemma 3.9, it follows that there exists a tally set T and sets $C_i \in R_c^p(T)$, $D_i \in R_d^p(T)$ such that $prefix(Left(A)) = \bigcup_{i=1}^{k}(C_i \cap D_i)$ and $C_1 \supseteq \overline{D_1} \supseteq C_2 \supseteq \overline{D_2} \ldots \supseteq C_k \supseteq \overline{D_k}$.

Let f_i be the conjunctive reduction that witnesses $C_i \in R_c^p(T)$ and g_i be the disjunctive reduction that witnesses $D_i \in R_d^p(T)$. Without loss of generality, we assume that these reductions are all computable in time bounded by a fixed polynomial p.

We first give an intuitive overview of the polynomial-time[5] algorithm recognizing A. As stated above, this algorithm performs a breadth-first search through the tree of witness prefixes for an input x. Let x be an element of A, and let $N = \{y_1, \ldots, y_t\}$ be a lexicographically ordered set of prefixes (all the same length) that includes the prefix of w_{max} of that length. We exploit some crucial properties of the Hausdorff representation $\bigcup_{i=1}^{k}(C_i \cap D_i)$ of $prefix(Left(A))$ for the design of a procedure pruning N to a polynomially size-bounded set that still includes the prefix of w_{max}.

Let y_m be the prefix of w_{max} in $\{y_1, \ldots, y_t\}$. Then, letting $d = 1$ and $l(0) = 1$, it holds that

$$\{\langle x, y_{l(d-1)} \rangle, \ldots, \langle x, y_m \rangle\} \subseteq C_d$$

Inductively, for $d = 1, \ldots, k$, let $r(d)$ be the largest index r such that $\{\langle x, y_{l(d-1)} \rangle, \ldots, \langle x, y_r \rangle\}$ is contained in C_d, and let $l(d)$ be the least index l such that $1 \leq l \leq r(d)+1$ and $\{\langle x, y_l \rangle, \ldots, \langle x, y_{r(d)} \rangle\} \subseteq \overline{D_d}$. Observe that since $\{\langle x, y_{l(d-1)} \rangle, \ldots, \langle x, y_m \rangle\} \subseteq C_d$ it follows that $r(d) \geq m$. Similarly, since $\{\langle x, y_{m+1} \rangle, \ldots, \langle x, y_{r(d)} \rangle\} \subseteq \overline{D_d}$, it holds that $l(d) \leq m+1$. We consider the following two cases separately.

---

[5] It is implicit in this section that polynomial time and polynomial size always mean polynomial in $|x|$.

1. $\langle x, y_m \rangle \in D_d$.

   Then $l(d) = m+1$ since $y_m \notin \{y_{l(d)}, \ldots, y_{r(d)}\}$, i.e., $l(d) > m$.

2. $\langle x, y_m \rangle \notin D_d$. (This case is possible only if $d < k$.)

   In this case, $y_m \in \{y_{l(d)}, \ldots, y_{r(d)}\}$. Since $\{\langle x, y_{l(d)} \rangle, \ldots, \langle x, y_m \rangle\} \subseteq prefix(Left(A))$ but $\{\langle x, y_{l(d)} \rangle, \ldots, \langle x, y_m \rangle\} \subseteq \overline{D_d}$, it follows that $\{\langle x, y_{l(d)} \rangle, \ldots, \langle x, y_m \rangle\} \subseteq C_{d+1}$, and the above analysis can be repeated.

If we could compute the prefixes $y_{l(d)}$ and $y_{r(d)}$ defined above in polynomial time, we could use the above properties in order to design a recursive procedure that collects all the prefixes $y_{l(d)-1}$ found in the recursive calls. This procedure would return a small subset of N containing y_m. Starting with $N = \{\epsilon\}$, the overall algorithm can repeatedly use such a pruning step at each level of the tree of possible witness prefixes by first expanding all the prefixes y in N to $y0$ and $y1$ (thus doubling N) and then pruning N back to a small subset. In that way, the algorithm finally computes a small subset of $\Sigma^{q(|x|)}$ that, if $x \in A$, contains w_{max}.

Although we cannot explicitly compute the required prefixes $y_{l(d)}$ and $y_{r(d)}$, instead we can compute, given $y_{l(d-1)}$, in polynomial time (polynomially size-bounded) sets $J_{right}(d)$ and $J_{left}(d)$ of prefixes such that $y_{r(d)} \in J_{right}(d)$ and $y_{l(d)} \in J_{left}(d)$. This suffices since for each prefix candidate $y \in J_{left}(d)$, the search for $y_{l(d+1)}$ can be done recursively. Since the depth of the recursion is a constant, namely k, the resulting sets $J_{left}(d)$ of candidates for $y_{l(d)}$ still have polynomially bounded cardinality.

We now describe the algorithm in detail. The algorithm calls a recursive pruning procedure PRUNE, which in turn calls two functions SEARCH-RIGHT and SEARCH-LEFT. SEARCH-RIGHT is used to search for candidates for $y_{r(d)}$ that are to the right of previously found candidates for $y_{l(d-1)}$, resulting in a polynomially size-bounded set $J_{right}(d)$ containing $y_{r(d)}$. SEARCH-LEFT is used to search to the left of the prefixes in $J_{right}(d)$, to form a polynomially size-bounded set

$J_{left}(d)$ containing $y_{l(d)}$.

SEARCH-RIGHT(d, N, y_l, x)
(* returns a set $J \subseteq N = \{y_1, \ldots, y_t\}$ that includes the largest prefix $y_r \in N$ such that $\{\langle x, y_l\rangle, \ldots, \langle x, y_r\rangle\} \subseteq C_d$ *)
begin
    $J := \{y_t\}$
    for $j := 0$ **to** $p(|x|)$ **do**
        $J := J \cup \{y_h \mid y_{h+1}$ is the smallest y in N
                s.t. $y \geq y_l$ and $0^j \in f_d(\langle x, y\rangle)\}$
    end
    return J
end

Claim 1 Function SEARCH-RIGHT(d, N, y_l, x), when called with parameter $y_l = y_{l(d-1)}$, returns a set J containing $y_{r(d)}$.

Proof of Claim 1: There are two cases. If $r(d) = t$, then $y_{r(d)}$ is clearly in the returned set J. Otherwise, since $\{\langle x, y_{l(d-1)}\rangle, \ldots, \langle x, y_{r(d)}\rangle\} \subseteq C_d$ and $\langle x, y_{r(d)+1}\rangle \notin C_d$, all the queries in the sets $f_d(\langle x, y_{l(d-1)}\rangle), \ldots, f_d(\langle x, y_{r(d)}\rangle)$ are in T but at least one query 0^j in $f_d(\langle x, y_{r(d)+1}\rangle)$ is not in T. Thus $y_{r(d)+1}$ is the smallest prefix y in N such that $y \geq y_{l(d-1)}$ and $0^j \in f_d(\langle x, y\rangle)$, i.e., $y_{r(d)}$ is included in J in the jth run of the for-loop. ∎

SEARCH-LEFT(d, N, y_r, x)
(* returns a set $J \subseteq N = \{y_1, \ldots, y_t\}$ that includes the smallest prefix $y_l \in N$ such that $\{\langle x, y_l\rangle, \ldots, \langle x, y_r\rangle\} \subseteq \overline{D_d}$ *)
begin
    $J := \{y_1\}$
    for $j := 0$ **to** $p(|x|)$ **do**
        $J := J \cup \{y_h \mid y_{h-1}$ is the largest y in N
                s.t. $y \leq y_r$ and $0^j \in g_d(\langle x, y\rangle)\}$
    end
    return J
end

Claim 2 Function SEARCH-LEFT(d, N, y_r, x), when called with parameter $y_r = y_{r(d)}$, returns a set J containing $y_{l(d)}$.

Proof of Claim 2: Again, there are two cases. If $l(d) = 1$, then $y_{l(d)}$ is clearly in the returned set J. Otherwise, since $\{\langle x, y_{l(d)}\rangle, \ldots, \langle x, y_{r(d)}\rangle\} \subseteq \overline{D_d}$ and $\langle x, y_{l(d)-1}\rangle \in D_d$, all the queries in the sets $g_d(\langle x, y_{l(d)}\rangle), \ldots, g_d(\langle x, y_{r(d)}\rangle)$ are outside of T but at least one query 0^j in $g_d(\langle x, y_{l(d)-1}\rangle)$ is in T. Thus $y_{l(d)-1}$ is the largest prefix y in N such that $y \leq y_{r(d)}$ and $0^j \in g_d(\langle x, y\rangle)$, i.e., $y_{l(d)}$ is included in J in the jth run of the for-loop. ∎

PRUNE(N, J'_{left}, d, x)
(* returns a subset of $N = \{y_1, \ldots, y_t\}$ that contains the prefix y_m of w_{max} if $y_m \in N \cap C_d$ and $\{y_1, \ldots, y_m\} \subseteq C_d$ for a $y_l \in J'_{left}$ with $l \leq m$ *)
begin
  if $d = k + 1$ then return $\emptyset$ end
  $J_{right} := \emptyset$
  for each $z \in J'_{left}$ do
    $J_{right} := J_{right} \cup$ SEARCH-RIGHT(d, N, z, x)
  end
  $J_{left} := \emptyset$
  for each $z \in J_{right}$ do
    $J_{left} := J_{left} \cup$ SEARCH-LEFT(d, N, z, x)
  end
  return $\{y_{l-1} \mid y_l \in J_{left}\} \cup$ PRUNE$(N, J_{left}, d+1, x)$
end

Claim 3 If $y_m \in N$, $\langle x, y_m\rangle \in C_d$, and $y_{l(d-1)} \in J'_{left}$ then function PRUNE(N, J'_{left}, d, x) returns a set I containing y_m.

Proof of Claim 3: If $y_m \in N$ and $\langle x, y_m\rangle \in C_d$ then $\langle x, y_m\rangle$ is also in the sets $\overline{D_{d-1}}, \ldots, \overline{D_1}$. By the above analysis (since case 2 always happens up to $d - 1$) it follows that $\{\langle x, y_{l(d-1)}\rangle, \ldots, \langle x, y_m\rangle\} \subseteq C_d$. Since $y_{l(d-1)} \in J'_{left}$, using Claim 1, $y_{r(d)}$ is included in J_{right} by the call of SEARCH-RIGHT$(d, N, y_{l(d-1)}, x)$. Using Claim 2, $y_{l(d)}$ is included in J_{left} by the call of SEARCH-LEFT$(d, N, y_{r(d)}, x)$. Now we can prove by induction that y_m is included in the set returned by PRUNE. If $\langle x, y_m\rangle \in D_d$ (which must be true in the base case $d = k$), then $y_m =$

$y_{l(d)-1}$ and y_m is included in the set returned by PRUNE. If $\langle x, y_m \rangle \notin D_d$ then $\langle x, y_m \rangle$ is in C_{d+1} and we can use the induction hypothesis. ∎

We complete the algorithm with a description of the main program.

    input x
    begin
        $N := \{\epsilon\}$
        for $i := 1$ **to** $q(|x|)$ **do**
            $N := \{y0 \mid y \in N\} \cup \{y1 \mid y \in N\}$ (* expand the prefixes
                                                      to length i *)
            $N := \text{PRUNE}(N, \{y_1\}, 1, x)$
        end
        (* N now includes w_{max} if $x \in A$ *)
        if there is a witness for x in N **then** *accept* **else** *reject* **end**
    end

In order to prove the correctness of the algorithm it suffices to observe that it follows from Claim 3 that the prefix y_m of w_{max} is included in the pruned set returned by $\text{PRUNE}(N, \{y_1\}, 1, x)$, provided that y_m is in N. Also, since the sets returned by SEARCH-RIGHT and SEARCH-LEFT are bounded in size by $p(|x|) + 2$, it follows inductively that the set J_{left} computed by PRUNE at level d is bounded in size by $(p(|x|) + 2)^{2d}$. Thus, since the depth of recursion of function PRUNE is bounded by a constant, the finally returned set—being the union of all the J_{left}'s—is polynomially size-bounded, and it is easy to see that the algorithm runs in polynomial time. ∎

The Hausdorff characterization of boolean closures of classes of sets has turned out to be also useful in proving related results concerning randomized reductions and nondeterministic reductions to sparse sets (see [AKM92]).

We now briefly discuss the application of the above results to the classes UP, PP and $C_=P$. Since for every set $A \in$ UP it holds that $Left(A)$ is in UP, it also follows that if UP is contained in $R_b^p(R_c^p(\text{SPARSE}))$ then P $=$ UP. This strengthens the results of Watanabe [Wat91], who showed that if P $\neq$ UP then there exists a set in UP that does not many-one polynomial-time reduce to any sparse set.

Consider the set $\{\langle x, m \rangle \mid$ there are at least m satisfying assignments for $x\}$, which has properties similar to left sets and is complete for PP.

Under the assumption that this set is in $R_b^p(R_c^p(\text{SPARSE}))$, we can use the algorithm described in the proof of Theorem 3.2 to compute in polynomial time a set of numbers that includes $\#\text{SAT}(x)$, the number of satisfying assignments of formula x. Now we can use the result of Cai and Hemachandra [CH91] and Toda (see [ABG90]) that P = PP if there is an FP function that computes on input x a set of numbers that includes $\#\text{SAT}(x)$. Alternatively, Theorem 3.10 could be proved along the same lines as Theorem 3.11.

Theorem 3.10 If PP is contained in $R_b^p(R_c^p(\text{SPARSE}))$, then P = PP.

We can also show that if $C_=P$ is contained in $R_b^p(R_c^p(\text{SPARSE}))$ then $P = C_=P$.

Theorem 3.11 If $C_=P$ is contained in $R_b^p(R_c^p(\text{SPARSE}))$ then $P = C_=P$.

Proof of Theorem 3.11:
There exist complete sets in $C_=P$ that are one word-decreasing self-reducible [OL]. Balcazár has shown that every one word-decreasing self-reducible set in $R_T^p(\text{SPARSE})$ is in Σ_2^p [Bal90]. So it follows from the assumption of the theorem that $C_=P \subseteq \Sigma_2^p$. Furthermore, since coNP $\subseteq C_=P$, if $C_=P \subseteq R_b^p(R_c^p(\text{SPARSE}))$ then also NP $\subseteq R_b^p(R_c^p(\text{SPARSE}))$, and it follows from Corollary 3.5 that $P = \Sigma_2^p$. ∎

As discussed previously, Ogiwara and Watanabe [OW91] showed that if for some k NP has a k-truth-table hard sparse set then P = NP. It is natural to ask whether the result of Ogiwara and Watanabe can be extended to truth-tables that have non-constant bounds on their number of queries. We show that, even with respect to the weakened conclusion that the boolean hierarchy [CGH+88,CGH+89] collapses, the Ogiwara-Watanabe result cannot be improved to $\omega(\log n)$-bounded truth-table reductions by any relativizable proof technique. Our result strengthens the work of Homer and Longpré [HL91], who independently of the work of this paper showed that there are relativized worlds in which there are sparse sets that are NP-complete with respect to such bounds and yet P $\neq$ NP. The strongest earlier result was the result of Kadin ([Kad89], see also [IM89,CGH+89]) that for every nice function $f(n) = o(\log n)$ there are relativized worlds in which the polynomial hierarchy does not collapse to $P^{\text{NP}[f(n)]}$ yet NP has sparse Turing complete sets. Since NP has sparse Turing complete sets if and only if NP has sparse truth-table complete sets—this can be seen either directly, via the parallel census technique discussed later in this subsection, or as a consequence of Hartmanis's sparse set that is truth-table complete for the sparse sets in NP ([Har83], only Turing completeness is stated, but Hartmanis's set is clearly truth-table complete)—Kadin's result applies equally well to the (seemingly stronger) truth-table case.

The boolean hierarchy [CGH+88,CGH+89] is the closure of NP under boolean operations; equivalently, it is the class of sets that can be accepted by finite amounts of hardware applied to NP predicates. Here, we define the hierarchy via one of its normal forms—namely, as the union of differences of NP sets.

Definition 3.12 [CGH+88]

1. For $k \geq 1$, the k-th level of the boolean hierarchy is defined by: $L \in \text{NP}(k)$ if and only if there exist $L_1, \ldots, L_k \in \text{NP}$ such that

$$L = \begin{cases} (L_1 - L_2) \cup \ldots \cup (L_{k-2} - L_{k-1}) \cup L_k & \text{if } k \text{ is odd} \\ (L_1 - L_2) \cup \ldots \cup (L_{k-1} - L_k) & \text{if } k \text{ is even.} \end{cases}$$

2. $\text{coNP}(k) = \{L \mid \overline{L} \in \text{NP}(k)\}$.

3. $\text{BH} = \bigcup_{k \geq 1} \text{NP}(k)$, defines the *boolean hierarchy*.

Like the polynomial hierarchy, the boolean hierarchy does have downward separation; if for some k_0 it holds that $\text{NP}(k_0) = \text{coNP}(k_0)$, then $\text{NP}(k_0) = \text{BH}$ [CGH+88]. In such a case, we say that the boolean hierarchy *collapses* (to level k_0).

Theorem 3.13 If f is a polynomial-time computable, nondecreasing function such that $f(n) = \omega(\log n)$, then there exist a set A and a tally set T such that BH^A does not collapse and T is $\leq_{f(n)-tt}^{p,A}$-complete for NP^A.

A detailed proof of Theorem 3.13 can be found in [AHH+92]. The coding used in that proof suggests an interesting issue: given a set S of low density (for example, a sparse set), can we find its elements via *parallel* access to some NP^S set (rather than using the obvious sequential prefix-searching algorithm)? The answer, perhaps surprisingly, is that parallel access suffices. Though the tools to note this have been implicit since the important sparse set research of Hartmanis, Immerman, and Sewelson [HIS85], it is noted most clearly—in slightly different form—in a recent paper of Selman ([Sel90], see that paper for a fuller discussion of the history of this notion). The following result states that one can tighten the bound on the number of queries needed slightly beyond that found in [Sel90], and can extend the range of applicability of Selman's technique beyond the sets in NP.

Theorem 3.14 Let S be a sparse set and let d be a polynomial-time computable function such that $d(n) = n^{\mathcal{O}(1)}$ and $(\forall n)[||S^{=n}|| \leq d(n)]$. There is an $\text{FP}^{\text{NP}_c^S \cap \text{TALLY}}_{\left(1 + n\binom{d(n)+1}{2}\right)\text{-}tt}$ algorithm[6] that, on input 1^n, outputs all length n strings belonging to S.[7]

A full proof of Theorem 3.14 can be found in [AHH+92].

Since a conjunctive reduction is a positive reduction, if S is in NP, then $\text{NP}_c^S = \text{NP}$. In this case, the above result states and generalizes Selman's result that all sparse NP sets are printable via parallel access to NP. Recall that a set L is P-printable [HY84] if there is a polynomial-time computable function f such that, for every n, on input 1^n the function f outputs a list of all strings in L of length at most n; relativized P-printability is defined analogously.

Corollary 3.15 If S is a sparse set, then S is $\text{P}_{tt}^{\text{NP}_c^S \cap \text{TALLY}}$-printable. In particular, all sparse sets are $\text{P}_{tt}^{\text{TALLY}}$-printable [Rub90], and all sparse NP sets are $\text{P}_{tt}^{\text{NP} \cap \text{TALLY}}$-printable [Sel90].

---

[6]That is, a polynomial-time machine given $1 + n\binom{d(n)+1}{2}$ parallel queries to a set in NP_c^S, the class of sets that nondeterministically conjunctively reduce [LLS75] to S.

[7]Clearly, the algorithm requires no queries for the case $d(n) = 0$, and it can be seen that there are relativized worlds in which for some set S having at most one string of each length it holds that $1 + n\binom{1+1}{2} = n + 1$ queries are actually required. We commend to the reader the open question of whether the $1 + n\binom{d(n)+1}{2}$ bound can be replaced in general by some tighter bound; we conjecture that it cannot. At issue here is the rather interesting question of the exact amount of parallel access needed to recover information about sparse sets.

3.2 Mod_kP and Nearly Near-Testable Sets

Mod_kP [CH90,BGH90] is the class of sets L for which there is a nondeterministic polynomial-time Turing machine M such that for every x, $x \in L$ if and only if the number of accepting computation paths of M on x is not a multiple of k.

By applying a different proof technique we obtain results similar to Corollary 3.4 for the classes Mod_kP, $k \geq 2$.

Definition 3.16 A set L is *rotatively one word-decreasing self-reducible* if there exist a deterministic polynomial-time Turing transducer M and a polynomial p satisfying the following conditions:

1. L is a set of strings of the form $\langle x, y, i \rangle$ with $i < p(|x|)$,

2. for every x and for every y, z, there is some $d < p(|x|)$ such that for every $i < p(|x|)$, $\chi_L(\langle x, y, i \rangle) = \chi_L(\langle x, z, i \circ d \rangle)$, where $i \circ d = (i + d) \bmod p(|x|)$, and

3. for every x and y, either

   (a) $M(x, y)$ outputs $\chi_L(\langle x, y, 0 \rangle) \cdots \chi_L(\langle x, y, p(|x|) - 1 \rangle) \in \Sigma^{p(|x|)}$, or

   (b) $M(x, y)$ outputs $d < p(|x|)$ such that for every $i < p(|x|)$, $\chi_L(\langle x, y, i \rangle) = \chi_L(\langle x, \text{pred}(y), i \circ d \rangle)$.

Theorem 3.17 Any rotatively one word-decreasing self-reducible set that conjunctively reduces to a sparse set is in P.

We prove Theorem 3.17 at the end of this section.

Since each set in Mod_kP, $k \geq 2$, is many-one reducible to a rotatively one word-decreasing self-reducible set in Mod_kP (in fact, these are essentially the strictly one word-decreasing self-reducible sets of [OL] that are complete for Mod_kP) we have the following corollary.

Corollary 3.18 For each $k \geq 2$: if Mod_kP has a sparse conjunctively-hard set then $P = Mod_kP$.

Proof of Corollary 3.18
Let L be an arbitrary set in Mod_kP. Let W be in P and let p be a polynomial such that for all x

$$x \in L \iff ||\{y \in \Sigma^{p(|x|)} \mid \langle x, y \rangle \in W\}|| \not\equiv 0 \pmod{k}.$$

Define A to be the set of strings of the form $\langle x, y, i \rangle$ such that z is not equivalent to i modulo k, where z is the number $y' \in \Sigma^{p(|x|)}$ such that $y' \leq y$ and $\langle x, y' \rangle \in W$. Note that

- A is rotatively one word-decreasing self-reducible and
- for every x, $x \in L$ iff $\langle x, 1^{m(|x|)}, 0 \rangle \in A$, and thus, L is many-one reducible to A.

So, if Mod_kP has a sparse conjunctively-hard set S, then $A \leq_c^p S$, and so $A \in \text{P}$. Thus $L \in \text{P}$. ∎

For nearly near-testable sets [HH91], the class of sets that have "implicit" polynomial-time membership tests, a result similar to Theorem 3.17 holds.

Definition 3.19 [HH91] A set A is *nearly near-testable* if there exists a polynomial-time function $N \colon \Sigma^* \to \{\mathit{true}, \mathit{false}, \leftrightarrow, \not\leftrightarrow\}$ such that for every x one of the following holds ($x - 1$ denotes the string lexicographically preceding x):

- $N(x) = \mathit{true}$ and $x \in A$
- $N(x) = \mathit{false}$ and $x \notin A$
- $x \neq \epsilon$ and $N(x) = \leftrightarrow$ and $(x \in A \iff x - 1 \in A)$
- $x \neq \epsilon$ and $N(x) = \not\leftrightarrow$ and $(x \in A \iff x - 1 \notin A)$

Theorem 3.20 Any nearly near-testable set that conjunctively reduces to a sparse set is in P.

A full proof of Theorem 3.20 can be found in [AHH+92]. Below we prove Theorem 3.17.

Proof of Theorem 3.17:
Let L be a rotatively one word-decreasing self-reducible set, as certified by machine M and polynomial p as in Definition 3.16. Suppose that L is R_c-reducible to a sparse set S via a function f. We will give a polynomial-time algorithm for L. Without loss of generality, we may assume that there exist polynomials q and r such that for every w, $f(w)$ is an encoding of a set in $\Sigma^{\leq q(|w|)}$ and for every n, $||S^{\leq q(n)}|| \leq r(n)$. Let $w_0 = \langle x_0, y_0, i_0 \rangle$ be a fixed input whose membership in L we are testing. As we have fixed the input, let p, q, and r denote $p(|x_0|)$,

$q(|w_0|)$, and $r(|w_0|)$, respectively. Let $I = \{0, \cdots, q-1\}$. For $a, b \in I$, let $a \circ b = (a+b) \bmod q$.

For a string y, let $\alpha(y)$ denote $\chi_L(\langle x, y, 0 \rangle) \cdots \chi_L(\langle x, y, p-1 \rangle)$. For a string $u \in \Sigma^p$ and $d \in I$, let $\rho(u, d)$ denote $u_{d+1} \cdots u_p u_1 \cdots u_d$. Note that, by definition, for every y, it holds that

(*) $M(x_0, y)$ is either $\alpha(y)$ or $d \in I$ such that $\alpha(y) = \rho(\alpha(\text{pred}(y)), d)$.

For a string y and $i \in I$, let $w(y, i)$ denote $\langle x_0, y, i \rangle$. An *argument sequence* is a sequence $(A_0, \cdots, A_{p-1})$ with each $A_i \subseteq \Sigma^{\leq q}$. An argument sequence $\mathcal{A} = (A_0, \cdots, A_{p-1})$ is said to be *correct* for a string y if for every $i \in I$, $w(y, i) \in L$ iff $A_i \subseteq S$. Note that if $\mathcal{A} = (A_0, \cdots, A_{p-1})$ is correct for y, then for every $i \in I$ with $||A_i|| > r$, $w(y, i) \notin L$ because $||S^{\leq q}|| \leq r$. For an argument sequence $\mathcal{A} = (A_0, \cdots, A_{p-1})$ and $d \in I$, $\mathcal{A}(d)$ denotes an argument sequence $(A_d, \cdots, A_{p-1}, A_0, \cdots, A_{d-1})$. Let $\mathcal{A} = (A_0, \cdots, A_{p-1})$ and $\mathcal{B} = (B_0, \cdots, B_{p-1})$ be given two argument sequences. We write $\mathcal{B} \subseteq_L \mathcal{A}$ to denote that for every $i \in I$ with $||A_i|| \leq r$, $B_i \subseteq A_i$. Also, $\mathcal{A} \cup \mathcal{B}$ denotes $(A_0 \cup B_0, \cdots, A_{p-1} \cup B_{p-1})$. For y, Φ_y denotes the argument sequence defined by $(f(w(y, 0)), \cdots, f(w(y, p-1)))$. Note that Φ_y is correct for y for every y.

Claim 1 Let $\mathcal{A} = (A_0, \cdots, A_{p-1})$ and $\mathcal{B} = (B_0, \cdots, B_{p-1})$ be argument sequences that are correct for y and z, respectively. Let $d \in I$ be such that $\mathcal{B}(d) \subseteq_L \mathcal{A}$. Then, $\rho(\alpha(z), d) = \alpha(y)$.

Proof of Claim 1: Let $\mathcal{A}, \mathcal{B}, y, z$, and d be as in the hypothesis. Since $\mathcal{A}$ is correct for y, for every $i \in I$, $w(y, i) \in L$ iff $(||A_i|| \leq r$ and $A_i \subseteq S)$. Since $\mathcal{B}(d) \subseteq_L \mathcal{A}$, for every $i \in I$ with $||A_i||$, $B_{i \circ d} \subseteq A_i$, and thus, for every $i \in I$ with $w(y, i) \in L$, $w(z, i \circ d) \in L$. Since $\mathcal{B}$ is correct for z, this implies that for every $i \in I$ with $w(y, i) \in L$, $w(z, i \circ d) \in L$. Since the number of i with $w(y, i) \in L$ is equal to the number of i with $w(z, i) \in L$, we have that for every $i \in I$ with $w(y, i) \notin L$, $w(z, i \circ d) \notin L$. Thus, $\rho(\alpha(z), d) = \alpha(y)$. ∎

Claim 2 Let $\mathcal{A} = (A_0, \cdots, A_{p-1})$ and $\mathcal{B} = (B_0, \cdots, B_{p-1})$ be argument sequences that are correct for y and z, respectively. Let $d \in I$ be such that $\mathcal{B}(d) \subseteq_L \mathcal{A}$. Then

1. if $M(x_0, z) = u$ for some $u \in \Sigma^p$, then $\rho(u, d) = \alpha(y)$, and

2. if $M(x_0, z) = e$ for some $e \in I$, then $\rho(\alpha(\text{pred}(z)), e \circ d) = \alpha(y)$, and $\mathcal{A} \cup \Phi_y(e \circ d) = (A_0 \cup f(w(\text{pred}(z), e \circ d)), \cdots,$

$A_{p-1} \cup f(w(\text{pred}(z), (p-1) \circ (e \circ d))))$ is correct for y.

Proof of Claim 2: Let $\mathcal{A}$, $\mathcal{B}$, y, z, and d be as in the hypothesis. From Claim 1, we have $\alpha(y) = \rho(\alpha(z), d)$. Suppose that $M(x_0, z) = u$ for some $u \in \Sigma^p$. By definition, $u = \alpha(z)$, and thus $\alpha(y) = \rho(u, d)$.

On the other hand, suppose that $M(x_0, z) = e$ for some $e \in I$. As discussed previously, it holds that $\alpha(z) = \rho(\alpha(\text{pred}(z)), e)$. By taking $\rho(\cdot, d)$ of both sides, we have $\rho(\alpha(z), d) = \rho(\rho(\alpha(\text{pred}(z)), e), d)$, and thus, $\alpha(y) = \rho(\alpha(\text{pred}(z)), e \circ d)$.

By definition, $(f(w(\text{pred}(z), 0)), \cdots, f(w(\text{pred}(z), p-1)))$ is correct for $\text{pred}(z)$. Thus, for every $i \in I$, $w(y, i) \in L$ iff $A_i \subseteq S$ iff $w(y, i \circ d \circ e) \in L$ iff $f(w(y, i \circ d \circ e)) \subseteq S$. So, $w(y, i) \in L$ iff $A_i \cup f(w(y, i \circ d \circ e)) \subseteq S$. This proves the claim. ∎

Now we define the algorithm. We operate on d, a string y, and an argument sequence $\mathcal{A} = (A_0, \cdots, A_{p-1})$. Initially, we set y to y_0, d to 0, and A_i to $f(w(y_0, i))$ for each $i \in I$, so that the following two conditions are satisfied:

(c1) $\mathcal{A}$ is correct for y_0, and

(c2) $\rho(\alpha(y), d) = \alpha(y_0)$.

The main part of the algorithm is the repetition of two steps defined below. We require that at the beginning of the first step both (c1) and (c2) hold.

First, we find $z \leq y$ such that

(d1) for some $c \in I$, $\Phi_z(c) \subseteq_L \mathcal{A}$, and

(d2) either $M(x_0, z) \in \Sigma^p$ or there is no e such that $\Phi_{\text{pred}(z)}(e) \subseteq_L \mathcal{A}$.

Note that, under the assumption that (c1) and (c2) hold at the beginning of this step, we have (i) every $z \leq y$ satisfies at least one of these conditions, (ii) $z = y$ satisfies (d1), and (iii) $z = \epsilon$ satisfies (d2). So, by executing a simple divide-and-conquer algorithm over $[\epsilon, y]$, we can easily find z for which (d1) and (d2) are satisfied. We set c to one of the values establishing (c1).

Next, we compute $M(x_0, z)$. If this is in Σ^p, from Claim 1, $\alpha(y_0) = \rho(\alpha(z), c) = \rho(M(x_0, z), c)$. So, $w_0 = w(y_0, i_0)$ is in L iff $w(z, i_0 \circ c) \in L$, and this is easily computed from the output of M. Hence, if this is the case, we obtain $\chi_L(w_0)$ and we accept w_0 iff it is 1. If $M(x_0, z)$ is not in

Σ^p, i.e., it is in I, let e be the value and we set A_i to $A_i \cup f(w(z, i \circ c \circ e))$ for every i, and set $\mathcal{A}$ to the resulting sequence. We claim that $\mathcal{A}$ is correct for y_0. This is seen as follows. Clearly, $\Phi_{\text{pred}(z)}$ is correct for pred(z). Since $M(x_0, z) = e$, $\Phi_{\text{pred}(z)}(e)$ is correct for z. Since $\Phi_z(c) \subseteq_L \mathcal{A}$, $\Phi_{\text{pred}(z)}(c \circ e)$ is correct for y_0. So $\Phi_{\text{pred}(z)}(c \circ e) \cup \mathcal{A}$ is correct for y_0.

After executing the above two steps, we set y to pred(z) and d to $c \circ e$. At this point, if A_{i_0} has more than r elements, then since $A_{i_0} \subseteq S$ iff $w_0 \in L$ and it is impossible that $A_{i_0} \subseteq S$, we reject w_0 and terminate the algorithm. If A_{i_0} has at most r elements, we go back to the start of the first step and repeat these two steps.

Note that each time $\mathcal{A}$ is updated there is some $i \in I$ such that A_i gets at least one new element. So the loop is executed at most $\mathcal{O}(pr)$ times, and this is bounded by some polynomial in $|w_0|$. It is not hard to see that all the other operations can be done in time polynomial in $|w_0|$, and the algorithm correctly decides whether $w_0 \in L$. So $L \in \text{P}$. ∎

4 Reductions to Simple Sparse Sets

In this section, we explore the second question mentioned in the introduction:

> If a set A reduces to a sparse set, does it follow that A is reducible to some sparse set that is "simple" relative to A?

Earlier work along these lines has been done both for the case (which is also the case of this paper) of reductions less flexible than Turing reductions [AHOW92] and for the case of Turing reductions [GW]. However, both these papers are concerned with "equivalence," and this saddles the results with weaknesses that are best illustrated by an example.

Theorem 4.1 [AHOW92] If P = NP and set A 2-truth-table reduces to a sparse set S, then there is another sparse set $\widehat{S}$ to which A is truth-table equivalent.

The key point to notice here is that no claim is being made that A 2-truth-table reduces to $\widehat{S}$; the "equivalence" claim hides a slippage (from 2-truth-table potentially to truth-table, though in fact [AHOW92] holds

the slippage to 5-truth-table) of the complexity of the reduction from
A. This is not a trivial point; the slippage is crucial to the structure
of earlier proofs. When reducing a set A to a sparse set via a certain
fixed reducing function (we speak now not of the reduction type, such
as 2-truth-table, but of the actual function that generates the queries),
there will in general be many possible sparse sets to which A reduces;
however, for the sparse set to be consistently defined by a reduction
back to A, exactly one such set must be selected. The slippage in
earlier results occurs exactly because of the cost of the disambiguating
down to a single sparse set.

We now show that one can obtain results that contain no slippage
at all. Very informally, to do this we avoid coding too much of the
disambiguating information into the sparse set. Using this type of
approach, we obtain the following results. (Note that Theorems 4.3
and 4.4 are incomparable.)

Theorem 4.2 If $A \leq^p_{bd} S$ for some sparse set S, then there is a sparse set $\widehat{S}$ such that $A \leq^p_{bd} \widehat{S}$ and $\widehat{S} \in \mathrm{P}^{\mathrm{NP}^A[\log]}$.

Theorem 4.3 If $A \leq^p_{2\text{-}tt} S$ for some sparse set S, then there is a sparse set $\widehat{S}$ such that $A \leq^p_{2\text{-}tt} \widehat{S}$ and $\widehat{S} \in \mathrm{P}^{\mathrm{NP}^A[\log]}$.

Theorem 4.4 For $k = 2$ and $k = 3$: If $A \leq^p_{k\text{-}d} S$ for some sparse set S, then there is a sparse set $\widehat{S}$ such that $A \leq_{k\text{-}d} \widehat{S}$ and $\widehat{S} \in \mathrm{P}^{\mathrm{NP} \oplus A}$.

Theorem 4.5 If $A \leq^p_d S$ for some sparse set S, then there is a sparse set $\widehat{S}$ such that $A \leq^p_d \widehat{S}$ and $\widehat{S} \in \mathrm{P}^{\mathrm{NP}^A}$.

Theorem 4.6 If $A \leq^p_c S$ for some sparse set S, then there is a sparse set $\widehat{S}$ such that $A \leq^p_c \widehat{S}$ and $\widehat{S} \in \mathrm{NP}^A$.

The rest of this section is devoted to proving Theorems 4.2 and 4.3.
Detailed proofs of Theorems 4.4, 4.5, and 4.6 can be found in [AHH+92].

We introduce some notations and lemmas that will be helpful in
proving Theorem 4.2. We say that $A \leq^p_{k\text{-}d} B$ via σ if σ is a polynomial-time function such that, for all x, $||\sigma(x)|| \leq k$ and $[x \in A \iff \sigma(x) \cap B \neq \emptyset]$. A collection of distinct sets $a_1, \ldots, a_h$ is called an h-sunflower if the intersection $a_i \cap a_j$ is the same for every pair of distinct indices; the common part $a_i \cap a_j$ is called the center of the sunflower. A collection W of sets is called h-compact if there are no subcollections

of W that are $(h+1)$-sunflowers. The following combinatorial lemma about sunflowers, due to Erdős and Rado [ER60] (see also [BS90]), will be used extensively.

Lemma 4.7 [ER60] If W is an h-compact collection of sets, each of cardinality at most k, then there are at most $h^k k!$ sets in W.

The proof of Theorem 4.2 is obtained from the following technical lemma.

Lemma 4.8 Let $k \geq 1$. Suppose that $A \leq^p_{k\text{-}d} S$ via σ and B, D are two sets such that:

1. S is a sparse set,
2. $D \in \text{NP}^{A \oplus B}$, $S \cap D = \emptyset$, and
3. σ satisfies the following "honesty" condition: a polynomial-time function r exists such that, for all z and y, $z \in \sigma(y) \Rightarrow |y| = r(0^{|z|})$, and for some polynomial p, $(\forall n)[r(0^n) \leq p(n)]$.

Then there is a sparse set $\widehat{S}$ such that $A \leq^p_{k\text{-}d} \widehat{S}$ via σ, $\widehat{S} \in \text{P}^{\text{NP}^{A \oplus B}[\log]}$, and $\widehat{S} \cap D = \emptyset$.

Proof of Lemma 4.8:
The idea can be expressed intuitively as follows. If many strings in A query the same strings in a set c of cardinality less than k, then a simple sparse set S' can be found by induction. Other queries can be proved by Lemma 4.7 to be covered by a sparse set S'' whose simplicity, relative to A, is shown by an algorithm for checking h-compactness.

The proof is by induction on k. If $k = 1$, define $\widehat{S} = \{z \mid (\exists y)[|y| = r(0^{|z|})$ and $y \in A$ and $z \in \sigma(y)]\}$. It is easy to verify that $\widehat{S}$ satisfies the thesis. Indeed, in this case it even holds that $\widehat{S} \in \text{NP}^A$.

Let $k > 1$, and suppose that $A \leq^p_{k\text{-}d} S$ via σ and B, D are two sets that satisfy conditions (1)-(3). Since σ is computable in polynomial time and S is a sparse set, there exists a polynomial p such that, for all y and z, $z \in \sigma(y) \Rightarrow |z| \leq p(|y|)$, and, for all n, $||S^{\leq n}|| \leq p(n)$. Let q be a polynomial such that $q(n) > p(p(n))$. The crucial fact that allows us to apply the inductive hypothesis is the following claim, which follows from the sparseness bound and is stated without proof.

Claim 1 If $y_1, \ldots, y_h \in A$ are such that $|y_1| = \cdots = |y_h|$, $h = q(|y_1|)$, and the collection of sets $\sigma(y_1), \ldots, \sigma(y_h)$ is an h-sunflower whose center is c, then $c \cap S \neq \emptyset$.

Observe that, in the case described in Claim 1, the cardinality of the center c certainly is strictly less than k. For each $m = 1, \ldots, k-1$, define A_m and σ_m as follows: $A_m = \{\langle y_1, \ldots, y_h\rangle \mid y_1, \ldots, y_h \in A$ and $|y_1| = \cdots = |y_h|$ and $h = q(|y_1|)$ and $\{\sigma(y_1), \ldots, \sigma(y_h)\}$ is an h-sunflower whose center has cardinality m $\}$, and

$$\sigma_m(y) = \begin{cases} c & \text{if } y = \langle y_1, \ldots, y_h\rangle \text{ and } |y_1| = \cdots = |y_h| \text{ and} \\ & h = q(|y_1|) \text{ and } \{\sigma(y_1), \ldots, \sigma(y_h)\} \text{ is an } h\text{-sunflower} \\ & \text{whose center } c \text{ has cardinality } m \\ \emptyset & \text{otherwise.} \end{cases}$$

Now, we prove that for all m, $1 \leq m \leq k-1$, it holds that $A_m \leq^p_{m\text{-}d} S$ via σ_m. Take m such that $1 \leq m \leq k-1$. Suppose that $y \in A_m$. Then there are $y_1, \ldots, y_h$ such that $y = \langle y_1, \ldots, y_h\rangle$, $h = q(|y_1|)$, $y_1, \ldots, y_h \in A$, $|y_1| = \cdots = |y_h|$, and $\{\sigma(y_1), \ldots, \sigma(y_h)\}$ is an h-sunflower whose center c has cardinality m; thus $\sigma_m(y) = c$, and from Claim 1 it holds that $c \cap S \neq \emptyset$, and so $\sigma_m(y) \cap S \neq \emptyset$. If $y = \langle y_1, \ldots, y_h\rangle \notin A_m$, then we have two cases: if there is a string y_i that does not belong to A, then $\sigma(y_i) \cap S = \emptyset$ and thus $\sigma_m(y) \cap S = \emptyset$ (since $\sigma_m(y) \subseteq \sigma(y_i)$); on the other hand, if every y_i belongs to A, then, since $y \notin A_m$, it must be the case that $\sigma_m(y) = \emptyset$.

Let $OUT = \{z \mid (\exists y)[|y| = r(0^{|z|}) \text{ and } y \notin A \text{ and } z \in \sigma(y)]\}$, i.e., the set of strings that are "provably" out of S. Let $D' = D \cup OUT$. It is easy to see that, for all m, $1 \leq m \leq k-1$, it holds that A_m, S, σ_m, $A \oplus B$, and D' satisfy conditions (1)-(3) (with $A_m \to A$, $S \to S$, $\sigma_m \to \sigma$, $A \oplus B \to B$, $D' \to D$), so we can apply the inductive hypothesis, which ensures, for every $1 \leq m \leq k-1$, the existence of a sparse set S_m such that $A_m \leq^p_{m\text{-}d} S_m$ via σ_m, $S_m \in \mathrm{P}^{\mathrm{NP}^{A_m \oplus (A \oplus B)}[\log]}$, and $S_m \cap D' = \emptyset$. Define $S' = S_1 \cup \cdots \cup S_{k-1}$; since $A_m \in \mathrm{P}^A$, it holds that $S' \in \mathrm{P}^{\mathrm{NP}^{A \oplus B}[\log]}$. Furthermore, if $y \notin A$ then $\sigma(y) \cap S' = \emptyset$ (since $S' \cap D' = \emptyset$). Unfortunately, if $y \in A$ we cannot prove that $\sigma(y) \cap S' \neq \emptyset$, thus we have to add other elements to S'. Consider the collection of sets that are not "covered" by S': $V : = \{a \mid (\exists y)[y \in A \text{ and } \sigma(y) = a]$ and $a \cap S' = \emptyset\}$. This collection of sets can be subdivided into subcollections: $V_n = \{a \mid a \in H_n$ and $a \cap S' = \emptyset\}$, where $H_n = \{a \mid (\exists y)[y \in A$ and $|y| = n$ and $\sigma(y) = a]\}$. Clearly $V = \bigcup_n V_n$. The collection V_n has the following important property.

Claim 2 *If $a \in V_n$ then there is no collection E such that $E \subseteq H_n$, $a \in E$, and E is a $q(n)$-sunflower.*

Proof of Claim 2: Let $a \in V_n$. Suppose that there exists a collection $E \subseteq H_n$ with $a \in E$, and E is a $q(n)$-sunflower. Then there exist $y_1, \ldots, y_h$ with $h = q(n)$, such that $y_1, \ldots y_h \in A$, $|y_1| = \cdots = |y_h| = n$, $\sigma(y_1) = a$, and $\{\sigma(y_1), \ldots, \sigma(y_h)\} = E$. Let c be the center of E, and $m = ||c|| < k$. Thus it holds that $\langle y_1, \ldots, y_h \rangle \in A_m$ and $\sigma_m(\langle y_1, \ldots, y_h \rangle) = c$, and, since $A_m \leq_{m-d}^p S_m$, it follows that $c \cap S_m \neq \emptyset$. Furthermore, $c \subseteq \sigma(y_1) = a$, and thus $a \cap S_m \neq \emptyset$, which contradicts the assumption that $a \in V_n$. ∎

Claim 2 suggests consideration of the following collection: $W_n = \{a \mid a \in H_n$ and there is no collection E such that $E \subseteq H_n$, $a \in E$, and E is a $q(n)$-sunflower $\}$. Define $S'' = \{z \mid (\exists a)[a \in W_{r(0|z|)}$ and $z \in a]\} - D'$, and $\widehat{S} = S' \cup S''$. From the definition of W_n it is clear that W_n is $(q(n) - 1)$-compact, thus by Lemma 4.7 S'' is a sparse set. Furthermore, it is not hard to verify that $A \leq_{k-d}^p \widehat{S}$ via σ. It remains to show that $S'' \in P^{NP^{A \oplus B}[\log]}$. A naïve algorithm, based directly on the definitions of S'' and W_n, yields only $S'' \in \Sigma_2^{p, A \oplus B}$. In order to accomplish our goal we need the following algorithm, which, given a collection of sets T and an integer h, "approximately" checks whether or not T is h-compact. We assume a total ordering of all finite sets of strings.

    APPROX(T,h)
    begin
        for each subset c of some set $a \in T$ **do**
            $I := \emptyset$
            $m := 0$
            while $m = 0$ **do**
                find (if such exists) the minimum (with respect to the total ordering of finite sets) set $a \in T$ such that:
                (1) $a \notin I$, (2) $c \subseteq a$, and (3) $(\forall b \in I)[b \cap a = c]$
                if such a exists **then** $I := I \cup \{a\}$
                else $m := ||I||$
            end
            end (*while*)
            if $m > h$ **then** *reject* **end**
        end (*for*)
        accept
    end

The above algorithm has the following properties.

1. If the cardinalities of the sets of the input collection T are bounded by a constant then APPROX runs in polynomial time.

2. If $(\forall a \in T)[||a|| \leq k]$ and APPROX(T,h) accepts then T is kh-compact.

3. If APPROX(T,h) rejects then T is not h-compact.

4. If APPROX(T,h) accepts and APPROX($T \cup \{a\}$,h) rejects then there is a collection $E \subseteq T \cup \{a\}$ such that $a \in E$ and E is a $(h+1)$-sunflower.

Properties (1), (3), and (4) are easy to prove. In order to prove property (2) we need the following.

Claim 3 Let E and F be two collections of sets with each set of cardinality at most k. If E is a m-sunflower and F is a ℓ-sunflower whose center is the same of that of E, then there exist at least $m - \ell k$ sets a in E such that $F \cup \{a\}$ is a $(\ell+1)$-sunflower.

Proof of Claim 3: Let $E = \{a_1, \ldots, a_m\}$, $F = \{b_1, \ldots, b_\ell\}$, and let c be the common center of E and F. Clearly, if $F \subseteq E$ then the assertion is true. Hence, in the following we suppose that $F \not\subseteq E$.

The proof is by induction on ℓ. Let $\ell = 1$, define, for each $i = 1, \ldots, m$, $d_i = (a_i \cap b_1) - c$. Since $\{a_1, \ldots, a_m\}$ is a m-sunflower whose center is c, it holds that, for all distinct i, j, $d_i \cap d_j = \emptyset$. Furthermore, for each $i = 1, \ldots, m$, $d_i \subseteq b_1$ and $||b_1|| \leq k$, thus there are at least $m - k$ sets d_i such that $d_i = \emptyset$. It follows that there are at least $m - k$ sets a in E such that $a \neq b_1$ and $a \cap b_1 = c$, that is, $\{b_1, a\}$ is a 2-sunflower.

Let $\ell > 1$, since $F \not\subseteq E$, we can assume that $b_\ell \notin E$. Applying the inductive hypothesis to $\{b_1, \ldots, b_{\ell-1}\}$ we obtain that there are at least $m - (\ell - 1)k$ sets a in E such that $\{b_1, \ldots, b_{\ell-1}, a\}$ is a ℓ-sunflower. Without loss of generality, we can assume that, for each $i = 1, \ldots, n$ with $n = m - (\ell-1)k$, $\{b_1, \ldots, b_{\ell-1}, a_i\}$ is a ℓ-sunflower. Define, for each $i = 1, \ldots, n$, $d_i = (a_i \cap b_\ell) - c$. Since $\{a_1, \ldots, a_n\}$ is a n-sunflower whose center is c, it must be the case that, for all distinct i, j, $d_i \cap d_j = \emptyset$. Furthermore, for each $i = 1, \ldots, n$, $d_i \subseteq b_\ell$ and $||b_\ell|| \leq k$. Thus, there are no less than $n - k$ sets d_i such that $d_i = \emptyset$. It follows that there are at least $m - \ell k$ sets a in E such that $F \cup \{a\}$ is a $(\ell+1)$-sunflower. ∎

Suppose that T is not kh-compact. This means that a subcollection E of T exists that is a $(kh + 1)$-sunflower. Let c be the center of E. Claim 3 implies that the while-loop of APPROX(T,h), relative to c, makes at least $h + 1$ iterations, and thus APPROX(T,h) rejects. This contradicts the assumption that APPROX(T,h) accepts.

A useful consequence of property (4) is the following.

Claim 4 If $T \subseteq H_n$ and APPROX(T,$q(n)$) accepts then APPROX($T \cup W_n$, $q(n)$) accepts.

Proof of Claim 4: By induction on the cardinality of W_n, using property (4) of APPROX. ∎

Let $m_n = \max\{||T|| \mid T \subseteq H_n$ and APPROX(T,$q(n)$) accepts$\}$. From Claim 4 we have that if $T \subseteq H_n$, APPROX(T,$q(n)$) accepts and $||T|| = m_n$ (T is maximal), then $W_n \subseteq T$. Thus, if we knew m_n and $||W_n||$, then given any string z with $r(0^{|z|}) = n$ we could check via one query to a suitable NP^A oracle whether or not there is a set $a \in W_n$ such that $z \in a$. In fact, an NP^A machine can guess a collection $T \subseteq H_n$ and verify that $||T|| = m_n$ and that APPROX(T,$q(n)$) accepts (observe that from property (2) of APPROX and Lemma 4.7 there is a polynomial that bounds m_n), subsequently (let $h = m_n - ||W_n||$) the NP^A machine guesses sets $a_1, \ldots, a_h$ and collections $E_1, \ldots, E_h$ such that $E_i \subseteq H_n$, $a_i \in E_i \cap T$, and $a_i \neq a_j$, and verifies that for every i it holds that E_i is a $q(n)$-sunflower; at this point, the computation accepts if and only if there is a set $a \in T - \{a_1, \ldots, a_h\}$ such that $z \in a$.

It is not hard to see that m_n and subsequently $||W_n||$ can be computed in polynomial time via $\mathcal{O}(\log n)$ queries to a suitable NP^A oracle. ∎

Proof of Theorem 4.2:
Let $A \leq^p_{k\text{-}d} S$ via σ, and let S be a sparse set. Define $S' = \{\langle 0^l, x \rangle_2 \mid x \in S$ and $l \geq 0\}$, and, for each x, $\sigma'(x) = \{\langle 0^{|x|p(|x|)}, y \rangle_2 \mid y \in \sigma(x)\}$, where p is a polynomial such that, for all y and z, $z \in \sigma(y) \Rightarrow |z| \leq p(|y|)$. It is easy to verify that $A \leq^p_{k\text{-}d} S'$ via σ', and that A, S', and σ' satisfy conditions (1)-(3) of Lemma 4.8 (with $B = D = \emptyset$). Thus, applying Lemma 4.8, we obtain a sparse set $\widehat{S}$ such that $A \leq^p_{k\text{-}d} \widehat{S}$ and $\widehat{S} \in \text{P}^{\text{NP}^A[\log]}$. ∎

We now turn to the proof of Theorem 4.3. Recall, as we discussed earlier, that when reducing a set A to a sparse set via a certain fixed reducing function, there will usually be many possible sparse sets to which A reduces. For the sparse set to be consistently defined by a reduction back to A, exactly one such set must be selected. The following proof tries to eliminate this ambiguity as follows. Via a binary search for census information, the machine to accept a sparse set (via access to the original set) obtains information about it. The search is "promise"-like (see, e.g., [Sel88]) in that it works only because the census is found to be sparse in the previous step. Thus, the machine for the sparse set has obtained census information about certain subsets of the strings in and out of itself; using this, it reduces (taking the case of Proposition 4.18 as an example) all remaining ambiguity to a feasible graph coloring problem. Crucially, all inputs (of the same connected component) will obtain the same coloring problem, and this will cause the sparse set to be consistently and correctly defined.

Proof of Theorem 4.3:
Without loss of generality, it is assumed that exactly two queries are asked in the 2-truth-table reductions. Given an input y, a 2-truth-table reduction generates a truth-table as well as a pair of queried strings; the acceptance of the input is decided by looking up the truth-table using the result of queries as an index. There are sixteen different truth-tables that can be generated; they are shown in Figures 2–7. (In the tables, entry 1 means acceptance, and 0 means rejection.) Let $\tau(y)$ denote the truth-table generated on the input y; and, let $\mathcal{T}$ be the set of sixteen truth-tables of arity two. For simplicity we will denote each member of $\mathcal{T}$ by its corresponding table number in Figures 2–7. For each $t \in \mathcal{T}$, let $B_t = \{y \in \Sigma^* \mid \tau(y) = t\}$. The following proposition is clear.

Proposition 4.9 Let $\tau(\cdot)$ be a polynomial-time computable function that on an input y, generates a truth-table in $\mathcal{T}$. Then the following holds.
(i) $(\forall t \in \mathcal{T})\ B_t \in \mathrm{P}$,
(ii) $(\forall t, t' \in \mathcal{T})\ [t \neq t' \Longrightarrow B_t \cap B_{t'} = \emptyset]$,
(iii) $\bigcup_{t \in \mathcal{T}} B_t = \Sigma^*$.

This proposition says that given a 2-truth-table reduction, the input string domain is completely covered by disjoint polynomial-time sets each of which is associated with a corresponding truth-table.

Henceforth, we will call $\{B_t \mid t \in \mathcal{T}\}$ the decomposition of the input string domain associated with the given 2-truth-table reduction.

In order to be able to discuss separately each of the truth-tables that may be used in a 2-truth-table reduction, we introduce the following definition.

Definition 4.10 Fixed truth-table reductions [Wec85] A truth-table reduction is called a fixed truth-table reduction if it uses the same truth-table for all its inputs. We denote such a reduction with $\leq_t^p$, where t denotes a truth-table: $A \leq_t^p B$ means $A \leq_{tt}^p B$ with the fixed truth-table t. We also denote such a reduction with $\leq_{ftt}^p$: i.e., $A \leq_{ftt}^p B$ with t means $A \leq_{tt}^p B$ with the fixed truth-table t. Similarly, $A \leq_{k\text{-}ftt}^p B$ with t means $A \leq_{k\text{-}tt}^p B$ with the fixed truth-table t.

The next two propositions, together with Proposition 4.9 enable us to find a corresponding sparse set individually for each fixed truth-table reduction, and combine the results together to form a full sparse set. The proofs are straightforward and we leave them as a exercise to the reader. Note that in Proposition 4.11, the *truth-table* 2 is given a special treatment since the only set that 2-truth-table reduces via the *truth-table* 2 is Σ^*.

Proposition 4.11 Let A be a set that 2-truth-table reduces to a sparse set. Let $\{B_t \mid t \in \mathcal{T}\}$ be the decomposition of the input string domain associated with the 2-truth-table reduction. Then the following holds.
(i) $(\forall t \in \mathcal{T}) \; [t \neq 2 \implies (\exists S_t \in \text{SPARSE}) \; B_t \cap A \leq_t^p S_t]$,
(ii) $t = 2 \implies (\exists S_t \in \text{SPARSE} \cap \text{P}) \; B_t \cap A \leq_{2\text{-}tt}^p S_t$.

Proposition 4.12 Let $\mathcal{C}$ be a complexity class such that for every oracle D (i) $\mathcal{C}^D$ is closed under $\leq_m^p$, (ii) $\mathcal{C}^D$ has a $\leq_m^p$-complete set, and (iii) for any $B \in \text{P}$, $\mathcal{C}^{B \cap D} \subseteq \mathcal{C}^D$. Let A be a set that 2-truth-table reduces to a sparse set. Let $\{B_t \mid t \in \mathcal{T}\}$ be the decomposition of the input string domain associated with the 2-truth-table reduction. Suppose that for all $t \in \mathcal{T}$, there exists a sparse $S_t \in \mathcal{C}^{B_t \cap A}$ such that $B_t \cap A \leq_{2\text{-}tt}^p S_t$. Then there exists a sparse set $\widehat{S} \in \mathcal{C}^A$ such that $A \leq_{2\text{-}tt}^p \widehat{S}$.

Clearly, for every oracle D, $\text{P}^{\text{NP}^D[\log]}$ qualifies as the complexity class $\mathcal{C}^D$ in Proposition 4.12. To establish the theorem, it remains to show that Proposition 4.11 implies that for all $t \in \mathcal{T}$, there exists a sparse

$S_t \in \mathrm{P}^{\mathrm{NP}^{B_t \cap A}[\log]}$ such that $B_t \cap A \leq_{2\text{-}tt}^p S_t$. Propositions 4.13–4.18 below accomplish this task. In the following propositions, on input y to the reduction $A \leq_{2\text{-}ftt}^p S$, $g_1(y)$ and $g_2(y)$ denote the first and the second queried strings, respectively. $g(y)$ denotes the set of queried strings on input y, i.e., $\{g_1(y), g_2(y)\}$. We say that $A \leq_{2\text{-}ftt}^p S$ via g.

Proofs of the following two propositions are immediate, and thus are omitted.

Proposition 4.13 If $A \leq_{2\text{-}ftt}^p S$ via g with any of the truth-tables of Figure 2, and S is sparse, then there exists a sparse set S' in P^A such that $A \leq_{2\text{-}ftt}^p S'$ via g with the same truth-table.

Table	First Query Answered **yes**		First Query Answered **no**	
Number	2nd Ans. **yes**	2nd Ans. **no**	2nd Ans. **yes**	2nd Ans. **no**
1	0	0	0	0
2	1	1	1	1

Figure 2: Trivial truth-tables of arity two.

Proposition 4.14 If $A \leq_{2\text{-}ftt}^p S$ via g with any of the truth-tables of Figure 3, and S is sparse, then there exists a sparse set S' in NP^A such that $A \leq_{2\text{-}ftt}^p S'$ via g with the same truth-table.

Table	First Query Answered **yes**		First Query Answered **no**	
Number	2nd Ans. **yes**	2nd Ans. **no**	2nd Ans. **yes**	2nd Ans. **no**
3	0	0	1	1
4	1	1	0	0
5	0	1	0	1
6	1	0	1	0

Figure 3: 1-tt-related truth-tables of arity two.

Proposition 4.15 If $A \leq_{2\text{-}ftt}^p S$ via g with any of the truth-tables of Figure 4, and S is sparse, then there exists a sparse set S' in NP^A such that $A \leq_{2\text{-}ftt}^p S'$ via g with the same truth-table.

Table	First Query Answered **yes**		First Query Answered **no**	
Number	2nd Ans. yes	2nd Ans. no	2nd Ans. yes	2nd Ans. no
7	0	0	1	0
8	1	1	0	1
9	0	1	0	0
10	1	0	1	1

Figure 4: Implication-related truth-tables of arity two.

Proof of Proposition 4.15: These tables are variations of the truth-table for implication. We prove for the case of *truth-table* 10. The other cases can be proved similarly.

Without loss of generality, we assume that g_1 and g_2 are honest and that for all y, $g_1(y)$ and $g_2(y)$ begin with 0 and 1, respectively. Let $S_1 = \{0x \mid (\exists y)[y \in \overline{A} \land 0x \in g(y)]\}$. S_1 has the strings in S that guarantee all the strings in $\overline{A}$ to be rejected. However, some of the strings in A may also get rejected if S_1 is used instead of S in the reduction. In order to remedy this problem, we add S_2 to S', where S_2 is given by
$$S_2 = \{1x \mid (\exists x', y_1, y_2)$$
$$[y_1 \in \overline{A} \land y_2 \in A \land 0x' \in g(y_1) \land g(y_2) = \{0x', 1x\}]\}.$$
Since S_2 contains only second queries, there is no more chaining of side effects. Note that, if the first and second queries were not separated, some strings in S_2 might be used as first queries, potentially leading to a chain of side effects. Let $S' = S_1 \cup S_2$. It is easy to verify that $S' \subseteq S$ and S' satisfies the theorem. ∎

The following proposition is essentially a special case of Theorem 4.6.

Proposition 4.16 If $A \leq^p_{2\text{-}ftt} S$ via g with any of the truth-tables of Figure 5, and S is sparse, then there exists a sparse set S' in NP^A such that $A \leq^p_{2\text{-}ftt} S'$ via g with the same truth-table.

Proof of Proposition 4.16: These tables are variations of the truth-table for conjunction. Without loss of generality, we assume that g_1 and g_2 are honest. It is easy to see that, for the case of truth-table 11, $S' = \{x \mid (\exists y)[y \in A \land x \in g(y)]\}$ satisfies the condition. The other case can be proved similarly. ∎

Reductions to Sets of Low Information Content 33

Table	First Query Answered **yes**		First Query Answered **no**	
Number	2nd Ans. **yes**	2nd Ans. **no**	2nd Ans. **yes**	2nd Ans. **no**
11	1	0	0	0
12	0	1	1	1

Figure 5: Conjunctive-related truth-tables of arity two.

The following proposition is essentially a special case of Theorem 4.2. We omit its proof.

Proposition 4.17 If $A \leq^p_{2\text{-}ftt} S$ via g with any of the truth-tables of Figure 6, and S is sparse, then there exists a sparse set S' in $\text{P}^{\text{NP}^A[\log]}$ such that $A \leq^p_{2\text{-}ftt} S'$ via g with the same truth-table.

Table	First Query Answered **yes**		First Query Answered **no**	
Number	2nd Ans. **yes**	2nd Ans. **no**	2nd Ans. **yes**	2nd Ans. **no**
13	1	1	1	0
14	0	0	0	1

Figure 6: Disjunctive-related truth-tables of arity two.

Proposition 4.18 If $A \leq^p_{2\text{-}ftt} S$ via g with any of the truth-tables of Figure 7, and S is sparse, then there exists a sparse set S' in $\text{P}^{\text{NP}^A[\log]}$ such that $A \leq^p_{2\text{-}ftt} S'$ via g with the same truth-table.

Table	First Query Answered **yes**		First Query Answered **no**	
Number	2nd Ans. **yes**	2nd Ans. **no**	2nd Ans. **yes**	2nd Ans. **no**
15	0	1	1	0
16	1	0	0	1

Figure 7: Exclusive-or-related truth-tables of arity two.

Proof of Proposition 4.18: These tables are variations of the truth-table for exclusive-or. We prove for the case of *truth-table* 15. The other case can be proved similarly.

We will show that, given a string x, we can determine whether $x \in S'$ for some unambiguously determined S', which is a well-behaved approximation to S. Although we cannot guarantee $S' \subseteq S$, we can make sure that S' is sparse and satisfies the requirements of the proposition.

Without loss of generality, assume that there exists a polynomial-time computable and invertible one-to-one function $h : 0^* \to 0^*$ such that for all $y \in \Sigma^*$, it holds that $0^{|g_1(y)|} = 0^{|g_2(y)|} = h(0^{|y|})$. (This condition makes the sparseness of S' very simple to argue.) Let $D_x = \{y \in \Sigma^* \mid 0^{|y|} = h^{-1}(0^{|x|})\}$, i.e., the set of strings, when reduced, might query x. Let $g(A) = \bigcup_{y \in A} g(y)$. Clearly, $\|g(D_x) \cap S\|$ is bounded by a polynomial in $|x|$; let $s(|x|)$ denote this polynomial bound.

Let $G_1 = \{g(y) \mid y \in D_x \cap A\}$, $G_2 = \{g(y) \mid y \in D_x \cap \overline{A}\}$, and $G = G_1 \cup G_2$. The problem can be easily visualized using the graph defined by the set of edges G. G consists of two different types of edges: exclusive-or type and coexistence type. The edges in G_1 are of exclusive-or type: of the two nodes of each edge in G_1, one is in S while the other is in $\overline{S}$. The edges in G_2 are of coexistence type: the two nodes of each edge in G_2 either both are in S or both are in $\overline{S}$.

Suppose that a sparse set S' is chosen, and that all the elements in S' are given the same color while those in $\overline{S'}$ are given another color. It is easy to see that, in order to satisfy the requirement of the above proposition, it suffices to show that the two endpoints of each edge in G_1 are colored differently while those of each edge in G_2 have the same color. Thus, the problem of choosing S' can be considered as a two-coloring problem in which the sets G_1 and G_2 has to be preserved. We now describe a $P^{NP^A[\log]}$ algorithm to choose such S'.

(1) Check if there exists a connected component of G that contains x and at least $2s(|x|)$ other nodes. This can be checked by asking a single query (with $m = 2s(|x|) + 1$) to an NP^A oracle defined by the following machine description.

Machine M^A on input $\langle 0^m, x \rangle$
Guess a graph G' with m distinct vertices.
If G' contains x, and forms a connected subgraph of G, then accept.

If such a connected component (G') does not exist, proceed to step (2). Otherwise, accept x if and only if $x \in S$. This can be checked by asking a single query to an NP^A oracle similar to the above one since $x \in S$ if and only if x has the less populous color in a two-coloring of G'.

(2) Find G_L, the set of edges in the maximal connected subgraph of G that contains x. Obviously, G_L is unique. Moreover, since G_L has no more than $2s(|x|)$ nodes, it has no more than $s(|x|)(2s(|x|)-1)$ edges. Hence, G_L can be found along some computation path of an NP^A machine, once $||G_L||$ is known. Accept x if and only if x has the less populous color in a two-coloring of G_L.

Note that $||G_L||$ can be obtained by a binary search using an NP^A oracle defined by the following machine description.

Machine $M_{G_L}^A$ on input $\langle 0^m, x \rangle$
Guess a graph G' with m distinct edges.
If G' contains x, and forms a connected subgraph of G, then accept.

Clearly, both the steps (1) and (2) maintain $||S'^{=|x|}|| \leq ||S^{=|x|}||$, thereby keeping S' sparse. It is easy to see that S', via the reduction g, preserves G_1 and G_2; this proves the correctness of the algorithm. It is not hard to fill in the details of the algorithm in a way that guarantees that $S' \in \text{P}^{\text{NP}^A[\log]}$. ∎

The above Propositions 4.13–4.18, together with Proposition 4.11 and Proposition 4.12, prove Theorem 4.3. ∎

5 Low Instance Complexity and Polynomial-Time Reductions

Instance complexity, as defined by Ko, Orponen, Schöning, and Watanabe [KOSW86,Orp90,KOSW90], is a notion of the complexity of specific instances of a problem—a topic that standard complexity theory is ill-suited to study (as any single instance is trivial). In this paper, we are primarily concerned with sets of "low" instance complexity: IC[log,poly] (introduced in [KOSW86]).

Definition 5.1 We say that a set A is in IC[log,poly] if there exist a constant $c > 0$, a polynomial t and a set $\Pi \subseteq \Sigma^*$ of programs[8] such that for every $x \in \Sigma^*$

1. there exists a $p \in \Pi^{\leq c\log(|x|)+c}$ such that p decides x in time $t(|x|)$ according to A, and

2. for every $p \in \Pi$ it holds that if p decides x in time $t(|x|)$ then p decides x according to A.

We show that the sets of low instance complexity are intimately related to the study of reductions to sets of low information content. In particular, a set A is of low instance complexity if and only if it both conjunctively and disjunctively reduces to tally sets (say T_0 and T_1, respectively). Note that this implies that A reduces disjunctively and conjunctively to the single tally set: $\{0^{2i} \mid 0^i \in T_0\} \cup \{0^{2i+1} \mid 0^i \in T_1\}$.

Theorem 5.2 A is in IC[log,poly] if and only if there exist tally sets T_0 and T_1 such that $A \leq_c^p T_0$ and $A \leq_d^p T_1$.

Proof of Theorem 5.2:
For $A \in$ IC[log,poly], let c be a constant, t be a polynomial and Π be a set of programs as in Definition 5.1. We denote by $ord(p)$ the position of the string p in the lexicographical enumeration of Σ^*. We can encode Π into the tally set $T = \{0^{ord(p)} \mid p \in \Pi\}$. Let g be a truth-table condition generator that on input x computes the disjunction of the encodings $0^{ord(p)}$ of length less than or equal to $c\log(|x|) + c$ for all programs p that accept x in time $t(|x|)$. Then $A \leq_d^p T$ via g since x is in A if and only if there exists a program in $\Pi^{\leq c\log(|x|)+c}$ that accepts x in time $t(|x|)$.

Since IC[log,poly] is closed under complementation, this proves also the inclusion IC[log,poly] $\subseteq R_c^p(\text{TALLY})$.

For the reverse inclusion, let $A \leq_c^p C$ and $A \leq_d^p D$ where C and D are tally sets, and g_c and g_d are the respective polynomial-time truth-table condition generators. We assume that all generated queries are in 0^*. For every $0^i \in 0^*$ it is easy to construct a program p_i^c that on input x computes $g_c(x) = y_1 \wedge \ldots \wedge y_m$ and rejects if $0^i \in \{y_j \mid 1 \leq j \leq m\}$.

---

[8]For a fixed efficient universal machine. In fact, "p decides x in time $\cdots$" in this definition refers to the run of the fixed universal machine on $\langle p, x \rangle$. We refer the reader to [Orp90, p. 21] for details of the universal machine scheme.

Otherwise p_i^c goes into an infinite loop. It is clear that the running time of p_i^c is polynomially bounded on all inputs that it rejects, and that the size of p_i^c is $\mathcal{O}(\log(i))$. For every $x \notin A$ the conjunction $g_c(x)$ contains a query $0^j \notin C$. Thus for every $x \notin A$ there exists an index j, polynomially bounded in $|x|$, such that $0^j \notin C$ and p_j^c on input x rejects.

Similarly, for every $0^i \in 0^*$ there is a program p_i^d that on input x computes $g_d(x) = z_1 \vee \ldots \vee z_m$ and accepts if $0^i \in \{z_j \mid 1 \leq j \leq m\}$. Otherwise it goes into an infinite loop. The programs p_i^d are also $\mathcal{O}(\log(i))$ in size and have polynomial running time on all inputs that are accepted. For every $x \in A$ the disjunction $g_d(x)$ contains a query $0^j \in D$. Thus for every $x \in A$ there exists an index j, polynomially bounded in $|x|$, such that $0^j \in D$ and p_j^d on input x accepts.

Hence, taking $\Pi = \{p_j^d \mid 0^j \in D\} \cup \{p_j^c \mid 0^j \notin C\}$ as the set of programs, it follows that $A \in \text{IC}[\log,\text{poly}]$. ∎

Using the above characterization of IC[log,poly] it is easy to see that IC[log,poly] is closed under bounded truth-table reductions (a different proof that explicitly constructs programs appears in [KOSW90]):

Theorem 5.3 [KOSW90] IC[log,poly] is closed downward under $\leq_b^p$-reductions.

Proof of Theorem 5.3:
It is easy to see that $\text{coR}_d^p(\text{TALLY}) = \text{R}_c^p(\text{TALLY})$, which immediately implies the closure of $\text{R}_d^p(\text{TALLY}) \cap \text{R}_c^p(\text{TALLY})$ under complementation. In fact, for every set A in $\text{R}_d^p(\text{TALLY}) \cap \text{R}_c^p(\text{TALLY})$ there exists a single tally set T such that A and $\overline{A}$ are in $\text{R}_d^p(T) \cap \text{R}_c^p(T)$ and thus $\text{R}_d^p(\text{TALLY}) \cap \text{R}_c^p(\text{TALLY})$ is also closed under one truth-table reductions.

The second observation is that $\text{R}_{bc}^p(\text{R}_d^p(\text{TALLY})) \subseteq \text{R}_d^p(\text{R}_{bc}^p(\text{TALLY})) \subseteq \text{R}_d^p(\text{R}_m^p(\text{TALLY})) \subseteq \text{R}_d^p(\text{TALLY})$. Here we use that $\text{R}_{bc}^p(\text{TALLY}) = \text{R}_m^p(\text{TALLY})$ (see [Ko89]).

Similarly, $\text{R}_{bd}^p(\text{R}_c^p(\text{TALLY})) \subseteq \text{R}_c^p(\text{TALLY})$, and thus it follows that $\text{R}_d^p(\text{TALLY}) \cap \text{R}_c^p(\text{TALLY})$ is closed under bounded conjunctive and bounded disjunctive reducibilities. Combining this, we get

$$\begin{aligned}
\text{R}_b^p(\text{IC}[\log,\text{poly}]) &= \text{R}_b^p(\text{R}_d^p(\text{TALLY}) \cap \text{R}_c^p(\text{TALLY})) \\
&\subseteq \text{R}_{bc}^p(\text{R}_{bd}^p(\text{R}_{1-tt}^p(\text{R}_d^p(\text{TALLY}) \cap \text{R}_c^p(\text{TALLY})))) \\
&\subseteq \text{R}_d^p(\text{TALLY}) \cap \text{R}_c^p(\text{TALLY}) = \text{IC}[\log,\text{poly}]. \quad \blacksquare
\end{aligned}$$

Corollary 5.4 [KOSW90] If $P \neq NP$ and A is $\leq_b^p$-hard for NP, then $A \notin IC[\log, poly]$.

Proof of Corollary 5.4:
Suppose that a bounded truth-table hard set for NP is in IC[log,poly]. It follows by Theorem 5.3 that $NP \subseteq IC[\log, poly]$. From the result of [KOSW86] that if a set of low instance complexity is many-one hard for NP then P = NP, it follows that P = NP. ■

Using Theorem 5.2 and the fact that NP has neither $\leq_c^p$-hard tally sets nor $\leq_d^p$-hard tally sets unless P = NP, it follows that NP has neither $\leq_c^p$-hard sets nor $\leq_d^p$-hard sets in IC[log,poly] unless P = NP.

Corollary 5.5 If $P \neq NP$ and A is $\leq_d^p$-hard for NP, then $A \notin IC[\log, poly]$.

Proof of Corollary 5.5:
Suppose that a disjunctive truth-table hard set for NP is in IC[log,poly]. Since $R_d^p(IC[\log,poly]) \subseteq R_d^p(TALLY) \subseteq R_d^p(coSPARSE)$, it follows from Ukkonen's result [Ukk83] that P = NP. ■

Corollary 5.6 If $P \neq NP$ and A is $\leq_c^p$-hard for NP, then $A \notin IC[\log, poly]$.

Proof of Corollary 5.6:
Suppose that a conjunctive truth-table hard set for NP is in IC[log,poly]. Since $R_c^p(IC[\log,poly]) \subseteq R_c^p(SPARSE)$, it follows from Corollary 3.4 that P = NP. ■

Finally, using Theorem 3.13 and the fact that every tally set is in IC[log, poly], we can conclude that for truth-tables of size $\omega(\log n)$, no analog of Corollary 5.4 can be proven by any relativizable proof technique.

Acknowledgments

For helpful conversations, comments, suggestions, and literature pointers, we are grateful to E. Allender, L. Fortnow, W. Gasarch, A. Hoene, R. Rubinstein, R. Schuler, A. Selman, E. Spaan, J. Torán, S. Wahl, and O. Watanabe. We thank J. Balcázar, M. Hermo, and R. Schuler for discussions from which Theorem 5.2 resulted.

References

[ABG90] A. Amir, R. Beigel, and W. Gasarch. Some connections between bounded query classes and non-uniform complexity. In *Proceedings of the 5th Structure in Complexity Theory Conference*, pages 232–243. IEEE Computer Society Press, July 1990.

[AHH+92] V. Arvind, Y. Han, L. Hemachandra, J. Köbler, A. Lozano, M. Mundhenk, M. Ogiwara, U. Schöning, R. Silvestri, and T. Thierauf. Reductions to sets of low information content. Technical Report TR-417, University of Rochester, Department of Computer Science, Rochester, NY, 1992.

[AHH+92a] V. Arvind, Y. Han, L. Hemachandra, J. Köbler, A. Lozano, M. Mundhenk, M. Ogiwara, U. Schöning, R. Silvestri, and T. Thierauf. Reductions to sets of low information content. In *Proceedings of the 5th International Colloquium on Automata, Languages, and Programming*, pages 162–173. Springer-Verlag *Lecture Notes in Computer Science #623*, 1992.

[AHOW92] E. Allender, L. Hemachandra, M. Ogiwara, and O. Watanabe. Relating equivalence and reducibility to sparse sets. *SIAM Journal on Computing*, 21(3):521–539, 1992.

[AKM92] V. Arvind, J. Köbler, and M. Mundhenk. On bounded truth-table, conjunctive, and randomized reductions to sparse sets. In *Proceedings of the 12th Conference on Foundations of Software Technology and Theoretical Computer Science*, pages 140–151. Springer-Verlag *Lecture Notes in Computer Science #652*, 1992.

[Bal90] J. Balcázar. Self-reducibility. *Journal of Computer and System Sciences*, 41(3):367–388, 1990.

[Ber78] P. Berman. Relationship between density and deterministic complexity of NP-complete languages. In *Proceedings of the 5th International Colloquium on Automata, Languages, and*

Programming, pages 63–71. Springer-Verlag *Lecture Notes in Computer Science #62*, 1978.

[BGH90] R. Beigel, J. Gill, and U. Hertrampf. Counting classes: Thresholds, parity, mods, and fewness. In *Proceedings of the 7th Annual Symposium on Theoretical Aspects of Computer Science*, pages 49–57. Springer-Verlag *Lecture Notes in Computer Science #415*, February 1990.

[BH77] L. Berman and J. Hartmanis. On isomorphisms and density of NP and other complete sets. *SIAM Journal on Computing*, 6(2):305–322, 1977.

[BK88] R. Book and K. Ko. On sets truth-table reducible to sparse sets. *SIAM Journal on Computing*, 17(5):903–919, 1988.

[BLS92] H. Buhrman, L. Longpré, and E. Spaan. SPARSE reduces conjunctively to TALLY. Technical Report NU-CCS-92-08, Northeastern University College of Computer Science, Boston, MA, 1992. To appear in *Proceedings of the 8th Structure in Complexity Theory Conference*.

[BS90] R. Boppana and M. Sipser. The complexity of finite functions. In J. Van Leeuwen, editor, *Handbook of Theoretical Computer Science*, chapter 14, pages 757–804. MIT Press/Elsevier, 1990.

[CGH+88] J. Cai, T. Gundermann, J. Hartmanis, L. Hemachandra, V. Sewelson, K. Wagner, and G. Wechsung. The boolean hierarchy I: Structural properties. *SIAM Journal on Computing*, 17(6):1232–1252, 1988.

[CGH+89] J. Cai, T. Gundermann, J. Hartmanis, L. Hemachandra, V. Sewelson, K. Wagner, and G. Wechsung. The boolean hierarchy II: Applications. *SIAM Journal on Computing*, 18(1):95–111, 1989.

[CH90] J. Cai and L. Hemachandra. On the power of parity polynomial time. *Mathematical Systems Theory*, 23(2):95–106, 1990.

[CH91] J. Cai and L. Hemachandra. A note on enumerative counting. *Information Processing Letters*, 38(4):215–219, 1991.

[ER60] P. Erdős and R. Rado. Intersection theorems for systems of sets. *Journal of the London Mathematical Society*, 35:85–90, 1960.

[For79] S. Fortune. A note on sparse complete sets. *SIAM Journal on Computing*, 8(3):431–433, 1979.

[Gav] R. Gavaldà. On conjunctive and disjunctive reductions to sparse sets. Manuscript, January 1992.

[GW] R. Gavaldà and O. Watanabe. On the computational complexity of small descriptions. *SIAM Journal on Computing*. To appear. Preliminary versions appear as [GW91] and [Gav].

[GW91] R. Gavaldà and O. Watanabe. On the computational complexity of small descriptions. In *Proceedings of the 6th Structure in Complexity Theory Conference*, pages 89–101. IEEE Computer Society Press, June/July 1991.

[Har83] J. Hartmanis. Generalized Kolmogorov complexity and the structure of feasible computations. In *Proceedings of the 24th IEEE Symposium on Foundations of Computer Science*, pages 439–445. IEEE Computer Society Press, 1983.

[Hau14] F. Hausdorff. *Grundzüge der Mengenlehre*. Leipzig, 1914.

[Hel86] H. Heller. On relativized exponential and probabilistic complexity classes. *Information and Control*, 71:231–243, 1986.

[HH91] L. Hemachandra and A. Hoene. On sets with efficient implicit membership tests. *SIAM Journal on Computing*, 20(6):1148–1156, 1991.

[HIS85] J. Hartmanis, N. Immerman, and V. Sewelson. Sparse sets in NP-P: EXPTIME versus NEXPTIME. *Information and Control*, 65(2/3):159–181, 1985.

[HL91] S. Homer and L. Longpré. On reductions of NP sets to sparse sets. In *Proceedings of the 6th Structure in Complexity Theory Conference*, pages 79–88. IEEE Computer Society Press, June/July 1991.

[HOW92] L. Hemachandra, M. Ogiwara, and O. Watanabe. How hard are sparse sets? In *Proceedings of the 7th Structure in Complexity Theory Conference*, pages 222–238. IEEE Computer Society Press, June 1992.

[HY84] J. Hartmanis and Y. Yesha. Computation times of NP sets of different densities. *Theoretical Computer Science*, 34:17–32, 1984.

[IM89] N. Immerman and S. Mahaney. Relativizing relativized computations. *Theoretical Computer Science*, 68:267–276, 1989.

[JY85] D. Joseph and P. Young. Some remarks on witness functions for non-polynomial and non-complete sets in NP. *Theoretical Computer Science*, 39:225–237, 1985.

[JY90] D. Joseph and P. Young. Self-reducibility: Effects of internal structure on computational complexity. In A. Selman, editor, *Complexity Theory Retrospective*, pages 82–107. Springer-Verlag, 1990.

[Kad89] J. Kadin. $P^{NP[\log n]}$ and sparse Turing-complete sets for NP. *Journal of Computer and System Sciences*, 39(3):282–298, 1989.

[KL80] R. Karp and R. Lipton. Some connections between nonuniform and uniform complexity classes. In *Proceedings of the 12th ACM Symposium on Theory of Computing*, pages 302–309, April 1980.

[KMR89] S. Kurtz, S. Mahaney, and J. Royer. The isomorphism conjecture fails relative to a random oracle. In *Proceedings of the 21st ACM Symposium on Theory of Computing*, pages 157–166. ACM Press, May 1989.

[Ko89] K. Ko. Distinguishing conjunctive and disjunctive reducibilities by sparse sets. *Information and Computation*, 81(1):62–87, 1989.

[KOSW86] K. Ko, P. Orponen, U. Schöning, and O. Watanabe. What is a hard instance of a computational problem. In *Proceedings of the 1st Structure in Complexity Theory*

Conference, pages 197–217. Springer-Verlag *Lecture Notes in Computer Science #223*, June 1986.

[KOSW90] K. Ko, P. Orponen, U. Schöning, and O. Watanabe. Instance complexity. Technical Report A-1990-6, University of Helsinki Department of Computer Science, Helsinki, Finland, September 1990. To appear in *Journal of the ACM*.

[KSW87] J. Köbler, U. Schöning, and K. Wagner. The difference and truth-table hierarchies of NP. *R.A.I.R.O. Informatique théorique et Applications*, 21(4):419–435, 1987.

[LLS75] R. Ladner, N. Lynch, and A. Selman. A comparison of polynomial time reducibilities. *Theoretical Computer Science*, 1(2):103–124, 1975.

[Mah82] S. Mahaney. Sparse complete sets for NP: Solution of a conjecture of Berman and Hartmanis. *Journal of Computer and System Sciences*, 25(2):130–143, 1982.

[Mah86] S. Mahaney. Sparse sets and reducibilities. In R. Book, editor, *Studies in Complexity Theory*, pages 63–118. John Wiley and Sons, 1986.

[Mah89] S. Mahaney. The isomorphism conjecture and sparse sets. In J. Hartmanis, editor, *Computational Complexity Theory*, pages 18–46. American Mathematical Society, 1989. Proceedings of Symposia in Applied Mathematics #38.

[OL] M. Ogiwara and A. Lozano. On one-query self-reducible sets. *Theoretical Computer Science*. To appear. Preliminary version appears as [OL91].

[OL91] M. Ogiwara and A. Lozano. On one-query self-reducible sets. In *Proceedings of the 6th Structure in Complexity Theory Conference*, pages 139–151. IEEE Computer Society Press, June/July 1991.

[Orp90] P. Orponen. On the instance complexity of NP-hard problems. In *Proceedings of the 5th Structure in Complexity Theory Conference*, pages 20–27. IEEE Computer Society Press, July 1990.

[OW91] M. Ogiwara and O. Watanabe. On polynomial-time bounded truth-table reducibility of NP sets to sparse sets. *SIAM Journal on Computing*, 20(3):471–483, 1991.

[RR92] D. Ranjan and P. Rohatgi. On randomized reductions to sparse sets. In *Proceedings of the 7th Structure in Complexity Theory Conference*, pages 239–242. IEEE Computer Society Press, June 1992.

[Rub90] R. Rubinstein. Relativizations of the P-printable sets and the sets with small generalized Kolmogorov complexity. Technical Report WPI-CS-TR-90-3, Worcester Polytechnic Institute, Worcester, MA, March 1990.

[Sel88] A. Selman. Promise problems complete for complexity classes. *Information and Computation*, 78:87–98, 1988.

[Sel90] A. Selman. A note on adaptive vs. nonadaptive reductions to NP. Technical Report 90-20, State University of New York at Buffalo Department of Computer Science, Buffalo, NY, September 1990.

[Ukk83] E. Ukkonen. Two results on polynomial time truth-table reductions to sparse sets. *SIAM Journal on Computing*, 12(3):580–587, 1983.

[Wag90] K. Wagner. Bounded query classes. *SIAM Journal on Computing*, 19(5):833–846, 1990.

[Wat88] O. Watanabe. On $\leq^p_{1-tt}$ sparseness and nondeterministic complexity classes. In *Proceedings of the 15th International Colloquium on Automata, Languages, and Programming*, pages 697–709. Springer-Verlag *Lecture Notes in Computer Science #317*, July 1988.

[Wat91] O. Watanabe. On intractability of the class UP. *Mathematical Systems Theory*, 24:1–10, 1991.

[Wec85] G. Wechsung. On the boolean closure of NP. In *Proceedings of the 5th Conference on Fundamentals of Computation Theory*, pages 485–493. Springer-Verlag *Lecture Notes in Computer Science #199*, 1985. (An unpublished precursor of this paper was coauthored by K. Wagner).

[Yap83] C. Yap. Some consequences of non-uniform conditions on uniform classes. *Theoretical Computer Science*, 26:287–300, 1983.

[Yes83] Y. Yesha. On certain polynomial-time truth-table reducibilities of complete sets to sparse sets. *SIAM Journal on Computing*, 12(3):411–425, 1983.

On Average P vs. Average NP*

Jie Wang †   Jay Belanger
Wilkes University ‡

Abstract

This paper studies average-case complexity classes. Structures of polynomial-time computable distributions and many-one reductions on randomized decision problems are investigated. We widen the scope from the most studied class DNP (Distributional-NP) to the class ANP (Average-NP) which consists of randomized decision problems accepted by nondeterministic Turing machines in average polynomial time. Our results include: 1) P $\neq$ NP if and only if there exists a randomized decision problem $(D, \mu) \in$ ANP$-$AP (Average-P) such that it is hard on positive instances and D is almost in NP with respect to μ. 2) All polynomial-time many-one complete problems for DNP are average polynomial-time many-one complete for ANP with respect to polynomial-time computable distributions. We also prove that there is a randomized decision problem which is average polynomial-time many-one complete for ANP with respect to polynomial-time computable distributions but not contained in DNP. 3) AP and ANP with polynomial-time computable distributions, and DNP are not closed under polynomial-time many-one reductions. So these classes are not closed under any weaker reductions. 4) AP and ANP with respect to arbitrary distributions do not have complete problems under polynomial-time one-one reductions.

---

*The preliminary version of this paper has been presented in the 7th IEEE Conference on Structure in Complexity Theory, June 1992, Boston.
†Supported in part by NSF grant CCR-9108899.
‡Mathematics and Computer Science Dept., Wilkes University, Wilkes-Barre, PA 18766, U.S.A. Internet: jwang@wilkes1.wilkes.edu and belanger@wilkes1.wilkes.edu.

1 Introduction

NP-completeness is a worst-case concept. Although they are most likely difficult in general, many NP-complete problems have been shown to be easily solvable on average, i.e. when the instances are chosen at random. For example, the Hamiltonian Circuit Problem has been shown to have a polynomial on average time algorithm [GS87] and the Graph 3-Colorability Problem has been shown to have a constant on average time algorithm [Wil84], where the instances in each case are taken randomly according to a natural distribution. In both cases, the hard instances are extremely rare, and add very little to the expected running time of the algorithms. Thus, it seems natural to study the difficulty of solving problems when the instances are chosen at random. This motivated the recent study of average-case complexity, for example, see [Lev84, BCGL89, Gur91, and Ven91]. To state a theorem about behavior "on average," one needs to be able to describe mathematically the probability distribution on instances over which the average is taken. So an average-case complexity class consists of pairs of a decision problem [1] and a probability distribution on problem instances. Such a pair is called a randomized decision problem. A randomized decision problem (D, μ) is considered easy if it is solvable in average polynomial-time with respect to the probability distribution μ. AP (Average-P) is the class of all easy randomized decision problems.

Most of the previous work deals with the class DNP (Distributional-NP) of pairs of an NP problem and a probability distribution which has a polynomial-time computable distribution function. One major open problem in average-case complexity theory is whether DNP is included in AP.

As in the worse-case complexity theory, reductions are an important concept in studying randomized decision problems. Reductions between randomized decision problems are defined in a way guaranteeing that they are closed for AP, i.e., if (D_1, μ_1) is reducible to (D_2, μ_2), and $(D_2, \mu_2) \in$ AP, then $(D_1, \mu_1) \in$ AP. Polynomial-time many-one re-

---

[1] We may consider randomized search problems as well. Ben-David, Chor, Goldreich, and Luby [BCGL89] proved that randomized search problems are equivalent to randomized decision problems with respect to randomized polynomial-time reductions. For more information about randomized reductions of search problems, we refer the reader to Blass and Gurevich [BG91].

ductions on randomized decision problems were first defined in [Lev84] (see also [Gur91]) and it was proved that the randomized tiling problem [Lev84], the randomized halting problem [Gur91], and the randomized Post Correspondence Problem [Gur91] are polynomial-time many-one (symbolically, $\leq_m^p$) complete for DNP. Venkatesan and Levin [VL88] proved that a randomized graph-coloring problem is complete for DNP with respect to coin-flipping polynomial-time reductions but it was shown in [Gur91] that it can not be $\leq_m^p$-complete for DNP unless $\text{DTIME}(2^{\text{poly}}) = \text{NTIME}(2^{\text{poly}})$. Recently, some other problems have been shown to be complete for DNP with respect to coin-flipping polynomial-time reductions [Gur90, VR91]. However, unlike in the worst-case complexity, only a small number of complete randomized decision problems are known so far. Is it by nature that proving a randomized decision problem to be hard on average is difficult? To answer this question, we investigate the structures of polynomial-time computable distribution functions and the polynomial-time many-one reductions on randomized decision problems. Our results indicate that the structures of average-case complexity classes are very different from their counterparts in worst-case complexity due to the complex structures of probability distributions and distribution functions. These results are based on the construction of certain randomized decision problems. Whether the structure of "natural" randomized decision problems is decidedly different from the structure of "natural" decision problems is still open.

We first consider the relationship between average-case complexity classes and the worst-case complexity classes. It is easy to see that P = NP implies that DNP is included in AP. Ben-David et. al. [BCGL89] proved that if $\text{DTIME}(2^{O(n)}) \neq \text{NTIME}(2^{O(n)})$, then DNP is not included in AP. Does P $\neq$ NP imply that DNP is not included in AP? We observe that unlike worst-case complexity where NP includes P, DNP does not include AP even when distribution functions are polynomial-time computable. This makes it harder to study the relationship between DNP and AP. To resolve this problem, we consider a more general average-case complexity class. Along the definition of AP, it is both logical and natural to define a new class ANP (Average-NP) to be the class of randomized decision problems (D, μ) such that D is acceptable by a (nondeterministic) Turing machine in polynomial time on μ-average. This has also been noted in [BCGL89] and [Gur91].

ANP is a natural analog of NP and ANP includes AP. We prove that the average-case complexity classes do indeed have a very strong connection with the worst-case complexity classes. In particular, we prove that $P \neq NP$ if and only if there is a randomized decision problem $(D, \mu) \in ANP - AP$ such that D is almost in NP with respect to μ and (D, μ) is hard on positive instances.

In the definition of AP, there is no requirement on the computability of the probability distribution μ. Hence, μ does not even have to be computable. Levin hypothesizes [Jo84] that any natural probability distribution either has a polynomial-time computable distribution function, or else is dominated by a function that does. So we consider a natural subclass AP_P of AP to be the class of randomized decision problems (D, μ) in AP such that μ has a polynomial-time computable distribution. Similarly, we can define the natural subclass ANP_P of ANP. Ben-David et. al. [BCGL89] argued that requiring probability distributions to have polynomial-time computable distributions may seem too restricting. They presented a wider family of distributions, P-samplable, which consists of distributions that can be sampled by probabilistic algorithms working in time polynomial in the length of the sample generated. However, they also noted that the distributions in P-samplable seem to be too complicated. Impagliazzo and Levin [IL90] recently proved that every DNP problem complete for polynomial-time computable distributions is also complete for all P-samplable distributions. Therefore, using P-samplable distributions does not generate harder random instances than using polynomial-time computable distributions. So in this paper we will focus on polynomial-time computable distributions. We will focus on many-one reductions as well. We show that ANP_P properly includes DNP. We prove that every $\leq_m^p$-complete randomized decision problem for DNP is average polynomial-time complete (symbolically, $\leq_m^{ap}$-complete) for ANP_P. They are also $\leq_m^{ap}$-complete for the class of randomized decision problems in ANP whose probability distributions are dominated by functions that have polynomial-time computable distributions. Moreover, we prove that there is a randomized decision problem which is $\leq_m^{ap}$-complete for ANP_P but not contained in DNP. So this problem is not complete for DNP under any reductions. All these known many-one complete problems are in fact one-one complete.

We use the polynomial-time computability of distribution functions in proving our completeness theorems. However, one could lose some nice properties by restricting to such simple distributions. For example, we prove that AP$_P$, DNP, and ANP$_P$ are not closed under $\leq_m^p$-reductions. So these classes are not closed under any reductions weaker than $\leq_m^p$-reductions. On the other hand, if there is no restriction at all on the computability of distribution functions, we can prove that neither AP nor ANP contains hardest randomized decision problems with respect to polynomial-time one-one reductions.

2 Preliminaries

In the worst-case time complexity, the size of inputs is the length. In the average-case time complexity, one may allow an algorithm to run longer on "rare" inputs. So one uses $|x|r(x)$ rather than $|x|$ as the size of instance x, where $r(x)$ is a measure of rareness which satisfies a randomness test, i.e., its expectation $E_x r(x) = O(1)$. This is Levin's definition of a randomness test in Martin Löf's sense. A running time $t(x)$ is polynomial on average if $t(x) = (|x|r(x))^k$ for some $k > 0$ and some r as above. Let $\epsilon = 1/k$, then $E_x t^\epsilon(x)/|x| = E_x r(x) = O(1)$. This is equivalent to the following definition.

Let μ be a probability distribution [2] on Σ^*, i.e., $\mu(x) \geq 0$ for any $x \in \Sigma^*$ and $\sum_{x \in \Sigma^*} \mu(x) = 1$. In general, we only require that $\sum_{x \in \Sigma^*} \mu(x)$ converges. We also require that $\mu(x) > 0$ for infinitely many x to avoid trivial results. The distribution function (or distribution, in short) of a probability distribution μ is defined by $\mu^*(x) = \sum_{y \leq x} \mu(y)$, where $\leq$ is the standard lexicographical order on Σ^*. So $\mu(x) = \mu^*(x) - \mu^*(x-1)$.

Definition 1 [Lev84] Let μ be a probability distribution. A function f is polynomial on μ-average if there is an $\epsilon > 0$ such that

$$\sum_{|x|>0} \frac{f^\epsilon(x)}{|x|} \cdot \mu(x) = O(1).$$

For more information and discussion about this definition, the reader is referred to Gurevich [Gur91].

---

[2] Probability distribution is also called "probability function" in [Gur91] and "density function" in [BCGL89].

Denote by AP the class of randomized decision problems (D,μ) such that there is a deterministic Turing machine M accepting D in time T which is polynomial on μ-average. We assume that T is time constructible. Denote by PDF the class of probability distributions μ such that its distribution function μ^* is computable in polynomial time. Let DNP denote the class of all randomized decision problems (D,μ) such that $D \in$ NP and $\mu \in$ PDF; ANP the class of all (D,μ) such that D is accepted by a Turing machine in time T which is polynomial on μ-average. So AP is a subset of ANP. Let AP_P denote the class of $(D,\mu) \in$ AP such that $\mu \in$ PDF; and ANP_P the class of $(D,\mu) \in$ ANP such that $\mu \in$ PDF.

The following material is abstracted from Gurevich [Gur91].

A function f from Σ^* to the interval $[0,1]$ of reals is computable in polynomial time (cf. [Ko83]) if there exists an algorithm $A(x, 1^k)$ which runs in time polynomial in $|x|$ and k and such that, for every Σ-string x and every positive integer k, $A(x, 1^k)$ is a binary fraction and $|f(x) - A(x, 1^k)| < 2^{-k}$.

Notice that the polynomial-time computability of a function f from Σ^* to $[0,1]$ in the above definition does not guarantee the computability of the kth digit of $f(x)$ [Gur91]. It is easy to see that a probability distribution μ is computable in polynomial time if the corresponding distribution μ^* is polynomial-time computable. But the converse is not true unless P = NP (see [Gur91]).

We now define reductions from one randomized decision problem to another. Intuitively, such a reduction should be efficiently computable, and "preserve" the probability distribution. In other words, the reduction should not transform very likely instances of the first problem to rare instances of the second problem.

Let μ_1 and μ_2 be probability distributions on strings in the same alphabet Σ. μ_2 dominates (resp. weakly dominates) μ_1 if there is a function f from Σ^* to non-negative reals such that $\mu_1(x) \leq f(x)\mu_2(x)$ and f is polynomially bounded (resp. polynomial on μ_1-average). A function f transforms a probability distribution μ_1 into a probability distribution μ_2 if $\mu_2(y) = \sum_{x \in f^{-1}(y)} \mu_1(x)$. A function f transforms (D_1, μ_1) to (D_2, μ_2) if it transforms μ_1 to μ_2 and $D_1|\{x : \mu_1(x) > 0\}$ is many-one reducible to D_2 via function f.

Let $\mu_1 \preceq \mu_2$ (resp. $\mu_1 \preceq_w \mu_2$) denote that μ_1 is dominated (resp. weakly dominated) by μ_2. μ_2 *dominates* (resp. *weakly dominates*) μ_1

with respect to a function f, symbolically $\mu_1 \preceq^f \mu_2$ (resp. $\mu_1 \preceq^f_w \mu_2$), if there exists some $\nu \succeq \mu_1$ (resp. $\nu \succeq_w \mu_1$) such that f transforms ν into a restriction of μ_2.

Definition 2 [Gur91]

1. (D_1, μ_1) is polynomial-time many-one reducible (symbolically, $\leq^p_m$-reducible) to (D_2, μ_2) if there is a total polynomial-time computable function f such that $D_1|\{x : \mu_1(x) > 0\} \leq^p_m D_2$ via f and $\mu_1 \preceq^f \mu_2$.

2. (D_1, μ_1) is average polynomial-time many-one reducible (symbolically, $\leq^{ap}_m$-reducible) to (D_2, μ_2) if there is a total function f which is polynomial-time computable on μ_1-average such that $D_1|\{x : \mu_1(x) > 0\} \leq^p_m D_2$ via f and $\mu_1 \preceq^f_w \mu_2$.

In both of these reductions, if the function f is one-one, then they are called one-one reductions and they are denoted symbolically by $\leq^p_1$ and $\leq^{ap}_1$ respectively.

A class $\mathcal{C}$ is closed under a reduction $\leq_r$ if $x \leq_r y$ and $y \in \mathcal{C}$ implies $x \in \mathcal{C}$. Here $\mathcal{C}$ could be a class of sets or a class of randomized decision problems. It was proved in [Gur91] that both $\leq^p_m$ and $\leq^{ap}_m$ reductions for randomized decision problems are transitive and they are closed for AP. Clearly, if (D, μ) is $\leq^p_m$-complete for DNP, then D must be $\leq^p_m$-complete for NP.

One can easily prove that every randomized decision problem in AP (resp. DNP, ANP, AP$_P$, ANP$_P$) is polynomial-time transformable to some randomized decision problem in AP (resp. DNP, ANP, AP$_P$, ANP$_P$) over the binary alphabet [Gur91]. So without loss of generality, we use the binary alphabet in this paper.

Let $M_1, M_2, \ldots$ be a fixed standard enumeration of all (deterministic and nondeterministic) Turing machines in which i is a program which simply codes up the states, symbols, tuples, etc. of the ith Turing machine M_i.

3 Average-Case and Worst-Case Complexity

We now study the relationship between average-case complexity classes and the worst-case complexity classes. We first study the inclusion relations between average-case complexity classes. It is easy to see that AP is not included in DNP because there is no restriction on the computability of probability distributions in AP. However, even when probability distributions have polynomial-time computable distributions, AP_P is still not included in DNP. So DNP does not include all easy problems.

Theorem 1 AP_P *is not included in* DNP.

Proof. Let A be a set not in NP which is accepted by a deterministic Turing machine M. Let $D = 1A \cup \{0\}^*$. Then $D \notin$ NP. Let $\mu(x) = 1/n^2$ if $x = 0^n$, 0 otherwise. Clearly, μ is a probability distribution and μ^* is polynomial-time computable. We construct a deterministic Turing machine M' as follows: on any input x, if $x = 0^n$, then M' accepts; if $x = 1y$, then M' simulates M on input y; otherwise, M' rejects. Notice that $\mu(x) = 0$ when $x \neq 0^n$, so clearly, M' accepts D in polynomial time on μ-average. ∎

Remark 1 The probability distribution μ constructed above only generates strings 0^n. What is more interesting is to have a probability distribution μ such that it is positive everywhere. We can show that if $DTIME(2^{poly}) \neq$ NP, then there is a set D and a probability distribution μ such that $\mu(x) > 0$ for any x and $(D, \mu) \in ANP_P -$ DNP as follows. Let $D \in DTIME(2^{poly}) -$ NP, then there is a deterministic Turing machine accepting D in time $2^{p(n)}$ for some polynomial p. Let $\mu(x) = 2^{-p(|x|)-1}$, then $(D, \mu) \in AP_P -$ DNP. ∎

By definition, DNP is included in ANP_P. We can show that DNP is properly included in ANP_P.

Corollary 2 DNP *is properly included in* ANP_P.

Proof. That DNP is a subset of ANP_P follows directly from the definition. Now from the proof of Theorem 1, we know that there is a pair

(D, μ) in $\text{AP}_\text{P} - \text{DNP}$. Since $\text{AP}_\text{P} \subseteq \text{ANP}_\text{P}$, this completes the proof. ∎

Similar to the proof of Corollary 2, it is easy to prove that there are problems $(D, \mu) \in \text{ANP}$ such that $D \notin \text{NP}$. Given a set D and a probability distribution μ, we define a notion of D being almost in NP with respect to μ to indicate that hard instances of D beyond NP under the probability distribution μ are rare.

Definition 3 Let D be a set and μ be a probability distribution. D is *almost in NP with respect to μ* if there is a Turing machine that accepts D in time T, and D has a subset D' in NP such that

$$\sum_{x \in D - D'} \frac{2^{\delta T(x)}}{|x|} \mu(x) = O(1)$$

for some constant $\delta > 0$.

Definition 4 A randomized decision problem (D, μ) is *hard on positive instances* if for any deterministic Turing machine that accepts D in time $T(x)$ and for any $\epsilon > 0$

$$\sum_{x \in D} \frac{T^\epsilon(x)}{|x|} \mu(x) = \infty.$$

Notice that in general a randomized decision problem is hard (i.e. not in AP) does not imply that it must be hard on positive instances.

Clearly, P = NP implies that DNP is included in AP. Ben-David et. al. [BCGL89] proved that if $\text{DTIME}(2^{O(n)}) \neq \text{NTIME}(2^{O(n)})$, then DNP is not included in AP. Does P $\neq$ NP imply that DNP is not included in AP? Or in a more general setting, do average-case complexity classes have a strong connection with the worst-case complexity classes? We affirmatively answer this question. In particular, we prove that P $\neq$ NP if and only if there is a randomized problem (D, μ) in ANP $-$ AP such that D is almost in NP with respect to μ and (D, μ) is hard on positive instances. The proof uses the concept of polynomial complexity cores from worst-case complexity theory.

Let A be a set not in P. A polynomial complexity core of A is a set C such that for every machine M that accepts A and every polynomial p there are at most finitely many $x \in C$ on which the number of steps of M is bounded by $p(|x|)$. A polynomial complexity core C is proper if $C \subseteq A$. Du [Du85] (see also [DB89]) proved that if P $\neq$ NP, then many natural NP-complete problems contain nonsparse proper polynomial complexity cores in DTIME($2^{O(n)}$).

Theorem 3 P $\neq$ NP *if and only if there is a pair* $(D, \mu) \in$ ANP $-$ AP *such that D is almost in* NP *w.r.t.* μ *and* (D, μ) *is hard on positive instances.*

Proof. ($\Rightarrow$) Suppose P $\neq$ NP. Then there is an NP-complete set D in NP $-$ P such that D has a nonsparse proper polynomial complexity core C in DTIME($2^{O(n)}$) [Du85, DB89]. Therefore, $C \subseteq D$ and for every machine M that accepts D and every polynomial p there are at most finitely many $x \in C$ such that the number of steps of M on x is bounded by $p(|x|)$. Let $C^{=n} = \{x : |x| = n \text{ and } x \in C\}$. Define

$$\mu(x) = \begin{cases} \frac{1}{|x|^2 |C^{=|x|}|}, & \text{if } x \in C, \\ \frac{1}{|x|^2 2^{|x|}}, & \text{otherwise.} \end{cases}$$

Clearly, μ is a probability distribution and is DTIME($2^{O(n)}$) computable. Since C is a proper complexity core of D, for any deterministic Turing machine that accepts D in time T and any $\epsilon > 0$, there are at most finitely many $x \in C$ such that $T^\epsilon(x) \leq |x|^3$. Therefore, there is a constant $m > 0$, such that when $|x| \geq m$, $T^\epsilon(x) > |x|^3$ if $x \in C$. Hence,

$$\sum_{x \in D} \frac{T^\epsilon(x)}{|x|} \cdot \mu(x) \geq \sum_{x \in C, |x| \geq m} \frac{T^\epsilon(x)}{|x|^3 |C^{=|x|}|} \geq \sum_{i \geq m} 1 = \infty.$$

So (D, μ) is hard on positive instances. It is easy to see that D is almost in NP with respect to μ because $D \in$ NP.

($\Leftarrow$) Let (D, μ) be a pair in ANP $-$ AP such that D is almost in NP with respect to μ and (D, μ) is hard on positive instances. By definition, there is a Turing machine M that accepts D in polynomial time $T(x)$ on μ-average, and D has a subset $D' \in$ NP such that

$$\sum_{x \in D - D'} \frac{2^{\delta T(x)}}{|x|} \mu(x) < \infty$$

for some constant $\delta > 0$, and for any deterministic Turing machine that accepts D in time $t(x)$ and any $\epsilon > 0$,

$$\sum_{x \in D} \frac{t^\epsilon(x)}{|x|} \mu(x) = \infty.$$

We shall prove that $D' \notin$ P. Suppose that $D' \in$ P, then there is a deterministic Turing machine M' which accepts D' in polynomial time $|x|^k + k$ for some $k > 0$. Now construct a deterministic machine $\hat{M}$ as follows. On any input x, $\hat{M}$ first simulates M' on x. $\hat{M}$ accepts x if M' accepts x. If M' rejects x, then $\hat{M}$ deterministically simulates M on x. Clearly $\hat{M}$ accepts D within time $\hat{T}(x)$ which is equal to $|x|^k + k$ when $x \in D'$ and $2^{cT(x)}$ when $x \notin D'$ for some constant $c > 0$. Let $l \geq \max\{k, c\}$ such that $c/l \leq \delta$. Then

$$\sum_{x \in D} \frac{\hat{T}^{1/l}(x)}{|x|} \mu(x) \leq \sum_{x \in D'} O(\mu(x)) + \sum_{x \in D-D'} \frac{2^{(c/l)T(x)}}{|x|} \mu(x) < \infty.$$

This contradicts the assumption. ∎

4 Complete Problems for ANP$_P$

We know that the randomized tiling problem [Lev84], the randomized halting problem [Gur91], and the randomized Post Correspondence Problem [Gur91] are $\leq_m^p$-complete for DNP. We will show in this section that every $\leq_m^p$-complete problem for DNP is also $\leq_m^{ap}$-complete for ANP$_P$. We first show that the randomized halting problem is $\leq_m^{ap}$-complete for ANP$_P$.

Let $K = \{(i, x, 1^n) : M_i \text{ accepts } x \text{ within } n \text{ steps}\}$, and $\mu_K(i, x, 1^n)$ be a standard probability distribution defined by

$$\mu_K(i, x, 1^n) = \frac{2^{-|i|+1}}{|i|^2} \frac{2^{-|x|}}{|x|^2} \frac{1}{n^2}.$$

We define a version of the randomized halting problem RH by RH $= (K, \mu_K)$.

Theorem 4 RH is $\leq_m^{ap}$-complete for ANP$_P$.

Proof. The proof is a modification of [Gur91]. A simpler proof is presented here.

Let $(D, \mu) \in \text{ANP}_P$. Then μ has a polynomial-time computable distribution and there is a Turing machine M accepting D in time T which is polynomial on μ-average. It was proved in [Gur91] (Lemma 1.6) that for every probability distribution $\mu \in \text{PDF}$ there is a probability distribution $\mu_1 \in \text{PDF}$ such that every value of μ_1 has at most $4 + 2|x|$ binary digits and $\mu(x) < 4\mu_1(x)$. So $\mu \preceq \mu_1$. Therefore, without loss of generality we may assume that μ is such μ_1 for the purpose of constructing a reduction.

Let x' be the shortest binary string such that $\mu^*(x-1) < 0.x'1 \leq \mu^*(x)$. Then $0.x'1 - 2^{-|x'1|} \leq \mu^*(x-1) < \mu^*(x) < 0.x'1 + 2^{-|x'1|}$ because otherwise x' is not the shortest. Therefore, it is easy to see that $\mu(x) = \mu^*(x) - \mu^*(x-1) < 2^{-|x'1|+1}$. So $2^{-|x'|} > \mu(x)$. Given x, x' can be found in polynomial time of $|x|$ because μ^* is polynomial-time computable and $\mu(x)$ has at most $4 + 2|x|$ binary digits. So $|x'|$ is bounded by a polynomial of $|x|$.

Now we define the desired $\leq_m^{ap}$-reduction from (D, μ) to RH. Define a Turing machine M' as follows. On input w, find x (this x is unique) such that $\mu^*(x-1) < 0.w1 \leq \mu^*(x)$. This can be carried out in polynomial time of $|x|$ as follows: Find the smallest n such that $\mu^*(0^n) \geq 0.w1$, then binary search all strings of length $n-1$ to find such a unique x. M' then computes x' from x, simulates M on x if $w = x'$ and rejects otherwise.

Clearly M' accepts w if and only if $w = x'$ and M accepts x. Because of the time bound on M, it is easy to see that M' on input x' is bounded in time $g(x')$ such that $g(x')$ is polynomial of x on μ-average and g is time-constructible. Let i be a program such that $M' = M_i$. Let

$$f(x) = (i, x', 1^{g(x')}).$$

Clearly, f is computable in time polynomial on μ-average and $x \in D$ if and only if $f(x) \in K$ by the construction. It is easy to see that f is one-one. It is also easy to see that for a one-one function f, $\mu \preceq_w^f \nu$ if and only if $\mu(x) \preceq_w \nu(f(x))$ [Gur91]. We know that $\mu_K(f(x))$ is proportional to $g(x')^{-2}|x'|^{-2}2^{-|x'|}$ which exceeds $g(x')^{-2}|x'|^{-2}\mu(x)$. So $\mu \preceq_w^f \mu_K$. Hence, $(D, \mu) \leq_m^{ap}$ RH via f. ∎

Corollary 5 *Every $\leq_m^p$-complete problem for DNP is $\leq_m^{ap}$-complete for* ANP_P.

Proof. Similar to the proof of Theorem 4, we can prove that RH is $\leq_m^p$-complete for DNP since $K \in \text{NP}$ (see also [Gur91]). Thus, every $\leq_m^p$-complete problem for DNP is $\leq_m^{ap}$-complete for ANP_P. ∎

It would be interesting to know whether there are $\leq_m^p$-complete randomized decision problems for ANP_P.

Remark 2 Theorem 4 and Corollary 5 also hold for the class of randomized decision problems (D, μ) in ANP such that μ is dominated by a probability distribution which has polynomial-time computable distribution. ∎

We can similarly prove that AP_P has $\leq_m^{ap}$-complete problems. Let $K_d = \{(i, x, 1^n) : M_i \text{ is deterministic and accepts } x \text{ within } n \text{ steps}\}$, and μ_K be as above. Then (K_d, μ_K) is $\leq_m^{ap}$-complete for AP_P.

We know that K defined above is in NP. So (K, μ_K) is in DNP. Does there exist a randomized decision problem in $\text{ANP}_P -$ DNP which is complete for ANP_P? We affirmatively answer this question.

Let $K' = \{(i, x, n) : M_i \text{ accepts } x \text{ within } n \text{ steps}\}$, and

$$\mu_{K'}(i, x, n) = \frac{2^{-|i|+1}}{|i|^2} \frac{2^{-|x|}}{|x|^2} \frac{1}{n^3}.$$

Clearly, $\mu_{K'} \in \text{PDF}$. It is straightforward to prove that K' is $\leq_m^p$-complete for NEXP (NEXP = NTIME(2^{poly})): Let A be a set in NEXP, then there is a Turing machine M_i accepting A in time $2^{p(n)}$ for some polynomial p. Let $f(x) = (i, x, 2^{p(|x|)})$. Then $f(x)$ is polynomial-time computable. (Note that $2^{p(|x|)}$ can be written as 1 followed by $p(|x|)$ many 0's in the binary system.) Clearly, $x \in A$ if and only if $f(x) \in K'$.

Theorem 6 $(K', \mu_{K'}) \in \text{ANP}_P -$ DNP *and is $\leq_m^{ap}$-complete for* ANP_P.

Proof. It has been proved in Seiferas, Fischer, and Meyer [SFM78, Corollary 4.1] that NTIME(T_2)$-\cup\{$NTIME(T_1) : $T_1(n+1) \in o(T_2(n))\}$ contains a language over $\{0, 1\}$. So it is easy to see that NP $\neq$ NEXP. Hence, $K' \notin$ NP since K' is $\leq_m^p$-complete for NEXP and NP is closed under $\leq_m^p$-reduction. So $(K', \mu_{K'}) \notin$ DNP. Let M be a Turing machine that accepts (i, x, n) if and only if M_i accepts x within n steps. So the running time of M is bounded by $O(|(i, x, n)| + n)$. Hence,

$$\sum_{i,x,n} \frac{O(|(i, x, n)| + n)}{|(i, x, n)|} \mu_{K'}(i, x, n) \leq \sum_{i,x,n} O(n) \mu_{K'}(i, x, n) < \infty.$$

Therefore, $(K', \mu_{K'}) \in \text{ANP}_P$. Now we will show that $(K', \mu_{K'})$ is $\leq_m^{ap}$-complete for ANP_P by reducing RH to it. Let $f(i, x, 1^n) = (i, x, n)$. Then f is polynomial-time computable and f is one-one. Clearly, $(i, x, 1^n) \in K$ if and only if $f(i, x, 1^n) \in K'$. Since $\mu_{K'}(f(i, x, 1^n)) = n^{-1}\mu_K(i, x, 1^n)$ and $n < |(i, x, 1^n)|$, $\mu_K \preceq^f \mu_{K'}$. Hence, RH $\leq_m^p$ $(K', \mu_{K'})$. So $(K', \mu_{K'})$ is $\leq_m^{ap}$-complete for ANP_P. ∎

It is clear from the definitions that if (D, μ) is $\leq_m^p$-complete for DNP, then D must be $\leq_m^p$-complete for NP. We also have a partial converse.

Proposition 7 *If D is $\leq_m^p$-hard for NP, then there exists a probability distribution μ such that (D, μ) is $\leq_m^p$-hard for DNP.*

Proof. Suppose D is $\leq_m^p$-hard for NP. Let (D', μ') be any $\leq_m^p$-complete problem for DNP. Then, since $D' \in$ NP, there exists a polynomial-time computable function f such that $D' \leq_m^p D$ via f. Define $\mu(y) = \sum_{y=f(x)} \mu'(x)$. We then have $(D', \mu') \leq_m^p (D, \mu)$ via f, and so (D, μ) is $\leq_m^p$-hard for DNP. ∎

Notice that in the above proof, even if $D \in$ NP, (D, μ) is not necessarily $\leq_m^p$-complete for DNP. Since μ could be outside PDF, (D, μ) may not even be in DNP.

5 Non-Closure Properties

In this section, we will prove that by restricting to simple distributions, AP_P, DNP, and ANP_P are not closed under $\leq_m^p$-reductions. Therefore, these classes are not closed under any reductions weaker than $\leq_m^p$-reductions. We first show the following lemma.

Lemma 8 *There are probability distributions $\mu \notin$ PDF and $\nu \in$ PDF, and a polynomial-time computable function f such that f transforms μ to a restriction of ν, where $\mu(x) > 0$ and $\nu(x) > 0$ for all $x \in \Sigma^*$.*

Proof. We first prove that there exists a set $H = \{x_1, x_2, ...\} \subseteq \{0,1\}^*$ such that $|x_n| = n$ for all n and $H \notin $ P. Let $P_1, P_2, ...$ be an enumeration of all sets in P. For any n, define x_n by

$$x_n = \begin{cases} 0^n, & \text{if } 0^n \notin P_n, \\ 1^n, & \text{otherwise.} \end{cases}$$

Let $H = \{x_1, x_2, ...\}$. Then $H \neq P_n$ for any n. So $H \notin $ P.

Let $p(n) = n(n+1)/2$ and define μ by

$$\mu(x) = \begin{cases} 2^{-p(n)}, & \text{if } x = x_n, \\ 2^{-(p(n)+n)}, & \text{if } |x| = n \text{ and } x \neq x_n. \end{cases}$$

Then $\sum_{|x|=n} \mu(x) = 2^{-p(n)+1} - 2^{-(p(n)+n)}$. Since $p(n) = p(n-1) + n$, it is easy to show that $\sum_{|x| \leq n} \mu(x) = 1 - 2^{-(p(n)+n)}$. So μ is a probability distribution.

We claim that μ is not polynomial-time computable. Suppose, to the contrary, that μ is polynomial-time computable. Then there exists a polynomial-time algorithm $\mathcal{A}(x, 1^k)$ such that for all strings x and all positive integers k, $|\mathcal{A}(x, 1^k) - \mu(x)| < 2^{-k}$. Let $\alpha(x) = \mathcal{A}(x, 1^{p(|x|)+2})$. Then α is a polynomial-time algorithm, and for all x, $|\alpha(x) - \mu(x)| < 2^{-(p(|x|)+2)}$. Let $n = |x|$. If $x = x_n$, then $\mu(x) = 2^{-p(|x|)}$, and so $|\alpha(x) - 2^{-p(|x|)}| < 2^{-(p(|x|)+2)}$. This implies $\alpha(x) > 2^{-p(|x|)} - 2^{-(p(|x|)+2)} > 2^{-(p(|x|)+1)}$, which in turn means that the binary expansion of $\alpha(x)$ has a 1 by the $p(|x|)+1$ position to the right of the binary point. On the other hand, if $x \neq x_n$, then $\mu(x) = 2^{-(p(|x|)+|x|)}$, and so $|\alpha(x) - 2^{-(p(|x|)+|x|)}| < 2^{-(p(|x|)+2)}$. This implies $\alpha(x) < 2^{-(p(|x|)+2)} + 2^{-(p(|x|)+|x|)} \leq 2^{-(p(|x|)+1)}$ for $|x| \geq 2$, and so the binary expansion of $\alpha(x)$ does not have a 1 in the first $p(|x|)+1$ positions to the right of the binary point. Therefore,

$H = \{x : x = x_1 \text{ or } (|x| > 1 \text{ and the binary expansion of } \alpha(x)$
has a 1 by the $p(|x|)+1$ position)$\}$.

Therefore, $H \in $ P. It contradicts the fact that $H \notin $ P. So μ cannot be polynomial-time computable.

Define $f : \{0,1\}^* \to \{0\}^*$ by $f(x) = 0^{|x|}$. Define ν by

$$\nu(y) = \begin{cases} 2^{-p(n)+1} - 2^{-(p(n)+n)}, & \text{if } y = 0^n, \\ |y|^{-2} 2^{-|y|}, & \text{otherwise.} \end{cases}$$

Then ν^* is polynomial-time computable. Since for any $y = 0^n$ in the image of f, $\nu(y) = \sum_{y=f(x)} \mu(x)$, f transforms μ to a restriction of ν. ∎

Remark 3 Lemma 8 can also be proved such that μ is not computable for any super-polynomial time complexity. ∎

Theorem 9 AP_P, ANP_P, and DNP are not closed under $\leq_m^p$-reductions.

Proof. Let μ, ν, and f be the functions constructed in the proof of Lemma 8. Notice that $\mu(x) > 0$ for any $x \in \{0,1\}^*$. Let $B \subseteq \{0\}^*$ and $B \in P$. Let $A = \{x : 0^{|x|} \in B\}$. Then $A \in P$ and $A \leq_m^p B$ via f. So $(A, \mu) \leq_m^p (B, \nu)$ via f. However, (B, ν) is in AP_P, ANP_P and DNP, but (A, μ) is not in AP_P, ANP_P or DNP since $\mu \notin PDF$. Hence, AP_P, ANP_P and DNP are not closed under $\leq_m^p$-reductions. ∎

Remark 4 In fact, for each $C \in NP$, we can construct a pair of randomized decision problems witnessing that DNP is not closed under $\leq_m^p$-reductions. Let μ, ν, and f be the functions constructed in the proof of Lemma 8. For any $C \in NP$, let $A = \{x : (\exists y)[|y| = |x|$ and $y \in C]\}$, and $B = \{0^{|x|} : x \in C\}$. Then both A and B are in NP and $A \leq_m^p B$ via f. Therefore, $(A, \mu) \leq_m^p (B, \nu)$ via f. Since (B, ν) is in DNP but (A, μ) is not in DNP, (A, μ) and (B, ν) witness that DNP is not closed under $\leq_m^p$-reductions. ∎

6 Non-Completeness Results

From the proof of Theorem 4, we can see that RH is actually average polynomial-time one-one complete for ANP_P. In fact, all the many-one complete problems for DNP known so far [Lev84, Gur91] are one-one complete.

We use the polynomial-time computability of the distribution functions in proving completeness theorems. In general, proving completeness theorems needs certain assumptions on the distribution functions. If there is no requirement at all, then we can prove that neither AP nor ANP contains hardest randomized decision problems with respect to polynomial-time one-one reductions. Gurevich [Gur91] proved that many NP-complete problems with natural probability distributions cannot be $\leq_m^P$-complete for DNP unless EXP = NEXP, where EXP = DTIME(2^{poly}). We start with the following lemmas.

Lemma 10 *Let ν be a probability distribution on Σ^*, and $f : \Sigma^* \to \Sigma^*$ any one-one, total function. Then $\nu(f(x)) < 1/2^{|x|}$ for infinitely many x.*

Proof. Let ν and f be above. Suppose, to the contrary, that for some n, $\nu(f(x)) \geq 1/2^{|x|}$ for all x with $|x| \geq n$. Then

$$\infty > \sum_y \nu(y) \geq \sum_x \nu(f(x)) \geq \sum_{|x| \geq n} \nu(f(x)) \geq \sum_{|x| \geq n} \frac{1}{2^{|x|}},$$

but this last sum diverges. ∎

Lemma 11 *Let ν be a probability distribution on Σ^*. Then there exists a probability distribution μ on Σ^* such that for any one-one, total, polynomial-time computable function f, μ is not dominated by ν with respect to f.*

Proof. Let $f_0, f_1, f_2, \ldots$ be a listing of all total, one-one, polynomial-time computable functions from Σ^* to Σ^* such that each function is included in this sequence an infinite number of times. (Notice that we do not require that such a sequence to be recursively enumerable.) Choose a sequence $\langle x_k \rangle \subset \Sigma^*$ such that $|x_k| \geq 2^k$ and $\nu(f_k(x_k)) < 1/2^{|x_k|}$ for all k. This is possible by Lemma 10. Define the function μ on Σ^* by

$$\mu(x) = \begin{cases} \frac{|x_k|^k}{2^{|x_k|}}, & \text{if } x = x_k \\ \frac{1}{|x|^2 2^{|x|}}, & \text{otherwise.} \end{cases}$$

Then

$$\sum_x \mu(x) < \sum_x \frac{1}{|x|^2 2^{|x|}} + \sum_k \mu(x_k) = \sum_n \frac{1}{n^2} + \sum_k \frac{|x_k|^k}{2^{|x_k|}}.$$

Since $x^k/2^x$ is a decreasing function of x for $x > k/\ln 2$ and $|x_k| \geq 2^k$, we have

$$\frac{|x_k|^k}{2^{|x_k|}} \leq \frac{(2^k)^k}{2^{2^k}} = \frac{2^{k^2}}{2^{2^k}},$$

and so
$$\sum_x \mu(x) \le \sum_n \frac{1}{n^2} + \sum_k \frac{2^{k^2}}{2^{2^k}} < \infty.$$

Hence, μ is a probability distribution on Σ^*.

Suppose $\mu \preceq^f \nu$ for some total, one-one, polynomial-time computable function f. This means that for some probability distribution μ_1 on Σ^*, $\mu \preceq \mu_1$ and f transforms μ_1 to ν. Since $\mu \preceq \mu_1$, for some polynomially bounded function g, $\mu(x) \le g(x)\mu_1(x)$ for all x. This implies that there exists an m such that $\mu(x) \le |x|^m \mu_1(x)$, and so $\mu_1(x) \ge \mu(x)/|x|^m$ for all x. Since f transforms μ_1 to ν, we have

$$\nu(y) = \sum_{y=f(x)} \mu_1(x).$$

By the definition of sequence $f_0, f_1, \ldots$, we know that $f = f_k$ for infinitely many k, and for these values of k,

$$\begin{aligned}
\nu(f_k(x_k)) &= \sum_{f_k(x)=f_k(x_k)} \mu_1(x) \\
&\ge \mu_1(x_k) \\
&\ge \frac{\mu(x_k)}{|x_k|^m} \\
&= \frac{|x_k|^k}{2^{|x_k|}} \frac{1}{|x_k|^m} \\
&= \frac{|x_k|^{k-m}}{2^{|x_k|}}.
\end{aligned}$$

Since $\nu(f_k(x_k)) < 1/2^{|x_k|}$, this cannot happen for arbitrarily large k. So μ cannot be dominated by ν with respect to any total, one-one, polynomial-time computable function. ∎

Theorem 12 AP *(resp. ANP) does not contain polynomial-time one-one complete problems.*

Proof. Suppose that (E, ν) is $\le_1^p$-complete for AP. Let $D \in$ P, and let μ be as in the proof of Lemma 11. Then $(D, \mu) \in$ AP, but since μ cannot be dominated by ν with respect to any total, one-one, polynomial-time computable function, (D, μ) cannot be $\le_1^p$-reducible to (E, ν). This is a contradiction.

The proof for ANP is similar. ∎

It is easy to see that Theorem 12 also holds for functions f if $|\{y : f(y) = f(x)\}|$ is bounded by a fixed constant for any x. It is open, however, whether Theorem 12 is true for other polynomial-time reductions.

It should be clear that for any $D \in$ NP, there exists a distribution μ such that $(D, \mu) \in$ AP, (or even AP$_P$). We can show, however, that such a (D, μ) cannot be $\leq_m^p$-complete for AP (or AP$_P$) unless EXP = NP.

Proposition 13 *Let (D, μ) be a randomized decision problem with $D \in$ NP. If (D, μ) is $\leq_m^p$-complete for AP (or AP$_P$), then EXP = NP.*

Proof. Assume that (D, μ), $D \in$ NP, is $\leq_m^p$-complete for AP. The proof for AP$_P$ is the same.

Let B be an arbitrary set in EXP. Then there is a DTM M which decides B within $2^{p(n)}$ steps, where p is a polynomial. We turn B into a randomized decision problem (B, ν) by setting $\nu(x) = |x|^{-2} 2^{-|x|-p(|x|)}$. Clearly, $\nu \in$ AP$_P$. So $(B, \nu) \leq_m^p (A, \mu)$. Hence, $B \leq_m^p A$. Since $A \in$ NP, $B \in$ NP. This completes the proof. ∎

Acknowledgments. We are grateful to Yuri Gurevich and R. Venkatesan for their comments.

References

[BCGL89] S. Ben-David, B. Chor, O. Goldreich, and M. Luby, *On the theory of average case complexity*, Proc. 21st Annual ACM Symposium on Theory of Computing, 1989, pp.204-216.

[BG91] A. Blass and Y. Gurevich, *Randomizing reductions of search problems*, Proceedings of Foundations of Software Technology and Theoretical Computer Science, New Delhi, India, 1991, pp. 10-24.

[Du85] D.-Z. Du, *Generalized Complexity Cores and Levelability of Intractable Sets*, Ph.D. thesis, Department of Mathematics, University of California, Santa Barbara, 1985.

[DB89] D.-Z. Du and R. Book, *On inefficient special case of NP-complete problems*, Theoretical Computer Science, 63(1989), pp. 239-252.

[Fel68] W. Feller, *An Introduction to Probability Theory and Its Applications*, Vol. 1, 3rd Edition, Wiley, 1968.

[Gur87] Y. Gurevich, *Complete and incomplete randomized NP problems*, Proc. 28th IEEE Symposium on Foundations of Computer Science, 1987, pp. 111-117.

[Gur90] Y. Gurevich, *Matrix decomposition is complete for the average case*, Proc. 31th IEEE Symposium on Foundations of Computer Science, 1990, pp. 802-811.

[Gur91] Y. Gurevich, *Average case completeness*, J. of Computer and System Sciences, 42(1991), pp. 346-398.

[GS87] Y. Gurevich and S. Shelah, *Expected Computation Time for Hamiltonian Path Problem*, SIAM J. on Computing, 16:3(1987), pp. 486-502.

[IL90] R. Impagliazzo and L. Levin, *No better ways to generate hard NP instances than picking uniformly at random*, Proc. 31th IEEE Symposium on Foundations of Computer Science, 1990, pp. 812-821.

[Jo84] D. Johnson, *The NP-completeness column: an ongoing guide*, J. of Algorithms, 5(1984), pp. 284-299.

[Ko83] K. Ko, *On the definition of some complexity classes of real numbers*, Math. Systems Theory, 16(1983), pp.95-109.

[Lev84] L. Levin, *Average case complete problems*, SIAM J. on Computing, 15(1986), pp.285-286. Extended abstract appeared in Proc. 16th ACM Symposium on Theory of Computing, 1984, p. 465.

[VL88] R. Venkatesan and L. Levin, *Random instances of a graph coloring problem are hard*, Proc. 20th ACM Symposium on Theory of Computing, 1988, pp. 217-222.

[SFM78] J. Seiferas, M. Fischer, and A. Meyer, *Separating nondeterministic time complexity classes*, J. ACM, 25(1978), pp. 146-167.

[Ven91] R. Venkatesan, *Average Case Intractability*, Ph.D. thesis, Computer Science Department, Boston University, 1991.

[VR91] R. Venkatesan and S. Rajagopalan, *Average case intractability of Diophantine and matrix problems*, Proc. of the 7th IEEE Conference on Structure in Complexity Theory, 1992, pp. 318-326.

[Wil84] H. S. Wilf, *Backtracking: An O(1) expected time algorithm for the graph coloring problem*, Information Processing Letters, 18(1984), pp. 119-122.

Additional Queries and Algorithmically Random Languages[*]

Ronald V. Book
Department of Mathematics
University of California
Santa Barbara, CA 93106, USA
e-mail: book@math.ucsb.edu

Abstract

A language A has the "additional query property relative to a function f," denoted AQ_f, if more decision problems are solvable with $f(n)$ queries to A than with $f(n)-1$ queries to A. It is shown that for suitable functions f, every algorithmically random language has the AQ_f.

Key words: relativizations, additional queries, algorithmically random languages, random oracles.

AMS (MOS) classification: 68Q30, 68Q15

---

[*]This work was suported in part by the National Science Foundation under Grant CCR-8913584 and by the Alexander von Humboldt Stiftung while the author visited the Facultät für Informatik, Universität Ulm, Germany. Some of these results were reported at the 17th International Colloquium on Automata, Languages, and Programming, July 1990, and an extended abstract appears in the proceedings of that colloquium.

1 Introduction

The study of resource-bounded reducibilities plays an important part in computational complexity theory, particularly in structural complexity theory. Essentially, a reduction of a language B, denoted $A \leq_\mathcal{R} B$, provides technical machinery that allows one to determine bounds on the computational complexity of membership in language A in terms of the computational complexity of membership in language B. For different types of reducibilities and different resource bounds, the capacities for exploiting the information content of language B may vary. Examples of results showing that additional queries increase the appropriate class include the following:

(i) Ladner, Lynch, and Selman [LLS75] proved that for every $k > 1$, there exist languages A and B such that $A \leq_{k\text{-}tt}^P B$ but $A \not\leq_{(k-1)\text{-}tt}^P B$.

(ii) Watanabe [Wat87] proved that for every $k > 1$, there exists a language that is complete for $\text{DTIME}(2^{\text{lin}})(= \cup_{c>0} \text{DTIME}(2^{cn}))$ with respect to $\leq_{k\text{-}tt}^P$ but is not complete for $\text{DTIME}(2^{\text{lin}})$ with respect to $\leq_{(k-1)\text{-}tt}^P$.

A great deal is known about the capacity of a reducibility to exploit the information content of a language when the information content is low. For example, there are the following:

(iii) Book and Ko [BK88] proved that for every $k > 1$, there exists a language A and a sparse language S such that $A \leq_{k\text{-}tt}^P S$ but $A \not\leq_{(k-1)\text{-}tt}^P S'$ for every sparse S'.

(iv) Tang and Book [TB91] showed that for suitable functions f, for almost every tally language T there exist a language A such that $A \leq_{k\text{-}tt}^P T$ but $A \not\leq_{(k-1)\text{-}tt}^P T$.

In addition, a result in the spirit of (iv) can be interpreted as a result about languages with high information content.

(v) Cai [Cai87] showed that for almost every language B, the Boolean hierarchy on $\text{NP}(B)$ is a properly infinite hierarchy. From results of Köbler, Schöning, and Wagner [KSW87], this means that for almost every language B, for every $k > 1$, there exists a language A such that $A \leq_{k\text{-}tt}^P K(B)$ but $A \not\leq_{(k-1)\text{-}tt}^P K(B)$, where $K(B)$ is $\leq_m^P$-complete for $\text{NP}(B)$.

The reader should recall that bounded Turing reducibilities, sometimes called "bounded sequential" reducibilities, are no more powerful than bounded truth-table reducibilities, sometimes called "bounded parallel" reducibilities, in the following sense: for all languages A, B and every integer $k > 0$, if $A \leq_{k\text{-tt}} B$, then $A \leq_{k\text{-T}} B$, while if $A \leq_{k\text{-T}} B$, then $A \leq_{(2^k-1)\text{-tt}} B$. Also, these relationships remain true if the reducibilities are computed in polynomial time. Hence, the reader may consider the above results to results about "bounded" reducibilities.

In the present paper we investigate properties of reductions to languages with high information content. The sets with maximum information content are those that are "algorithmically random" as defined by Martin-Löf [Mar66]; let RAND denote the class of algorithmically random languages. We wish to compare the capacities of various reducibilities to exploit the high information content of algorithmically random languages. The emphasis is on reducibilities specified by machines with bounds on certain computational resources where the number of oracle queries allowed in a computation is also bounded. Intuitively, a language A has the "additional query property relative to a function f," denoted AQ_f, if more decision problems are solvable with $f(n)$ queries to A than with $f(n) - 1$ queries to A.

We show that for suitable functions f, with probability one a randomly selected language B has AQ_f property. From this result and a recent result of Book, Lutz, and Wagner [BLW93], we conclude that *every* language in RAND has this property.

2 Preliminaries

For the most part our notation is standard, following that used by Balcázar, Díaz, and Gabarró [BDG88]. We assume that the reader is familiar with the standard recursive reducibilities and the variants obtained by imposing resource bounds such as time or space of the algorithms that computed these reducibilities. For a relativizable complexity class C, a reducibility R, and a language B, $C_R(B)$ denotes $\{A \mid A \leq_R^C B\}$.

A word (string) is an element of $\{0,1\}^*$. The length of a word $w \in \{0,1\}^*$ is denoted $|w|$. We assume a fixed lexicographical enumeration of $\{0,1\}^*$.

The characteristic sequence of a language A is a (one-way) infinite sequence ξ_A on $\{0,1\}$. We freely identify a language with its characteristic sequence and the class of all languages on the fixed finite alphabet $\{0,1\}$ with the set $\{0,1\}^\omega$ of all such infinite sequences; the usage is based on context so that there should be no ambiguity on the part of the reader.

If X is a set of strings (that is, a language) and C is a set of sequences (that is, a class of languages), then $X \cdot C$ denotes the set $\{w\xi \mid w \in X, \xi \in C\}$. (Note that $X \cdot C$ is itself a class of languages under our identification of languages with sequences.)

For each string w, $C_w = \{w\} \cdot \{0,1\}^\omega$ is the *basic open set* defined by w. Note that C_w is the set of all sequences ξ such that w is a prefix of every string in ξ. An *open set* is a (finite or infinite) union of basic open sets, that is, a set of the form $X \cdot \{0,1\}^\omega$ where $X \subseteq \{0,1\}^*$. (This definition gives the usual product topology, also known as the Cantor topology, on $\{0,1\}^\omega$.) A *closed* set is the complement of an open set. A class of languages is *recursively open* if it has the form $X \cdot \{0,1\}^\omega$ for some recursively enumerable set $X \subseteq \{0,1\}^*$. A class of languages is *recursively closed* if it is the complement of some recursively open set.

For a class C of languages, we write Prob[C] for the probability that $A \in C$ when A is chosen by a random experiment in which an independent toss of a fair coin is used to decide whether each string is in A. This probability is defined whenever C is measurable in the usual product topology of $\{0,1\}^\omega$. In particular, if C is countable union or intersection of (recursively) open or closed sets, then C is measurable so Prob[C] is defined. Note that there are only countably many recursively open sets, so every intersection of recursively open sets can be expressed as a countable intersection of such sets, and hence is measurable; similarly every union of recursively closed sets is measurable. We say that "almost every language" is in C if and only if Prob[C] = 1.

A class C is *closed under finite variation* if $A \in C$ whenever $B \in C$ and A and B have finite symmetric difference. The Kolmogorov 0–1 Law says that for every measurable set $C \subseteq \{0,1\}^\omega$ which is closed under finite variation, either Prob[C] = 0 or Prob[C] = 1.

Assume an effective enumeration of the recursively enumerable languages as $w_0, w_1, \ldots$.

A *constructive null cover* of a class C of languages is a sequence of recursively open sets $W_{g(0)} \cdot \{0,1\}^\omega$, $W_{g(1))} \cdot \{0,1\}^\omega, \ldots$ specified by a total recursive function g with the properties that for every k,

(i) $C \subseteq W_{g(k)} \cdot \{0,1\}^\omega$, and

(ii) Prob[$W_{g(k)} \cdot \{0,1\}^\omega$] $\leq 2^{-k}$.

If a class C has a constructive null cover, then C is a *constructive null set*, so that Prob[C] = 0. Let NULL be the union of all constructive null sets, so that Prob[NULL] = 0 since NULL is a countable union of sets with probability 0. Let RAND be defined as the complement of NULL (with respect to $\{0,1\}^\omega$). Then Prob[RAND] = 1 since Prob[NULL] = 0.

A language is *algorithmically random* if and only if its characteristic sequence is in RAND. Since we identify a language with its characteristic sequence, we introduce no ambiguity by denoting the class of all algorithmically random languages by RAND.

The definition of the class RAND is due to Martin-Löf [Mar66].

3 Main Result

The notion of "the additional query property" was introduced in [BLT90].

A language B has the *additional query property relative to a function* f, denoted AQ_f, if there exists a language A such that for every n, $A \leq_{f(n)-tt} B$ but $A \not\leq_{f(n)-1)-T} B$.

If f is the constant function $f(n) = k$ for some integer $k > 1$, then we say that B has the AQ_k property if, in addition, the $k - tt$ reduction from A to B can be carried out in polynomial-time. Notice that in the case of $A \leq_{f(n)-tt} B$, the reduction is nonadaptive (sometimes called "parallel"), while in the case of $A \leq_{f(n)-T} B$ (sometimes called "sequential"), the reduction is adaptive.

The result is not surprising. But we give a careful proof of a basic lemma in order to be able to make a number of conclusions about reducibilities computed by machines with restrictions on different computational resources.

Lemma 1 *Let $k > 1$ be an integer. For almost every language $B \subseteq \Sigma^*$, B has the AQ_k property. That is, for almost every language $B \subseteq \Sigma^*$, there exists a language A such that $A \leq^P_{k-tt} B$ but $A \not\leq_{(k-1)-T} B$.*

Proof Recall that we have assumed an enumeration $w_0 < w_1 < w_2 < \ldots$ of Σ^*. For every language $B \subseteq \{0,1\}^*$, let $ODD_B = \{w_m \mid m \geq 0$ and $\| \{w_{m+1}, \ldots, w_{m+k}\} \cap B \|$ is ODD$\}$. (The reader should note that ODD_B is a simple variation on the notion of the parity function.) For every $x \in \Sigma^*$ there is a unique m (but it is not necessary to compute m) such that $w_m = x$ so, from x the language $\{w_{m+1}, \ldots, w_{m+k}\}$ can be computed in time polynomial in $|w_m|$. Then with k queries to an oracle for B, it can be determined whether w_m is in ODD_B. Thus, it is clear that for every language $B \subseteq \{0,1\}^*$, $\mathrm{ODD}_B \leq_{k-tt} B$, and this reduction can be implemented by some program that runs in time polynomial in $|w_m|$. (In fact the referee has observed this reduction needs at most $\log n$ workspace; however, the oracle tape needs linear space since the length of each query string is $O(|w_m|)$.)

The work to be done is to show that for almost every B, $\mathrm{ODD}_B \not\leq_{(k-1)-T} B$.

For any $j > 0$ and any fixed program Π that witnesses a $\leq_{j-T}$ reduction, there is a total recursive function r_Π (that depends only on Π) with the property that for every language B and every input x, every string queried in Π's computation on x relative to B has length at most $r_\Pi(|x|)$.

It is sufficient to show the following.

(*) For any fixed program Π that witnesses a $(k-1)-T$ reduction,
$\mathrm{Prob}[\{B \subseteq \{0,1\}^* \mid \Pi \text{ witnesses } \mathrm{ODD}_B \leq_{(k-1)-T} B\}] = 0$.

Let Π be a program that witnesses a $(k-1)-T$ reduction. Since Π makes at most $k-1$ oracle queries in any computation, the question of whether ODD_B and $L(\Pi, B)$ agree on some x depends on only a finite subset of B. Let q be a function with the properties that for all x, $r_\Pi(|x|) \leq q(|x|)$ and for each $m \geq 0$, $|w_{m+k}| \leq q(|w_m|)$. Thus, $\{B \subseteq \{0,1\}^* \mid \mathrm{ODD}_B \text{ and } L(\Pi, B) \text{ agree on } x\} = R_{y(1)} \cup \ldots \cup R_{y(j)}$, where for each i, $R_{y(i)}$ is the basic open set $y(i) \cdot \{0,1\}^\omega$, $j \geq 1$, the $y(i)$'s are distinct, and $|y(1)| = \ldots = |y(j)| = q(|x|)$. Hence, for any finite language $S \subset \{0,1\}^*$, $\{B \subseteq \{0,1\}^* \mid \mathrm{ODD}_B \text{ and } L(\Pi, B) \text{ agree on each word in } S\}$ is a union of finitely many basic open sets that are pairwise disjoint.

Claim Let m be fixed and let y be a string such that $y < w_m$. Then $\mathrm{Prob}[\{B \subseteq \{0,1\}^* \mid \mathrm{ODD}_B \text{ and } L(\Pi, B) \text{ agree on } w_m\} \cap R_y] = \mathrm{Prob}[R_y]/2$.

Proof Let $R_y = R_{yz(1)} \cup \ldots \cup R_{yz(t)}$, where $|yz(1)| = \ldots = |yz(t)| = q(|w_m|)$ and $i \neq j$ implies $R_{yz(i)} \cap R_{yz(j)} = \emptyset$; notice that for each i, $1 \leq i \leq t$, $\mu(R_{yz(i)}) = \mu(R_y)/t$. For any language $A \in R_y$, both $\mathrm{ODD}_A(w_m)$ and Π's computation on w_m relative to A depend only on $\{x \in A \mid |x| \leq q(|w_m|)\}$, so that for each i, either $R_{yz(i)} \subseteq \{B \mid \mathrm{ODD}_B \text{ and } L(\Pi, B) \text{ agree on } w_m\}$ or $R_{yz(i)} \cap \{B \mid \mathrm{ODD}_B \text{ and } L(\Pi, B) \text{ agree on } w_m\} = \emptyset$.

For any language $A_1 \in R_y$, recall that $w_m \in \text{ODD}_{A_1}$ if and only if $\| \{w_{m+1}, \ldots, w_{m+k}\} \cap A_1 \|$ is odd. Because Π's computation on w_m relative to A_1 queries the oracle about at most $k-1$ words, there exists some word in $\{w_{m+1}, \ldots, w_{m+k}\}$ that is not queried; let w_p be such a word. If A_1 and A_2 differ only on w_p (i.e., $A_2 = A_1 \Delta \{w_p\}$), then $A_2 \in R_y$, and $L(\Pi, A_1)$ and $L(\Pi, A_2)$ agree on w_m but ODD_{A_1} and ODD_{A_2} do not agree on w_m. Hence, exactly one of A_1 and A_2 is in $\{B \subseteq \{0,1\}^* \mid \text{ODD}_B$ and $L(\Pi, B)$ agree on $w_m\} \cap R_y$.

Suppose that for all j, $1 \leq j \leq t/2$, $yz(2j)$ and $yz(2j-1)$ differ only on the bit corresponding to w_p (so that $B \in R_{yz(2j-1)}$ if and only if $B \Delta \{w_p\} \in R_{yz(2j)}$). By the analysis above we have $R_{yz(2j-1)} \subseteq \{B \subseteq \{0,1\}^* \mid \text{ODD}_B$ and $L(\Pi, B)$ agree on $w_m\}$ if and only if $R_{yz(2j)} \cap \{B \subseteq \{0,1\}^* \mid \text{ODD}_B$ and $L(\Pi, B)$ agree on $w_m\} = \emptyset$. Thus, among the cylinders $R_{yz(1)}, \ldots, R_{yz(t)}$, exactly one half are included in $\{B \subseteq \{0,1\}^* \mid \text{ODD}_B$ and $L(\Pi, B)$ agree on $w_m\}$; each of the others is disjoint from that class. □Claim.

Continuing with the proof of (*), choose $n_1 < n_2 < \ldots$ such that for each i, $|w_{n_{i+1}}| > q(|w_{n_i}|)$. Let $\mathbf{C}_i = \{B \subseteq \{0,1\}^* \mid \text{ODD}_B$ and $L(\Pi, B)$ agree on each of $w_{n_1}, w_{n_2}, \ldots, w_{n_i}\}$. This implies that $\mathbf{C}_1 \supseteq \mathbf{C}_2 \supseteq \ldots \supseteq \{B \subseteq \{0,1\}^* \mid \Pi$ witnesses $\text{ODD}_B \leq_{(k-1)\text{-T}} B\}$. By the Claim, we have $\text{Prob}[\mathbf{C}_{i+1}] = \text{Prob}[\mathbf{C}_i]/2$. Thus, $\text{Prob}[\{B \subseteq \{0,1\}^* \mid \Pi$ witnesses $\text{ODD}_B \leq_{(k-1)\text{-T}} B\}] \leq 1/2^i$ for all $i > 0$, and so $\text{Prob}[\{B \subseteq \{0,1\}^* \mid \Pi$ witnesses $\text{ODD}_B \leq_{(k-1)\text{-T}} B\}] = 0$.

Thus, (*) is established.

Notice that $\{B \subseteq \{0,1\}^* \mid \text{ODD}_B \leq_{(k-1)\text{-T}} B\} = \cup \{B \subseteq \{0,1\}^* \mid \Pi$ witnesses $\text{ODD}_B \leq_{(k-1)\text{-T}} B\}$ where the union is taken over all programs Π that witness a $(k-1)-T$ reduction. From (*) we see that each language on the right hand side of the equation has probability 0. Since there are only countably many programs Π, $\text{Prob}[\{B \subseteq \{0,1\}^* \mid \text{ODD}_B \leq_{(k-1)\text{-T}} B\}] = 0$. □ Lemma 1.

The reader should observe that the proof involves more than a formalization of an argument about "coin-tossing" since it shows limitation of Turing (i.e., adaptive) reducibilities, but the proof does not involve the details of any specific formalization of the notion of computable function.

The main result is a simple generalization of Lemma 1.

A function f will be called "nice" if there is a (deterministic) program that on input of size n, n a non-negative integer, will output $n + f(n)$ in unary notation while running for at most $(n + f(n))^c$ steps for some integer $c > 0$ (which depends only on the program). Examples of nice functions include the following: $f(n) = j$, $f(n) = \log_2 n$, $f(n) = n^j$, $f(n) = 2^{n^j}$, where $j > 0$ is an integer. Every time-constructible function is nice.

Given B, let $A = \mathrm{ODD}_B = \{w_m \mid m \geq 1 \text{ and } \| \{w_{m+1}, \ldots, w_{f(|w_m|)+m}\} \cap B \|$ is odd$\}$. Then proof of Lemma 1 is easily altered to yield the following fact.

Theorem 2 *Let f be a nice function. For almost every language $B \subseteq \Sigma^*$, B has the AQ_f property. That is, for almost every language $B \subseteq \Sigma^*$, there exists a language A such that $A \leq_{f(n)-tt} B$ but $A \not\leq_{f(n)-1)-T} B$. The reduction $A \leq_{f(n)-tt} B$ can be implemented in time polynomial in $(n + f(n))$.*

From a recent result of Book, Lutz, and Wagner [BLW93], Theorem 2 yields the following fact.

Theorem 3 *For every nice function f, every language in RAND has the AQ_f property.*

It should be clear that the proof in Lemma 1 that $A \not\leq_{(k-1)-T} B$ can be applied widely. The only part of the proof that depends on the model of computation are the facts that in each accepting computation, there are at most $k - 1$ queries to the oracle and that there is a total function f_Π (that depends only on Π) with the property that for every language B and every input x, if Π's computation on x relative to B halts, then every string queried in that computation has size at most $r_\Pi(|x|)$. Thus, there are several corollaries.

Corollary 4 *Let $k > 1$ be an integer.*

a. *For almost every language B, $\mathrm{P}_{k-tt}(B) - \mathrm{NP}_{(k-1)-T}(B) \neq \emptyset$.*

b. *For almost every language B, $\mathrm{NP}_{k-T}(B) - \mathrm{NP}_{(k-1)-T}(B) \neq \emptyset$.*

c. *For almost every language B, $\mathrm{P}_{k-tt}(B) - \mathrm{PSPACE}_{(k-1)-T}(B) \neq \emptyset$.*

It is known that when a restriction on the number of oracle queries is placed on the computations of space-bounded machines, Savitch's Theorem no longer holds (for example, see [Boo81]). Thus, when considering polynomial space-bounded oracle machines, the class specified by nondeterminstic machines may be larger than the class specified by deterministic machines.

Corollary 5 *Let $k > 1$ be an integer. For almost every language B, $P_{k-tt}(B) - \text{NPSPACE}_{(k-1)-T}(B) \neq \emptyset$.*

For every set B, let $K(B)$ denote any language that is $\leq_m^P$-complete for $NP(B)$, and for each $j > 0$, let $NP_{j-T}(B)$ denote the class of languages A such that there is a nondeterministic oracle machine M such that $L(M, B)$ and M has the property that in each computation there are at most j oracle queries allowed. How do $NP_{j-T}(B)$ and $P_{j-T}(k(B))$ compare? Notice that for every B and every j, $NP_{j-T}(B) \subseteq NP(B) = \{A \mid A \leq_m^P K(B)\} \subseteq P_{j-tt}(K(B)) \subseteq P_{j-T}(K(B))$.

Corollary 6 *Let $k > 0$ be an integer. For almost every language B, $P_{k-T}(K(B)) - NP_{k-T}(B) \neq \emptyset$.*

The reader should note that the observation made by the reference in the proof of Lemma 1 also applies to Corollaries 4, 5, and 6.

Each of Corollaries 4–6 hold for every language in RAND.

Lemma 1 and each of Corollaries 4–6 hold for every language in RAND. In the case of Lemma 1, one can say something about the languages that serve as witnesses to the separation.

Proposition 7 *Let B in RAND. For any k, if L has the property that $L \leq_{k-tt}^P B$ and $L \not\leq_{(k-1)-tt}^P B$, then L is not recursive.*

Proof Suppose to the contrary that L is recursive. Then $L \in \text{REC} \cap P_{btt}(\text{RAND})$. As noted in [BLW93], this means that L is in P so that L can be recognized in polynomial time with no queries to any oracle. Hence, for every A and every j, $L \leq_{j-tt}^P A$, contradicting $L \not\leq_{k-tt}^P B$. □

Note that no r.e. language (hence, no recursive language) is in RAND [Mar66]. It would be interesting to find a characterization of all languages (or all recursive languages) that have the AQ_f property when f is a polynomial.

4 Additional Considerations

Amir, Beigel, and Gasarch [ABG90] studied the idea of "terse" languages.

For each language A, and each integer $k > 0$, define $F_k^A(x_1, \ldots, x_k) = \langle \chi_A(x_1), \ldots, \chi_A(x_k) \rangle$, where χ_A is the characteristic function of A. For each language A and each integer $k > 0$, define $FQ(k, A)$ as the set of all functions g such that g can be computed by an algorithm that makes at most k queries to χ_A, that is, $g \leq_{k\text{-T}} \chi_A$.

We will abuse the notation and write $h \leq_{k\text{-T}} \chi_A$ when $h \in FQ(k, A)$.

Language A is k-terse if the function $F_k^A \notin FQ(k-1, A)$. Language A is terse if it is k-terse for all $k > 0$.

The notion of terseness can be extended so that the number of queries allowed is not bounded by a constant. We do this in the following way.

Let g be a "nice" function. For every language A, define the function E_g^A as follows: for every $x \in \Sigma^*$, $E_g^A(x) = \langle \chi_Z(s_{i_1}), \ldots, \chi_A(s_{i_{g(|x|)}}) \rangle$, where $s_{i_1}, \ldots, s_{i_{g(|x|)}}$ is the (increasing) sequence of $g(|x|)$ consecutive (in the lexicographic order) strings in Σ^* that begins with $s_{i_1} = x$. Language A is g-terse if E_g^A cannot be computed by an algorithm that makes at most $g(|x|) - 1$ (sequential) queries to χ_A.

We will write $E_g^A \not\leq_{g(n)-1)\text{-T}} \chi_A$ if A is g-terse.

Proposition 8 *For every nice function g, almost every language is g-terse.*

Proof Consider a language A that is not g-terse, so that $E_g^A \leq_{(g(n)-1)-T} A$. This means that there is an algorithm that for any $x \in \Sigma^*$ that does the following:

(i) computes $\langle \chi_A(s_{i_1}), \ldots, \chi_A(s_{i_{g(|x|)}}) \rangle$ (where $s_{i_1}, \ldots, s_{i_{g(|x|)}}$ is the (increasing) sequence of $g(|x|)$ consecutive (in the lexicographic order) strings in Σ^* that begins with $s_{i_1} = x$) by making at most $g(x) - 1$ queries to an oracle for A; and then

(ii) determines the parity of $\langle \chi_A(s_{i_1}), \ldots, \chi_A(s_{i_{g(|x|)}}) \rangle$; and then

(iii) halts and gives output 1 if the parity is odd and 0 otherwise.

Thus, the algorithm shows that A does not have the AQ_g property. But Theorem 2 asserts that since g is taken to be a nice function, almost every language has the AQ_g property. □ Proposition 8

Dr. Richard Beigel (personal communication) has observed that the results in Section 3 regarding the AQ property do not follow from the results in [ABG90].

The question of whether the class NP is closed under complementation led in a natural way to the study of the Boolean hierarchy on NP and on relativizations of NP [CGHHSWW88]. Then Cai [Cai87] showed that for almost every language A, the classes making up the Boolean hierarchy on NP(A) form an infinite hierarchy.

The Boolean hierarchy based on NP can be described in several technically different ways. While the classes in the hierarchy may vary depending on the description, the union of the hierarchy does not change. For our purposes, the description of the Boolean hierarchy based on any language at all will depend on bounded truth-table reducibilities computed in polynomial time.

Let A be an arbitrary language. Then the Boolean hierarchy on A is the structure made up of the following classes:

$$P_{1-tt}(A),\ P_{2-tt}(A),\ldots,P_{k-tt}(A),\ldots.$$

The class BH(A) is defined to be the union of all such classes, that is, BH(A) = $\cup_{k>0} P_{k-tt}(A) = P_{btt}(A)$. This means that BH($A$) is the boolean closure of $\{B \mid B \leq_m^P A\}$. From Lemma 1, it is clear that for almost every language A, $P_{1-tt}(A),\ P_{2-tt}(A),\ldots$ forms an infinite hierarchy, that is, for every k, $P_{k-tt}(A) \neq P_{(k+1)-tt}(A)$.

Caution: The reader should note that in [CGHHSWW88] and [Cai87], BH(A) denotes BH(NP(A)).

The Boolean hierarchy on NP (often referred to as BH) is BH(SAT), where SAT denotes the first NP-complete set (conjunctive normal form formulas that are satisfiable); thus, BH(SAT) = P_{btt}(SAT) = P_{btt}(NP). Since it is not known whether NP is closed under complementation, it is not known whether BH = NP or BH extends to some finite number (greater than one) of levels or BH extends to infinitely many levels. Cai considered relativizations of NP, say NP(A), and showed that for almost every language A, $P_{1-tt}(K(A)),\ P_{2-tt}(K(A)),\ldots$ forms an infinite hierarchy, that is, for every $k > 1$, $P_{(k-1)-tt}(K(A)) \neq P_{k-tt}(K(A))$; here, $K(A)$ denotes any set that is $\leq_m^P$-complete for NP(A).

Thus, for almost every language A, the Boolean hierarchy on A and the Boolean hierarchy on $K(A)$ are infinite. These facts suggest the question of whether there is a relationship between the Boolean hierarchy on a language A being infinite and the Boolean hierarchy on the language $K(A)$ being infinite. More formally,

Question: Does either of the following hold:

(a) for any language A, if the Boolean hierarchy on A is infinite, then the Boolean hierarchy on $K(A)$ is infinite;

(b) for any language A, if the Boolean hierarchy on $K(A)$ is infinite, then the Boolean hierarchy on A is infinite.

There is reason to believe that the Boolean hierarchy on NP is infinite (since otherwise, the polynomial-time hierarchy collapses [Kad88]) and this suggests that (b) is false for some languages since for any language A in P, the Boolean hierarchy on A collapses to P, while the Boolean hierarchy on $K(A)$ is the Boolean hierarchy on NP.

One difficulty in answering this question is the fact that if the Boolean hierarchy on a language A is infinite, then for each k, any witness L to $P_{(k-1)-tt}(A) \neq P_{k-tt}(A)$ is in $P_{k-tt}(A) \subseteq P(A) \subseteq NP(A) = \{B \mid B \leq_m^P K(A)\} \subseteq P_{1-tt}(K(A))$, and the latter class is the first level of the Boolean hierarchy on $K(A)$. Thus the language L cannot be a witness to $P_{(k-1)-tt}(K(A)) \neq P_{k-tt}(K(A))$. Conversely, if the Boolean hierarchy on $K(A)$ is infinite, then for each k, a witness for the separation of the $(k-1)$st and kth levels of the Boolean hierarchy on $K(A)$ cannot be a witness for the separation of the $(k-1)$st and kth levels of the Boolean hierarchy on A.

There is an additional consideration. Let A = $\{A \mid$ the Boolean hierarchy on A is infinite$\}$ and B = $\{A \mid$ the Boolean hierarchy on $K(A)$ is infinite$\}$. Question (a) asks whether A $\subseteq$ B. By Lemma 1, Prob[A] = 1, and Cai's result shows that Prob[B] = 1; hence, Prob[A $\cap$ B] = 1. From a result in [BLW93], it is clear that RAND $\subseteq$ A $\cap$ B so that every language A in RAND has the property that both the Boolean hierarchy on A and the Boolean hierarchy on $K(A)$ are infinite.

These facts suggest that further investigation be carried out. Since languages that are $\leq_m^P$-complete for classes of the form $NP(A)$ appear to have a great deal of structure, it seems unlikely that there exist A such that $K(A) \in$ RAND; even more unlikely is the existence of a language $A \in$ RAND such that $K(A) \in$ RAND.

There are two additional quetions of interest here. Which recursive sets A have the property that the Boolean hierarchy on A is infinite? Which recursive sets A have the property that the Boolean hierarchy on $K(A)$ is infinite?

References

[ABG90]   A. Amir, R. Beigel, and W. Gasarch. Some connections between bounded query classes and non-uniform complexity, *Proc. 5th IEEE Conference on Structure in Complexity Theory* (1990), 232–243.

[BDG88] J. Balcázar, J. Díaz, and J. Gabarró. *Structural Complexity I*, Springer-Verlag, 1988.

[Boo81] R. Book. Bounded query machines: on NP() and NPQUERY(), *Theoret. Comput. Sci.* (1981), 27–39.

[BK88] R. Book and K-I. Ko. On sets truth-table reducible to sparse sets, *SIAM J. Comput.* 17 (1988), 903–919.

[BLT90] R. Book, J. Lutz, and S. Tang. Additional queries to random and pseudorandom oracles, M. Paterson (ed.) *ICALP 90 - Automata, Languages and Programming*, Lecture Notes in Computer Science 443 (1990), Springer-Verlag Publishing Co., 283–293.

[BLW93] R. Book, J. Lutz, and K. Wagner. An observation on probability versus randomness with applications to complexity classes, *Math. Systems Theory* 27 (1993), to appear. Also see, *Proc. STACS 92, Lecture Notes in Computer Sci.* 577, Springer-Verlag, 319–328.

[Cai87] J.-Y. Cai. Probability one separation of the boolean hierarchy. *STACS 87*, Lecture Notes in Computer Sci., 247, Springer-Verlag, (1987) 148–158.

[CGHHSWW88] J.-Y Cai, T. Gundermann, J. Hartmanis, L. Hemachandra, V. Sewelson, K. Wagner, and G. Wechsung. The Boolean hierarchy I: structural properties, *SIAM J. Comput.* 17 (1988), 1232–1252.

[Kad88] J. Kadin. The polynomial time hierarchy collapses if the Boolean hierarchy collapses, *SIAM J. Comput.* 17 (1988), 1263–1282.

[KSW87] J. Köbler, U. Schöning, and K. Wagner. The difference and truth-table hierarchies for NP. *RAIRO-Inform. Theorique et Appl.* 21 (1987), 419–435.

[LLS75] R. Ladner, N. Lynch, and A. Selman. A comparison of polynomial-time reducibilities, *Theoret. Comput. Sci.* 1 (1975), 103–123.

[Mar66]   P. Martin-Löf. On the definition of random sequences, *Info. and Control* 9 (1966), 602–619.

[TB91]   S. Tang and R. Book. Polynomial-time reducibilities and "almost-all" oracle sets, *Theoret. Comput. Sci.* 81 (1991), 35–47.

[Wat87]   O. Watanabe, A comparison of polynomial-time completeness notions, *Theoret. Comput. Sci.* 54 (1987), 249–265.

Bounded Reductions

Harry Buhrman   Edith Spaan
Leen Torenvliet
Departments of Mathematic and Computer Science
University of Amsterdam
Plantage Muidergracht 24 1018 TV Amsterdam
The Netherlands

Abstract

We study properties of resource– and otherwise bounded reductions and corresponding completeness notions on nondeterministic time classes which contain exponential time. As it turns out, most of these reductions can be separated in the sense that their corresponding completeness notions are different. There is one notable exception. On nondeterministic exponential time, 1-truth table and many-one completeness is the same notion.

1 Introduction

Efficient reducibilities and completeness are two of the central concepts of complexity theory. Since the first use of polynomial time bounded Turing reductions by Cook [4] and the introduction of polynomial time bounded many-one reductions by Karp[6], considerable effort has been put in the investigation of properties and the relative strengths of different reductions and corresponding completeness notions. In 1975, Ladner, Lynch and Selman [8] gave an extensive survey of different types of reductions and differences between these reductions on $E (= \cup_{c \in I\!N} \mathrm{DTIME}(2^{cn}))$. However, they did not present any conclusions concerning any differences in complete sets for these various reductions. In particular, they left open the question of whether these different reductions yield different complete sets. In 1987, Watanabe [10] building upon earlier work of L. Berman [1], proved almost all possible differences between the polynomial-time completeness notions on E and larger deterministic time classes.

The question of differentiating between complete sets for nondeterministic time classes with respect to various bounded reductions was considered by Buhrman, Homer and Torenvliet in [2]. Their paper however, concentrates on differentiating between the notions of completeness defined by many-one, bounded truth-table and standard (unbounded) Turing reductions, but did not consider the case of *bounded* Turing reductions, in both the polynomial time and logarithmic space case on nondeterministic time and space classes.

In this paper, we concentrate on the remaining open problems between notions of bounded reducibilities, and the corresponding completeness notions on E, NE, EXP and $NEXP$ (and solve all of these). As all of the considered reductions are bounded by a *constant* number of queries, the proofs are independent of the specific model for truth-table reducibilities, which may present a problem in the case of unbounded-query truth table reducibilities (cf. [3]).

- In section 3, we prove that k-conjunctive and k-disjunctive truth-table completeness are incomparable.

- In section 4, we show that many-one completeness is the same as 1-truth table completeness.

- In section 5, we give a precise relation between k-Turing and ℓ-truth-table completeness: for $k > 1$, k-Turing completeness strictly contains k-truth-table completeness, and for $k < \ell < 2^k - 1$, k-Turing completeness and ℓ-truth-table completeness are incomparable.

2 Preliminaries

2.1 Machines and languages

Let $\Sigma = \{0,1\}$. Strings are elements of Σ^*, and are denoted by small letters $x, y, u, v, \ldots$. For any string x, the length of a string is denoted by $|x|$. Languages are subsets of Σ^*, and are denoted by capital letters $A, B, C, S, \ldots$. The complement of a language A in Σ^*, $\Sigma^* - A$, will be denoted by $\overline{A}$. For any set S, the cardinality of S is denoted by $\|S\|$. We fix a pairing function $\lambda xy.<x,y>$ computable in polynomial time from $\Sigma^* \times \Sigma^*$ to Σ^*. Without loss of generality, we assume for all y, y' and x with $|y|, |y'| \leq |x|$ that $|<y,x>| = |<y',x>|$. We assume that the reader is familiar with the standard Turing machine model. An *oracle* machine is a multi-tape Turing machine with an input tape, an output tape, work tapes, and a *query* tape.

Oracle machines have three distinguished states QUERY, YES and NO, which are explained as follows: at some stage(s) in the computation the machine may enter the state QUERY and then goes to the state YES or goes to the state NO depending on the membership of the string currently written on the query tape in a fixed *oracle* set.

Oracle machines appear in the paper in two flavors: adaptive and non-adaptive. For a non-adaptive machine, queries may not be interdependent, whereas an adaptive machine may compute the next query depending on the answer to previous queries.

Whenever it is obvious that a universal recognizing or transducing machine exists for a class of languages (i.e. the class is recursively presentable), we will assume an enumeration of the acceptors and/or transducers and denote this enumeration by $M_1, M_2, \ldots$. For a Turing machine M, let $L(M)$ denote the set of strings accepted by M. For an oracle machine M, let $L(M, A)$ denote the set of strings accepted by M with oracle A.

2.2 Time classes

Let $\text{DTIME}(f(n))$ be the class of sets such that $A \in \text{DTIME}(f(n))$ iff there exists a Turing machine M for which the running time is bounded by $f(n)$ as $n \to \infty$ (n is the length of the input) and $A = L(M)$. Let $\text{NTIME}(f(n))$ be the corresponding nondeterministic class. We define the following classes:

$$NEXP = \bigcup_{i=1}^{\infty} \text{NTIME}\left(2^{n^i}\right)$$

$$EXP = \bigcup_{i=1}^{\infty} \text{DTIME}\left(2^{n^i}\right)$$

$$NE = \bigcup_{c=1}^{\infty} \text{NTIME}(2^{cn})$$

$$E = \bigcup_{c=1}^{\infty} \text{DTIME}(2^{cn})$$

2.3 Truth tables

The ordered pair $<<a_1, \ldots, a_k>, \alpha>$ ($k > 0$) is called a *truth-table condition of norm k* if $<a_1, \ldots, a_k>$ is a k-tuple of strings, and α is a k-ary Boolean function [8]. The set $\{a_1, \ldots, a_k\}$ is called the *associated set* of the tt-condition. A function f is a *truth-table function* if f is total and $f(x)$ is a truth-table condition for every x in Σ^*. If, for all x, $f(x)$ has norm less than or equal to k, then f is called a k-truth-table (k-tt) function. We say that a tt-function f is a *disjunctive* (*conjunctive*) truth-table (d-tt (c-tt)) function if f is a truth-table condition and its Boolean function is always a disjunction (conjunction) of its arguments.

2.4 Reductions, reducibilities and completeness

Let $A_1, A_2 \subseteq \Sigma^*$. We say that:

1. A_1 is polynomial-time many-one reducible to A_2 ($\leq_m^p$-reducible) iff there exists a function f computable within polynomial-time such that $x \in A_1$ iff $f(x) \in A_2$.

2. A_1 is polynomial-time k-truth-table reducible to A_2 ($\leq_{k-tt}^p$-reducible) iff there exists a polynomial-time bounded k-tt-function f such that $\alpha(\chi_{A_2}(a_1), \ldots, \chi_{A_2}(a_k))$ evaluates to **true** iff $x \in A_1$, where $f(x)$ is $<<a_1, \ldots, a_k>, \alpha>$ and χ_{A_2} is the characteristic function of the set A_2.

3. A_1 is polynomial-time Turing reducible ($\leq_T^p$-reducible) to A_2 iff there exists a polynomial-time bounded deterministic oracle machine such that $A_1 = L(M, A_2)$.

4. A_1 is polynomial-time disjunctive (conjunctive) reducible ($\leq_d^p$ ($\leq_c^p$)-reducible) to A_2, iff $A_1 \leq_{tt}^p A_2$ by some d-tt(c-tt)-function. For $k \geq 0$, A_1 is k-disjunctive(conjunctive) reducible ($\leq_{k-d}^p$ ($\leq_{k-c}^p$)) to A_2, if $A_1 \leq_{tt}^p A_2$ by some d-tt(c-tt)-function of norm k.

Let $\leq_r^p$ be any of the above reductions

1. A set A is $\leq_r^p$-hard for some complexity class C iff for all $B \in C$, B is $\leq_r^p$-reducible to A.

2. A set A is $\leq_r^p$-complete for some complexity class C iff A is $\leq_r^p$-hard for C and $A \in C$.

For *NEXP* we use a standard many-one complete set K. $K = \{<i, x, l>|$ machine M_i has an accepting computation on input x within steps$\}$. Note that this set can be recognized in 2^n steps and is therefore also complete for NE. For *EXP* and E we use $K = \{<i, x, l>|$ machine M_i is deterministic and accepts x within l steps$\}$.

3 Disjunctive and Conjunctive Truth-table Reductions

Theorem 1 *there exists a set $A \in NEXP$ such that A is $\leq^p_{2-d}$-complete but not $\leq^p_{2-c}$-complete.*

Proof: Let K be the standard $\leq^p_m$-complete set for NE as defined above. To achieve the separation, we construct a set $W \in E$ and a set $A \in NEXP$ such that $W \not\leq^p_{2-c} A$, but $K \leq^p_{2-d} A$. We assume an enumeration of polynomial time 2-conjunctive truth-table reductions $M_1, M_2, \ldots$ where M_i runs in time n^i. We need a set of elements on which to diagonalize. To get such a set, we define a sequence of integers $\{b(n)\}_n$:

$$b(n) = \begin{cases} 1 & \text{if } n \leq 1 \\ 2^{b(n-1)^{n-1}} + 1 & \text{otherwise} \end{cases}$$

We construct A and W in stages. $A \subseteq \{0,1\} \times \Sigma^*$ and $W \subseteq \{0\}^*$. At stage n, we define A_n and decide whether $0^{b(n)} \in W$ or $0^{b(n)} \in \overline{W}$. We let $A_{<n} = \bigcup_{i=0}^{n-1} A_i$ and $A = \bigcup_{n=0}^{\infty} A_n$

Stage 0: $A_0 = W = \emptyset$.

stage n:

Let $A'_n = \{<i,z>|\ z \in K$ and $b(n-1)^{n-1} < |<i,z>| \leq b(n)^n$ and $i \in \{0,1\}\}$

Simulate M_n on input $0^{b(n)}$. M_n will query at most two strings x and y. Without loss of generality, let x be the largest (in lexicographic order) of the two. M_n accepts iff x and y are both in the oracle set.

There are two cases:

1. $|x| \leq b(n-1)^{n-1}$

2. $b(n-1)^{n-1} < |x| \leq b(n)^n$

In case 1, compute the answers relative to $A_{<n}$ of both x and y and put $0^{b(n)} \in W$ iff M_n rejects. Let $A_n = A'_n$

In case 2, put $0^{b(n)} \in W$ and let $A_n = A'_n - \{x\}$. This ensures that M_n^A rejects on input $0^{b(n)}$.

end of stage n

The remainder of the proof consists of four items. We show that $A \in NEXP$, that $W \in E$, that A is not $\leq^p_{2-c}$-complete, and finally that A is $\leq^p_{2-d}$-complete.

- We show first that $A \in NEXP$. To decide $<i,z> \in A$ ($i = 0, 1$), compute n such that $b(n)^n \geq |<i,z>| > b(n-1)^{n-1}$, which can be done in linear time. Simulate machine M_n on input $0^{b(n)}$ and compute x and y. If $<i,z> = x$ reject, else accept iff $z \in K$. All this can be done in nondeterministic exponential time, since simulation of machine M_n on input $0^{b(n)}$ takes time $b(n)^n \leq 2^{n(b(n-1)^{n-1}+1)} \leq 2^{|<i,z>|^2}$, as $b(n) \geq 1$, and $b(n) \geq 2n$, whenever $n \geq 3$.

- Next we show that $W \in E$. On input $0^{b(n)}$, simulate M_n on input $0^{b(n)}$ and compute the queries x and y. Assume again, that x is the larger query in lexicographical order. If $|x| > b(n-1)^{n-1}$ we accept, else we will decide membership of x and y in A, and accept iff M_n rejects. To compute if $x \in A$, determine $n' < n$ such that $b(n'-1)^{n'-1} < |x| \leq b(n')^{n'}$. $x \in A$ iff x is not the largest query asked by $M_{n'}$ and $x \in K$. This takes deterministic time $2^{2^{|x|}} < 2^{b(n)}$. Membership of y in A can be decided by the same algorithm.

- Now assume for a contradiction, that A is $\leq^p_{2-c}$-complete. Note that $0^{b(n)} \in W$ iff M_n rejects. Then there must be a 2-conjunctive truth-table reduction from W to A. Let M_j be the machine witnessing this reduction. Then, by construction, $0^{b(j)}$ is in W iff M_j on input $0^{b(j)}$ rejects. This contradicts the assumption that M_j reduces W to A. This proves that A is not $\leq^p_{2-c}$-complete.

- Finally, we give the $\leq^p_{2-d}$-reduction from K to A. Since in every step, only one of the pairs $<1, x>$ or $<0, x>$ can be deleted, $x \in K$ iff $<0, x> \in A$ or $<1, x> \in A$. Therefore, the following reduction reduces K to A:

$$g(x) = \{<0, x> \vee <1, x>\}$$

□

The same proof technique yields the following theorem.

Theorem 2 *There exists a set $A \in NEXP$ such that A is $\leq^p_{2-c}$-complete, but not $\leq^p_{2-d}$-complete.*

Bounded Reductions

Proof: The proof is almost the same as the previous one. It differs only in case 2 of the diagonalization. Here we put $0^{b(n)}$ not in W and *add x to A'_n*. In this way we ensure that $0^{b(n)} \notin W$ iff M_n^A accepts. Note that $x \in K$ iff $<0, x> \in A$ and $<1, x> \in A$. The $\leq_{2-c}^p$-reduction from K to A becomes:

$$g(x) = \{<0, x> \wedge <1, x>\}$$

$\square$

It is easy to see, that the proofs generalize to $\leq_{k-d}^p$-complete sets v.s. $\leq_{k-c}^p$-complete sets (for $k \geq 2$). The theorems solve an open problem from Watanabe [10].

Corollary 3 *For all $k \geq 2$, there exists a set A that is $\leq_{k-tt}^p$-complete for NEXP, but not $\leq_{k-d}^p$ ($\leq_{k-c}^p$)-complete for NEXP.*

This corollary can be strengthened. We are now able to construct a set that is $\leq_{k-tt}^p$-complete, but neither $\leq_{k-d}^p$-complete nor $\leq_{k-c}^p$-complete.

Corollary 4 *For all $k \geq 2$, there exists a set A that is $\leq_{k-tt}^p$-complete for NEXP, but neither $\leq_{k-d}^p$-complete, nor $\leq_{k-c}^p$-complete for NEXP*

Proof: To do this we use the constructions of Theorem 1 at the even stages and the constructions of Theorem 2 at the odd stages $\square$

Corollary 5 *For all $k \geq 2$, there exists a set A that is $\leq_{k-tt}^p$-complete for NEXP, but neither $\leq_d^p$-complete, nor $\leq_c^p$-complete.*

By standard padding techniques the results in this section go through for NE, and since we do not make use of any special properties of nondeterminism, they also hold for E and EXP. By complementation the results also hold for Co-NE and Co-$NEXP$.

4 1-Truth-Table versus Many-One

Another question concerning reductions on E and NE is whether the notions $\leq_{1-tt}^p$ and $\leq_m^p$ differ for complete sets. From recursion theory it is known (and easy to prove), that these two reductions are the same with respect to RE sets if the time bounds on the reductions are omitted. Recently Homer et. al. [5] showed, that these two notions are also the same for E-complete sets. They also showed the notions to be identical in the case of RE-complete sets under polynomial time bounded reductions. They left open however the question for NE. We solve this question here. The idea is to first prove that sets in $NE \cap co\text{-}NE$ that are $\leq^p_{1-tt}$-reducible to a complete set are also $\leq^p_m$-reducible to this set. This can be done using a similar technique as in [5]. Once this is done, we are able to reduce the general case (if this is necessary) to this special case.

Lemma 6 *Let T be a $\leq^p_{1-tt}$-complete set for NE. For every set $A \in NE \cap co\text{-}NE$, $A \leq^p_m T$.*

Proof: We assume an enumeration of polynomial time 1-truth-table reductions $M_1, M_2, \ldots$. Without loss of generality, we may assume that M_i runs in time n^i. Let A be any set in $NE \cap co\text{-}NE$. We will construct a set $D \in NE$ with the property that $A \leq^p_{1-tt} D \leq^p_{1-tt} T$ and the $\leq^p_m$-reduction from A to T can be computed from the $\leq^p_{1-tt}$-reduction from D to T. We simulate M_i on input $<i, x>$, and let z be the string queried by M_i. Now there are four possible cases that can occur:

1. M_i accepts iff z is in the oracle set.

2. M_i accepts iff z is *not* in the oracle set.

3. M_i accepts. (M_i is not a $\leq^p_{1-tt}$-reduction on this input)

4. M_i rejects. (M_i is not a $\leq^p_{1-tt}$-reduction on this input)

In case 1 we put the pair $<i, x>$ in D iff $x \in A$
In case 2 we put $<i, x>$ in D iff $x \notin A$
In case 3 we put $<i, x>$ not in D
In case 4 we put $<i, x>$ in D
We first note, that D is in NE. To decide whether $<i, x>$ is in D or not, simulate machine M_i on input $<i, x>$ and find out in which case M_i ends up. The only problem is case 2, but since A is in $NE \cap co\text{-}NE$, we can compute whether x is in the complement of A. Having shown that D is in NE, we infer that D is 1-truth-table reducible to T. Let machine M_h witness this reduction and let z be the string queried by machine M_h on input $<h, x>$. Now we can construct the many one reduction f from A to T:

$$f(x) = z$$

Since machine M_h runs in polynomial time this reduction also runs in polynomial time. Machine M_h can not end up in case 3 or 4 on *any* input $<h, x>$, since this would contradict the fact the M_h is a 1-truth-table reduction from D to T. The following two cases remain:

- Machine M_h is in case 1. Then $x \in A$ iff $<h,x> \in D$ iff M_h accepts iff $z \in T$.

- Machine M_h is in case 2. Then $x \in A$ iff $<h,x> \notin D$ iff M_h rejects iff $z \in T$.

So in both cases $x \in A$ iff $z \in T$. □

Now for all sets in NE if a set is 1-truth-table reducible to a complete set T via say machine M_j there are strings that are accepted if the query is in T. Those strings are already many-one reducible to T. The other strings (i.e. the strings that get accepted by a query in the complement of T) form a set that is in $NE \cap co\text{-}NE$ and by Lemma 6 they are many-one reducible to T via some other reduction. We state:

Theorem 7 *Every $\leq^p_{1-tt}$-complete set for NE is also $\leq^p_m$-complete.*

Proof: Let A be a set in NE, T a 1-truth-table complete set in NE and let M_j witness the reduction from A to T. On any input M_j can end up in one of the following four situations:

1. M_j queries z and accepts iff $z \in T$

2. M_j queries z and accepts iff $z \notin T$

3. M_j accepts

4. M_j rejects

We now split set A in two subsets A_1 and A_2.

$$A_1 = \{x \mid x \in A \text{ and machine } M_j \text{ is not in case 2}\}$$
$$A_2 = \{x \mid x \in A \text{ and machine } M_j \text{ is in case 2}\}$$

CLAIM 8 A_2 *is in $NE \cap co\text{-}NE$.*

Proof: We need to show that there is a NE predicate for A_2 and for the complement of A_2.

$x \in A_2$ iff machine M_j in case 2 and $x \in A$
$x \notin A_2$ iff machine M_j not in case 2 or $z \in T$

It is clear that both predicates are NE. □

Now we can construct the many-one reduction from A to T: On input x simulate machine M_j on input x. If M_j is in case 1, then output z. If M_j in case 2, then x is in A iff x is in A_2. Since A_2 is in $NE \cap co\text{-}NE$ there is, by Lemma 6, a many-one reduction from A_2 to T say g. Now output $g(x)$. If M_j is in case 3, output a fixed element $t_0 \in T$, and if M_j is in case 4, output a fixed element $t_1 \notin T$. The entire construction can be carried out in polynomial time. □

The construction can be generalized to a recursion theoretic setting. We relax the time bounds and end up with recursive reductions. We now have the following equivalent reductions $\leq_m^{rec}$ for a many-one reduction and $\leq_{1-tt}^{rec}$ for a 1-truth-table reduction in exactly the same way as the above theorem was proven we can prove the following:

Corollary 9 *let Σ_k be the kth level of the arithmetic hierarchy as defined in [9]. For all k, if A is $\leq_{1-tt}^{rec}$-complete for Σ_k, then A is $\leq_m^{rec}$-complete for Σ_k.*

It would be interesting to prove the same result for the class NP. The problem is that the technique used in lemma 6 is not applicable for sets in NP. Under the strong assumption that $P = NP \cap co\text{-}NP$ however, we can prove it.

Corollary 10 *If $P = NP \cap co\text{-}NP$ then every $\leq_{1-tt}^p$-complete set for NP is $\leq_m^p$-complete.*

5 Bounded Turing versus bounded Truth-Table

We now turn our attention to bounded Turing reductions. Informally, these are Turing reductions where for any input x, the number of queries asked is bounded by a constant k. Note that by definition, every k-truth table reduction is a k-Turing reduction. It is well known, that every k-Turing reduction can be simulated by $(2^k - 1)$-truth-table reduction. A natural question one can ask is: "What is the relation between k-Turing reductions

Bounded Reductions 93

versus ℓ-truth-table reductions?" In the previous section, it was proven that for nondeterministic-exponential-time complete sets: many-one = 1-truth-table = 1-Turing. In this section we prove that k-Turing reductions are more powerful than k-truth-table reductions for $k > 1$, and that for $k < \ell < 2^k - 1$, k-Turing and ℓ-truth table reductions are incomparable. These results hold also for the corresponding completeness notions on $NEXP$.

Definition 11 *Let $Q(M, x, A)$ be the set of strings, queried in the computation of polynomial time oracle machine M with oracle A on input x. We say that $B \leq^p_{k-T} A$ if there exists a polynomial time oracle machine M such that $B = L(M, A)$ and for all x, $\|Q(M, x, A)\| \leq k$.*

Theorem 12 *For every k, there exists a set D in NEXP that is $\leq^p_{k-T}$-complete, but not $\leq^p_{(2^k-2)-tt}$-complete.*

As an example of the techniques used, we first prove the degenerate case $k = 2$, i.e. we will construct a set $D \in NEXP$, such that D is $\leq^p_{2-T}$-complete but not $\leq^p_{2-tt}$-complete.
Proof: Let $M_1, M_2, \ldots$, be an enumeration of the 2-truth-table reductions, where M_i runs in time n^i. Let K be the standard $\leq^p_m$-complete set for NE and let $\{b(n)\}_n$ the sequence defined in the proof of Theorem 1. We will construct sets $D \in NEXP$ and $W \in E$, such that $W \not\leq^p_{2-tt} D$, and $K \leq^p_{2-T} D$. W and D will be constructed in stages. At each stage n, we will define a set D_n and we let $D = \bigcup_{n=0}^{\infty} D_n$.

To ensure that $K \leq^p_{2-T} D$, we have to exploit the fact that a 2-Turing reduction can ask 3 queries in its *entire* query tree, while a 2-truth-table reduction can ask at most 2 queries in its entire query tree. We will ensure that $D \subseteq \{0, 1, 2\} \times K$, and use the following 2-Turing reduction M_T to reduce K to D:

On input x, first query $<0, x>$. If the answer is YES, query $<1, x>$, and accept iff the answer is YES. If the answer to query $<0, x>$ is NO, query $<2, x>$ and accept iff the answer is YES.

For every 2-truth-table reduction, and for every x, at least one of the strings $<0, x>, <1, x>, <2, x>$ is not queried on input $0^{b(n)}$, where n is such that $<0, x> \leq b(n)^n$. This provides enough freedom to diagonalize against the 2-truth-table reductions, while still keeping $K \leq^p_{2-T} D$ by M_T.

At stage $0 : D_0 = W = \emptyset$

stage n:

Let $D_n = \{<i,x>| \ x \in K \text{ and } b(n-1)^{n-1} < |<i,x>| \leq b(n)^r \text{ and } 0 \leq i \leq 2\}$.

Simulate M_n on input $0^{b(n)}$. If M_n queries strings of length $\leq b(n-1)^{(n-1)}$, compute the answers to those strings. I.e., compute the membership of those strings in $\bigcup_{i=0}^{n-1} D_i$.

Let Q be the set of (at most two) queries $\in \{0,1,2\} \times \Sigma^*$ with length $> b(n-1)^{(n-1)}$. Let $i_0 \in \{0,1,2\}$ be a number such that Q contains no string of the form $<i_0,x>$. Now we take the following action, depending on the value of i_0:

$i_0 = 0$: For every y occurring as second member in a pair of Q do
$D_n := (D_n \setminus \{<2,y>\}) \cup \{<1,y>\}$

$i_0 = 1$: For every y occurring as second member in a pair of Q do
$D_n := (D_n \setminus \{<2,y>\}) \cup \{<0,y>\}$

$i_0 = 2$: For every y occurring as second member in a pair of Q do
$D_n := (D_n \setminus \{<0,y>\}) \cup \{<1,y>\}$

Now we are able to compute whether M_n accepts or rejects the input $0^{b(n)}$, when given the oracle $\bigcup_{i=0}^{n} D_i$. The fact that no strings of length $\leq b(n)^n$ (i.e. strings that can be queried by M_n on input $0^{b(n)}$) are defined inside or outside D at subsequent stages ensures, that the computation on this input is the same as with oracle D

Put $0^{b(n)}$ in W iff M_n rejects on input $0^{b(n)}$.

end of stage n

We can use an argument similar to the one used in the proof of Theorem 1, to prove that $D \in NEXP$, $W \in E$ and W is not $\leq_{2-tt}^p D$. It remains to prove, that D is $\leq_{2-T}^p$-hard for $NEXP$. Our 2-Turing reduction M_T accepts x iff either $(<0,x> \in D \land <1,x> \in D)$ or $(<2,x> \in D \land <0,x> \notin D)$.

We have the following possibilities for $D \cap \{<0,x>,<1,x>,<2,x>\}$

$x \in K$: $\{<0,x>,<1,x>,<2,x>\}$ or $\{<0,x>,<1,x>\}$ or $\{<1,x>,<2,x>\}$.

$x \notin K$: $\emptyset$ or $\{<0,x>\}$ or $\{<1,x>\}$.

Thus, M_T accepts x iff $x \in K$ as required. □

For this proof, it was essential that a 2-Turing reduction can ask more queries in its entire query tree than can a 2-truth-table reduction. Since a

Bounded Reductions 95

k-Turing reduction can ask $2^k - 1$ queries in its entire query tree, while a $2^k - 2$ truth-table reduction can ask at most $2^k - 2$ queries in its entire oracle tree, we can use a generalization of the previous construction, to obtain a set D that is $\leq^p_{k-T}$-complete, but not $\leq^p_{(2^k-2)-tt}$-complete, thus proving Theorem 12.

Proof: Let $M_1, M_2, \ldots$, be an enumeration of the $2^k - 2$-truth-table reductions, where M_i runs in time n^i. Let K be a standard $\leq^p_m$-complete set for NE and let $\{b(n)\}_n$ the sequence defined in the proof of Theorem 1. We construct set D and W in stages; $D = \bigcup_{n=0}^{\infty} D_n$. We will ensure that $D \subseteq \{0, \ldots, 2^k - 2\} \times \Sigma^*$, and use the following k-Turing reduction M_T to reduce K to D.

On input x, first query $<0, x>$. For the j-th query $<i, x>$, where $j < k$ do the following: if the answer is YES, query $<2i + 1, x>$, else query $<2i + 2, x>$. Accept iff the k-th query asked gets answer YES.

At stage $0 : D_0 = W = \emptyset$

stage n:

Let $D_n = \{<i, x> | \ x \in K \text{ and } b(n-1)^{n-1} < |<i, x>| \leq b(n)^n \text{ and } 0 \leq i \leq 2^k - 2\}$.

Simulate M_n on input $0^{b(n)}$. If M_n queries strings of length $\leq b(n-1)^{(n-1)}$, compute the answers to those strings.

Let Q be the set of queries $\in \{0, \ldots, 2^k - 2\} \times \Sigma^*$ with length $> b(n-1)^{(n-1)}$. Let $i_0 \in \{0, \ldots, 2^k - 2\}$ be such that Q contains no string of the form $<i_0, x>$.

Consider the following tree of depth k, where the nodes are labeled $0, \ldots, 2^k - 2$: The root has label 0, and for each node at depth $< k$ with label i, the left child has label $2i + 1$, and the right child label $2i + 2$. We will identify roots with their labels, and we will visualize this tree as a query tree, such that on input x the node labeled i corresponds to query $<i, x>$. If our Turing reduction receives a YES answer to a query represented at some node, then the next query computed is represented by its left child and if it receives a NO answer, then the next query computed is represented by its right child.

For every y that occurs as second member in a pair of Q and and for every $i \in \{0, \ldots 2^k - 2\}, i \neq i_0$, we take the following action:

1. if i occurs on the path from the root to i_0 then

if i_0 is in the left subtree of i then $D_n := D_n \cup \{<i,y>\}$
if i_0 is in the right subtree of i then $D_n := D_n \setminus \{<i,y>\}$

2. if i occurs to the left of the path from the root to i_0 then
$D_n := D_n \cup \{<i,y>\}$

3. if i occurs to the right of the path from the root to i_0 then
$D_n := D_n \setminus \{<i,y>\}$

4. if i is in the left subtree of i_0 then $D_n := D_n \cup \{<i,y>\}$

5. if i is in the right subtree of i_0 then $D_n := D_n \setminus \{<i,y>\}$

Now we are able to compute whether M_n accepts or rejects. Put $0^{b(n)}$ in W iff M_n rejects on input $0^{b(n)}$.

end of stage n

We can use a similar argument as in the proof of theorem 1, to prove that $D \in \mathit{NEXP}$, $W \in E$ and $W \not\leq^p_{(2^k-2)-tt} D$. It remains to prove that D is $\leq^p_{k-T}$-hard for NEXP.

Recall that our k-Turing reduction M_T works as follows: On input x, first query $<0,x>$. For each query $<i,x>$ at depth $< k$ do the following: if the answer is YES, query $<2i+1,x>$, else query $<2i+2,x>$. Accept iff the last query asked gets answer YES. View this reduction as a tree of depth k, where the nodes are labeled by the queries, and a YES (resp. NO) answer to a query corresponds to taking the left (resp. right) branch.

If $x \in K$, then M_T, on input x, takes either the leftmost path in its oracle tree (if $i_0 = 0$), or the leftmost path through the node labeled i_0 which corresponds to query $<i_0,x>$. In either case we accept.

If $x \notin K$, then M_T, on input x, takes either the rightmost path in its oracle tree, or the rightmost path through $<i_0,x>$. In either case we reject.

Thus, M_T is a reduction from K to D. □

Now we will construct a set D in NEXP that is $\leq^p_{(k+1)-tt}$-complete, but not $\leq^p_{k-T}$-complete. A $\leq^p_{k-T}$-reduction can be represented as a binary tree of depth k, where every node in the tree represents a query. The reduction starts with the query represented by the root and proceeds after obtaining an anwer to each query as follows. If the answer to the query is YES, we proceed to the left branch otherwise to the right branch. The leaves of the tree are labeled with the quality accept or reject. The idea of the oracle construction is to force the $\leq^p_{k-T}$-reduction into one branch by leaving out all the queries (if possible) of that branch. Since there are only k queries on one branch, there remains the freedom to code an extra pair of K into D that can be queried by a $\leq^p_{(k+1)-tt}$-reduction.

Theorem 13 *There exists a set D in $NEXP$ that is $\leq^p_{(k+1)-tt}$-complete, but not $\leq^p_{k-T}$-complete.*

Proof: We only give the construction for $k = 4$. Let $M_1, M_2, \ldots$ be an enumeration of $\leq^p_{4-T}$ reductions. Let K be the standard $\leq^p_m$-complete set for NE and $\{b(n)\}_n$ the sequence defined in the proof of theorem 1. Again we use a stage construction.

stage n:
$D'_n := \{<i,x> | \ x \in K \text{ and } b(n-1)^{(n-1)} < |<i,x>| \leq b(n)^n \text{ and } 0 \leq i \leq 4\}$
Simulate M_n on input $0^{b(n)}$, and compute the answers to the queries of length $< b(n-1)^{n-1}$. Now evaluate the branch where all the other queries receive the answer NO. Let Q' be the set of the queries that are large (of length $\geq b(n-1)^{(n-1)}$).
Put $0^{b(n)}$ in W iff M_n rejects
$D_n := D'_n \backslash Q'$
end of stage n

Note that for every $x : x \in K$ iff $<i,x> \in D$ for some i. The 5-truth-table reduction from K to D becomes:

$$g(x) = \{<0,x> \vee \ldots <4,x>\}$$

□

Corollary 14 *If $k < \ell < 2^k - 1$, then $\leq^p_{k-T}$ and $\leq^p_{\ell-tt}$ are incomparable with respect to complete sets for $NEXP$.*

As before the results go through for NE, E and EXP.

6 Conclusions

In the previous sections, we proved that almost all reductions on NE, E, EXP and $NEXP$ are incomparable except those where inclusion is trivial.

As a consequence, the extended Berman Hartmanis conjecture for those reductions fails. It follows, that for example the sets in the degree of 2-truth-table complete sets are not p-isomorphic. An interesting step would be to disprove the extended conjecture for the degree of many-one complete sets. Perhaps the techniques discussed here could lead towards results in that direction.

The proof of the non-separation of many-one and 1-truth-table reductions fails for NP. The problem is, that it is not known whether the universal polynomial time function is computable in NP. For all well behaved classes that contain the universal polynomial time function, this non-separation result is true.

One area of great interest would be to separate the various polynomial time reductions on classes between P and $PSPACE$, and in particular to do this for NP.

All the previous obtained results go through with respect to logspace reductions for nondeterministic and deterministic space classes that contain that universal logspace function. It would be interesting to prove a similar result for $NLOGSPACE$.

Acknowledgements We would like to thank Steven Homer and Peter van Emde Boas for fruitful discussions.

References

[1] Berman L. *On the structure of complete sets*. Proc. 17th IEEE conference on Foundations of Computer Science (1976) pp76-80.

[2] Buhrman H., S. Homer & L. Torenvliet. *Completeness for nondeterministic complexity classes* Mathematical Systems Theory **24** (1991) 179–200.

[3] Buhrman H., E. Spaan & L. Torenvliet. *The relative power of logspace and polynomial time reductions*. To appear.

[4] Cook, S. A. *The complexity of theorem-proving procedures*. Proc. 3d ACM Symp. on Theory of Computing, Assoc. for Computing Machinery, New York (1971) pp151–158.

[5] Homer S, S. Kurtz & J. Royer *A note on many-one and 1-truth table complete languages* To appear in Theoretical Computer Science.

[6] Karp, R.M. *Reducibility among combinatorial problems*. Complexity of Computer Computations, R.E. Miller & J.W. Thatcher eds. Plenum N.Y. pp85–103.

[7] Ladner R.E. & N.A. Lynch. *Relativization of Questions About Log Space Computability*. Mathematical Systems Theory 10 (1976) pp19–32.

[8] Ladner, R.E., N. Lynch & A.L. Selman. *A comparison of polynomial time reducibilities*. Theoretical Computer Science 1 (1975) 103–123.

[9] Soare, R.I. *Recursively Enumerable Sets and Degrees*. Perspectives in Mathematical Logic. Springer-Verlag (1987) pp60–61.

[10] Watanabe, O. *A comparison of polynomial time completeness notions*. Theoretical Computer Science 54 (1987) pp249–265.

Promises and Fault-Tolerant Database Access

Jin-yi Cai*
Department of Computer Science
Princeton University
Princeton, NJ 08544

Lane A. Hemachandra[†]
Department of Computer Science
University of Rochester
Rochester, NY 14627

Jozef Vyskoč[‡]
Institute for Informatics and Statistics
842 21 Bratislava, Slovakia

---

*Research supported in part by the National Science Foundation under research grant CCR-9057486 and by a grant from MITL.

[†]Research supported in part by the National Science Foundation under research grants NSF-CCR-8957604 and NSF-INT-9116781/JSPS-ENG-207. Research done in part during visits to Universität Ulm and Princeton University; the latter visit was supported in part by DIMACS (Center for Discrete Mathematics and Theoretical Computer Science), a National Science Foundation Science and Technology Center—NSF-STC88-09648.

[‡]Work done in part while at the University of Rochester. Research supported in part by the National Science Foundation under grant NSF-CCR-8957604 and the Slovak Academy of Sciences under the grant "Complexity of Sequential and Parallel Computations."

Abstract

This paper studies the power of access, especially fault-tolerant access, to probabilistic databases and to unambiguous databases.

We study fault-tolerant access to probabilistic computation, and completely characterize the complexity classes R and ZPP in terms of fault-tolerant database access. We also show that consistent and inconsistent failure are in general interchangeable.

We study the power of three types of access to unambiguous computation: nonadaptive access, fault-tolerant access, and guarded access. (1) Though for NP it is known that nonadaptive access has exponentially terse adaptive simulations, we show that UP has no such relativizable simulations: there are worlds in which $k+1$-truth-table access to UP is not subsumed by k-Turing access to UP, or even to NP machines that are unambiguous on the questions actually asked. (2) Though fault-tolerant access (i.e., "1-helping" access) to NP is known to be no more powerful than NP itself, we give both structural and relativized evidence that fault tolerant access to UP suffices to recognize even sets beyond UP. Furthermore, we completely characterize, in terms of locally positive reductions, the sets that fault-tolerantly reduce to UP. (3) In guarded access, Grollmann and Selman's natural notion of access to unambiguous computation, a deterministic polynomial-time Turing machine asks questions to a nondeterministic polynomial-time Turing machine in such a way that the nondeterministic machine never accepts ambiguously. In contrast to guarded access, the standard notion of access to unambiguous computation is that of access to a set that is *uniformly* unambiguous—even for queries that it never will be asked by its questioner, it must be unambiguous. We show that these notions, though the same for nonadaptive reductions, differ for Turing and strong nondeterministic reductions.

1 Introduction

We say that a nondeterministic Turing machine N is *categorical* if for every $x \in \Sigma^*$ it holds that $N(x)$ has at most one accepting computation path; unambiguous polynomial time (UP), introduced by Valiant [Val76], is defined by: UP = $\{L \mid$ there is a categorical nondeterministic polynomial-time Turing machine N whose language is $L\}$. Clearly, P $\subseteq$ UP $\subseteq$ NP.

UP is intimately connected to such diverse issues as cryptographic security, the Berman-Hartmanis Isomorphism Conjecture, and the closure properties of the counting version of NP. For example, it is known that P $\neq$ UP if and only if one-way functions exist [Ko85, GS88], that UP equals probabilistic polynomial time if and only if the counting version of NP (i.e., #P) is closed under all polynomial-time operations [OH91], and that, though the classic potential counterexamples to the Berman-Hartmanis Isomorphism Conjecture are conditioned on the hope that P $\neq$ UP [JY85], even if no one-way functions exist and P $\neq$ NP, it is still plausible that there are non-isomorphic NP-complete sets [HH91].

In this paper, we are interested in viewing unambiguous computation as a database. That is, we are interested in how broad a class one gets via various types of access to a "database" that is itself a machine that in some sense computes unambiguously.

Section 2 studies the relationships between guarded access [GS88] to unambiguous computation and (conventional) access to UP. We prove that if guarded Turing access to unambiguous computation is no more powerful than conventional access to UP, then UP and FewP [AR88] are Turing equivalent. Additionally, we provide relativized evidence that for Turing, strong nondeterministic, and coNP reductions, guarded access yields tremendous power: there is a relativized world in which two guarded queries to unambiguous computation allows one to accept sets that are not in coNP^{UP}.

It is well-known that (both in the real world and in every relativized world, within the natural model of relativization) any set accepted via truth-table queries to NP can be accepted via logarithmically many Turing queries to NP [Hem89,Bei91]. Nonetheless, Section 3 gives evidence that UP exhibits no such efficient simulations. There is a relativized world in which truth-table queries to UP cannot be simulated by even one fewer Turing query to UP. Indeed, we prove this via an even stronger result. We show that there is a relativized world in which truth-table queries to UP cannot be simulated by even one fewer Turing query to any NP machine that is forbidden to accept ambiguously on any question actually asked (though the machine perhaps accepts ambiguously on some strings that never happen to be asked). That is, we show that there is a world in which k-truth-table access to UP is not subsumed by $k - 1$-Turing guarded [GS88] access to unambiguous computation.

Section 4 addresses a question of Ko. Ko [Ko87] proved that all sets in UP can be accepted *fault-tolerantly* given access to UP. He asked whether the converse inclusion holds. We provide a relativized counterexample to the converse inclusion, and completely characterize the sets that fault-tolerantly reduce to UP as the sets that left locally positive reduce [HJ91] to UP.

Section 5 address a related question of Ko—whether ZPP [Gil77], expected polynomial time, is exactly the class of sets accepted via fault-tolerant access to BPP [Gil77], bounded probabilistic polynomial time. We show that, in an appropriate access model, equality holds, thus completely characterizing ZPP (and, similarly, R, random polynomial time [Gil77]) in terms of fault-tolerant database access. This section also studies the relationship between reducing to each solution of a promise problem and reducing to a promise problem in a way that is robust with respect to *inconsistent* failures; we prove that these notions are identical for deterministic Turing reductions.

2 Guarded Access to Unambiguous Computation

Consider P^{UP}, the class of sets Turing reducible to some UP set. Unarguably, every set in this class is accepted via access to unambiguous computation. However, we feel that P^{UP} is too restrictive to capture adequately one's intuitive notion of Turing access to unambiguous computation. For a set to be in P^{UP}, it must be accepted by a P machine, M, given access to some nondeterministic polynomial-time Turing machine N that is unambiguous on every input question—even those questions that M never asks. We feel that this extremely strong uniformity requirement should be relaxed.

Grollmann and Selman suggested as a natural model of access to unambiguous computation a notion that removes this uniformity constraint [GS88]. In particular, they suggested allowing a querying machine to ask questions to a nondeterministic polynomial-time Turing machine, under the rule that on every question that it asks, the nondeterministic polynomial-time Turing machine must have at most one accepting path. The point is that the nondeterministic polynomial-time Turing machine need not be categorical. It is fine if the machine would on some inputs have more than one accepting path, as long as those inputs are never asked the machine. Note that in this scheme, all computations *actually performed* are indeed unambiguous. One can formalize this model by refining previous approaches to accessing "promise problems."

Promise problems, first defined and studied by Even, Selman, and Yacobi [EY80,ESY84], are a notion of conditionally correct computation: if some promise is met (e.g., "this boolean formula has at most one solution") then some property is correctly tested (e.g., "this boolean formula is satisfiable").

Definition 2.1 [ESY84] A *promise problem* is a pair of predicates (Q, R). The predicate Q is called the *promise*.

Definition 2.2 We define $\mathcal{UP}$ ("promise UP") as the following class of promise problems:
$$\{(Q_i, R_i) \mid i \geq 1\},$$
where $N_1, N_2, \cdots$ is a standard enumeration of nondeterministic polynomial-time Turing machines,[1] $Q_i = \{x \mid ||N_i(x)|| \leq 1\}$, and $R_i = \{x \mid ||N_i(x)|| \geq 1\}$.

Other papers (see, e.g., [KST92]) have discussed the promise problem (1SAT, SAT) (see Definition 5.4). Since Cook's Theorem (i.e.,

---

[1] Notation: For nondeterministic Turing machines N, $||N(x)||$ denotes the number of accepting paths of N on input x. For sets A, $||A||$ denotes the cardinality of A. For strings s, $|s|$ denotes the length of s.

the NP-completeness of SAT) can be made parsimonious [Sim75], access via any guarded reducibility to (1SAT, SAT) can substitute for access via the same reducibility to any member of $\mathcal{UP}$; this is a completeness result of sorts. However, (1SAT, SAT) is not as transparently amenable to relativization as is $\mathcal{UP}$, and so, since we prove some relativized results, we prefer the canonical approach of Definition 2.2.

The crucial issue at this point is how to naturally access a promise problem. If a query z made to (A, B) satisfies $z \in A$, then the promise is met, and the reply should be "yes" if $z \in B$ and "no" if $z \notin B$. But how should the case of $z \notin A$ be handled? Clearly, the answer to this depends on what intuition we wish "access to promise problems" to capture.

Section 5 presents a resolution fulfilling the notion that promise-breaking queries, though allowed, should result in answers that are completely unreliable. Each time such a query is asked, an arbitrary answer is returned, and our query machine must "do the right thing" regardless of which answer is returned. Additionally, Section 5 critically discusses a resolution in the literature—a resolution we feel to be less natural in conception, but that we prove to be equivalent. Section 5 also discusses similarities between the class $\mathcal{UP}$ and (1SAT, SAT), a previously studied promise problem.

The current section analyzes Grollmann and Selman's resolution in which promise-breaking queries are forbidden. Any query made in an actual run must obey the promise. Taking the case of unambiguous computation as an example, this resolution requires that every query that is asked must be a query on which the oracle machine's computation is unambiguous; we feel that this is the most satisfying way of modeling "access to unambiguous computation," as all the computation that actually occurs is indeed unambiguous, and we suggest that the definition below—which we'll use in this paper—is a good way of notating the Grollmann-Selman guarded access notion.[2]

Definition 2.3 Let $A = (Q, R)$ be a promise problem. We say that

---

[2] In their paper, Grollmann and Selman [GS88] used the term "smart" rather than "guarded."

$L \in \mathrm{P}^A$ if there is a deterministic polynomial-time Turing machine M such that:

1. $L = L(M^R)$, and

2. For every string x, in the computation of $M^R(x)$ every query z made to the oracle satisfies $z \in Q$.

For other standard reduction types [LLS75,Sel78,Lon82], we now similarly define guarded access. In this paper we use the following notations for various polynomial-time reductions: $\leq_m^p$ for many-one reductions; $\leq_c^p$ for conjunctive truth-table reductions or, equivalently, conjunctive Turing reductions; $\leq_d^p$ for disjunctive truth-table reductions or, equivalently, disjunctive Turing reductions; $\leq_b^p$ for bounded truth-table reductions or, equivalently, bounded Turing reductions; $\leq_{k\text{-}tt}^p$ for k-truth-table reductions, $k \geq 1$; $\leq_{tt}^p$ for truth-table reductions; $\leq_{k\text{-}T}^p$ for k-Turing reductions, $k \geq 1$; $\leq_T^p$ for Turing reductions; $\leq_{SN}^p$ for strong nondeterministic reductions.

Definition 2.4

1. Let $A = (Q, R)$ be a promise problem. For $r \in \{m, c, d, b, tt, 1\text{-}tt, 2\text{-}tt, \cdots, k\text{-}tt, \cdots \}$, let $\mathrm{R}_r^p(A) = \{L \mid L \leq_r^p R$ via a reducing function that asks only questions y such that $y \in Q\}$.

2. Let $A = (Q, R)$ be a promise problem. For $k \in \{1, 2, 3, ...\}$, a set L is in $\mathrm{R}_{k\text{-}T}^p(A)$ if there is a deterministic polynomial-time Turing machine M such that:

   (a) $L = L(M^R)$, and

   (b) For every string x, in the computation of $M^R(x)$ every query z made to the oracle satisfies $z \in Q$, and

   (c) For every string x, in the computation of $M^R(x)$, at most k queries are made to the oracle.

3. Let $A = (Q, R)$ be a promise problem. A set L is in $\mathrm{R}_{SN}^p(A)$ if there is a strong nondeterministic polynomial-time Turing

machine,[3] M such that:

(a) $L = L(M^R)$, and

(b) For every string x, in the computation of $M^R(x)$ every query z made to the oracle (along any computation path) satisfies $z \in Q$.

The notion of reductions to promise problems can be naturally extended to the notion of reductions to classes of promise problems.

Definition 2.5 (Guarded Access)
Let $\mathcal{C}$ be any class of promise problems, and let r be any reducibility for which the notion of reductions to promise problems has been defined. Then $R_r^p(\mathcal{C})$ denotes $\bigcup_{C \in \mathcal{C}} R_r^p(C)$.

UP is sometimes referred to as a "promise class" [HR92], as the machine "promises" that for no input will it have more than one computation path. This same promise burdens the UP machine of a P^{UP} computation. In contrast, in guarded access to unambiguous computation (e.g., in $P^{\mathcal{UP}}$), the only thing that is promised is that all questions actually asked will be unambiguously answered, and it is clear that the burden for fulfilling this relatively low-key promise is partially shouldered by the querying machine. In fact, in the proof of Proposition 2.7 below, it will be clear that the querying machine explicitly shields the oracle machine from having multiple accepting paths.

For clarity, throughout this paper we assume that our finite alphabet, Σ, is $\{0, 1\}$; however, all our results also hold for any finite alphabet.

First, we note that for nonadaptive reductions, guarded access gives no more power than access to UP.

---

[3]A nondeterministic polynomial-time Turing machine is said to be strong if for every input x: either one or more paths finish in an "accept" state (and all others finish in a "don't know" state) or one or more paths finish in a "reject" state (and all others finish in a "don't know" state) [Sel78,Lon82]. Following Johnson and others [Joh90], we eschew the quite different (see, e.g., [BLS84]) use of the same term in a way that requires a machine to have such properties robustly—that is, with respect to every oracle (see [HH90,GB91,BI87,Tar89]).

Proposition 2.6 Let $r \in \{m, c, d, tt, btt, 1\text{-}tt, 2\text{-}tt, \cdots, k\text{-}tt, \cdots\}$. Then $\mathrm{R}_r^p(\mathrm{UP}) = \mathrm{R}_r^p(\mathcal{UP})$.

Proof of Proposition 2.6 Nothing that is actually asked to $\mathrm{R}_r^p(\mathcal{UP})$ can generate ambiguous acceptance, and thus the very asking provides a certificate of sorts of non-ambiguity. For $r = tt$, if $L \in \mathrm{R}_{tt}^p(\mathcal{UP})$ via deterministic machine M and nondeterministic machine N, then $L \in \mathrm{R}_{tt}^p(\mathrm{UP})$ via the deterministic machine that on input x asks the same questions as M except with the input paired on (i.e., $y \longrightarrow \langle x, y \rangle$), and the categorical machine for the UP set $\{\langle x, y \rangle \mid M(x)$ queries y and $y \in L(N)\}$. ∎

On the other hand, Proposition 2.7 and Theorem 2.9 provide structural and relativized evidence that, for adaptive reductions, guarded access to unambiguous computation is more powerful than the conventional access to UP.

FewP, a generalization of UP defined by Allender and Rubinstein [AR88], is the class of sets that are accepted by nondeterministic machines of polynomially bounded ambiguity—machines N such that there exists some polynomial $p(\cdot)$ such that for every x it holds that $N(x)$ has at most $p(|x|)$ accepting paths. Clearly, $\mathrm{P} \subseteq \mathrm{UP} \subseteq \mathrm{FewP} \subseteq \mathrm{NP}$, and it is also known that $\mathrm{FewP} \subseteq \oplus \mathrm{P}$ ([CH90]; $\oplus \mathrm{P}$ is parity polynomial time [PZ83,GP86]). Proposition 2.7 says that if guarded Turing access to unambiguous computation fails to provide additional power, then UP and FewP are Turing equivalent.

Proposition 2.7 $\mathrm{P}^{\mathrm{UP}} = \mathrm{P}^{\mathcal{UP}} \Rightarrow \mathrm{P}^{\mathrm{UP}} = \mathrm{P}^{\mathrm{FewP}}$.

Proof of Proposition 2.7 Let $A = L(M^B)$, with $B \in \mathrm{FewP}$ and M a deterministic polynomial-time Turing machine. Let N be a machine whose ambiguity is bounded by polynomial $p(\cdot)$, such that $L(N) = B$. Let $B' = \{\langle x, n \rangle \mid N(x)$ has at least n accepting paths$\}$. Clearly, $B' \in \mathrm{NP}$. Let M' be the deterministic polynomial-time oracle machine that, on input x, simulates $M(x)$, except that each time $M(x)$ makes an oracle query, say z, M' handles it as follows. M' queries, one at a time, "$\langle z, p(|z|) \rangle \in B'?$," "$\langle z, p(|z|) - 1 \rangle \in B'?$," ..., "$\langle z, 1 \rangle \in B'?$," stopping the querying as soon as it gets its first "yes" response. As soon as a "yes" is returned, M' continues the simulation of $M(x)$ as if M had received the answer "yes"; if all queries were out of B', M' continues the simulation of $M(x)$ as if M had received the answer "no." M' and the natural NP machine for B' certify that $A \in P^{\mathcal{UP}}$. Thus, with no assumption, it holds that $P^{UP} \subseteq P^{FewP} \subseteq P^{\mathcal{UP}}$ (indeed, by the same argument, $P^{\mathcal{F}ew\mathcal{P}} = P^{\mathcal{UP}}$, where $P^{\mathcal{F}ew\mathcal{P}}$ is the natural[4] guarded analog of P^{FewP}). So by our hypothesis that $P^{UP} = P^{\mathcal{UP}}$, we have $P^{UP} = P^{FewP}$. ∎

The previous proposition provided a structural consequence of $P^{UP} = P^{\mathcal{UP}}$. We now provide evidence that even the weakest form of guarded adaptive access to unambiguous computation, guarded 2-Turing access, is not subsumed by Turing, strong nondeterministic, or even $\leq_{coNP}^p$ access to UP.

Definition 2.8 For any set A, we define $\mathcal{UP}^A$ as the following class of promise problems:
$$\{(Q_i, R_i) \mid i \geq 1\},$$
where $N_1, N_2, \cdots$ is a standard enumeration of nondeterministic polynomial-time Turing machines, $Q_i = \{x \mid ||N_i^A(x)|| \leq 1\}$, and $R_i = \{x \mid ||N_i^A(x)|| \geq 1\}$.

Theorem 2.9 $(\exists A)[R_{2\text{-}T}^p(\mathcal{UP}^A) \not\subseteq coNP^{UP^A} \cup NP^{UP^A}]$.

Corollary 2.10 Let $r \in \{SN, T, 2\text{-}T, \cdots, k\text{-}T, \cdots\}$. Then $(\exists A)[R_r^p(UP^A) \neq R_r^p(\mathcal{UP}^A)]$.

Before proving Theorem 2.9, we state a combinatorial lemma that will be useful (see also [CGH+89, p. 104] and [HH88, Lemma 3.2]).

Lemma 2.11 Let $G = (S, T, E)$ be any directed bipartite graph with out-degree bounded by d for all vertices. Let $S' \subseteq S$ and $T' \subseteq T$ be subsets such that $S' \supseteq \{s \in S \mid (\exists t \in T)[\langle s, t \rangle \in E]\}$, and $T' \supseteq \{t \in T \mid (\exists s \in S)[\langle t, s \rangle \in E]\}$. Then either:
  1. $||S'|| \leq 2d$, or

---
[4]That is, $\mathcal{F}ew\mathcal{P} = \{(Q_i, R_i) \mid i \geq 1\}$, where $N_1, N_2, \cdots$ is a standard enumeration of nondeterministic polynomial-time Turing machines, $Q_{\langle j, k \rangle} = \{x \mid ||N_j(x)|| \leq |x|^k + k\}$, and $R_{\langle j, k \rangle} = \{x \mid ||N_j(x)|| \geq 1\}$.

2. $||T'|| \leq 2d$, or

3. $(\exists s \in S')(\exists t \in T')[\langle s,t \rangle \notin E \land \langle t,s \rangle \notin E]$.

Note that one consequence of this lemma, very informally stated, is that if a nondeterministic polynomial-time machine rejects x with some oracle yet accepts x with the same oracle augmented by any one of a large collection of strings, then there are two elements of the collection such that if the oracle is augmented by both the elements then the machine has at least two accepting paths (this holds by putting a node into $S = S'$ and into $T = T'$ for each string in the collection, and for each string y in the collection putting $\langle s_y, z \rangle$ and $\langle t_y, z \rangle$ into the edge set if and only if z is queried on the lexicographically first accepting path of the machine on input x with the oracle augmented by string y).

Proof of Lemma 2.11 Consider the set:

$$X = \{(s,t) \in S' \times T' \mid \langle s,t \rangle \in E \text{ or } \langle t,s \rangle \in E\}.$$

Since the out-degree of every vertex in G is bounded by d, the cardinality of X is bounded by:

$$||X|| \leq (||S'|| + ||T'||)d \leq 2d\max(||S'||, ||T'||).$$

Suppose $||S'|| > 2d$ and $||T'|| > 2d$, then:

$$||X|| < \min(||S'||, ||T'||) \cdot \max(||S'||, ||T'||) = ||S' \times T'||.$$

Thus, alternative (3) holds. ∎

Proof of Theorem 2.9 Since $\mathrm{R}^p_{2\text{-}T}(\mathcal{UP}^A)$ is closed under complementation, we need prove only that for some A, $\mathrm{R}^p_{2\text{-}T}(\mathcal{UP}^A) \not\subseteq \mathrm{coNP}^{\mathcal{UP}^A}$. We will maintain, in our construction of A, the invariant: $(\forall n)[||A \cap \Sigma^n|| \leq 2]$. Let $L_A = \{1^n \mid ||A \cap \Sigma^n|| \neq 0\}$. Note that, thanks to the invariant, $L_A \in \mathrm{R}^p_{2\text{-}T}(\mathcal{UP}^A)$.

Indeed, L_A is even in a smaller class. $\mathrm{UP}_{\leq 2}$ denotes the class of languages that can be accepted by an NP Turing machine that on no input has more than two accepting paths [Wat88,Bei89]; of course, $\mathrm{UP} \subseteq \mathrm{UP}_{\leq 2} \subseteq \mathrm{FewP} \subseteq \mathrm{NP}$, and this inclusion chain itself relativizes.

Clearly, L_A is in $UP_{\leq 2}$, and thus the current proof actually constructs an A such that $UP_{\leq 2}^A \not\subseteq coNP^{UP^A}$.

Let A_ℓ, $\ell \geq 1$ indicate all strings added to the oracle by the end of stage ℓ, and set $A_0 = \emptyset$. Let $A = \bigcup_{\ell \geq 1} A_\ell$. Without loss of generality, let $N_0, N_1, \cdots$ be a standard enumeration of nondeterministic polynomial-time Turing machines such that for all k it holds that N_k runs in time $n^k + k$.

In stage $i = \langle j, k \rangle$, we will ensure that either (1) N_k is noncategorical with respect to oracle A, or (2) $1^n \notin \overline{L_A} \iff 1^n \in L(N_j^{L(N_k^A)})$. Thus, overall, the construction establishes that $\overline{L_A} \notin NP^{UP^A}$.

Stage $i = \langle j, k \rangle$: Choose n so large that it is larger than any string that has been placed into A or reserved for $\overline{A}$, and so large that:

$$2((n^j + j)^k + k) < \frac{2^n - ((n^j + j)^k + k)(n^j + j)}{(n^j + j)}.$$

Suppose $N_j^{L(N_k^{A_{i-1}})}(1^n)$ rejects. Freeze in their current state all strings of length at most $(n^j + j)^k + k$. Note that we have ensured that $1^n \in \overline{L_A}$ yet $1^n \notin L(N_j^{L(N_k^A)})$. Go to stage $i + 1$.

The remaining case is that $N_j^{L(N_k^{A_{i-1}})}(1^n)$ accepts. Let us focus our attention completely on one accepting path of $N_j^{L(N_k^{A_{i-1}})}(1^n)$ (for specificity, let us say the leftmost such path). For each query q that both is made (along this one path) to $L(N_k^{A_{i-1}})$ and receives the answer "yes," freeze in their current state the (polynomial number of) elements along a single accepting path of $N_k^{A_{i-1}}(q)$. Note that over all q's, this freezes at most a polynomial number—$((n^j+j)^k+k)(n^j+j)$—of elements. Of these at most $((n^j+j)^k+k)(n^j+j)$ elements, let B_0 denote those that are of length n.

Let $r_1, r_2, \cdots, r_s$ denote the queries made to $L(N_k^{A_{i-1}})$ (along this accepting path of $N_j^{L(N_k^{A_{i-1}})}(1^n)$) that receive the answer "no." (If there are *no* such queries, we simply add some unfixed string of length n to the oracle, and then freeze in their current state all strings of length at most $(n^j+j)^k+k$ and proceed to stage $i+1$, having already ensured that our machine is accepting when it should reject.) Note that $s \leq (n^j + j)$. Let B denote all strings y in $\Sigma^n - B_0$ such that for some t, $1 \leq t \leq s$, it holds that $N_k^{A_{i-1} \bigcup \{y\}}(r_t)$ accepts. Note that for each such y, for the r_t such that $N_k^{A_{i-1} \bigcup \{y\}}(r_t)$ accepts, we have that $N_k^{A_{i-1} \bigcup \{y\}}(r_t)$ queries y along each accepting path.

There are two subcases. (1) If $(B \bigcup B_0) \neq \Sigma^n$, then add any string from $\Sigma^n - (B \bigcup B_0)$ to the oracle, freeze in their current state all strings of length at most $(n^j + j)^k + k$, and go to stage $i+1$. (2) If $(B \bigcup B_0) = \Sigma^n$, then for at least one t, $1 \leq t \leq s$, it must be the case that for at least $(2^n - ||B_0||)/s$ strings y, $N_k^{A_{i-1} \bigcup \{y\}}(r_t)$ accepts. It follows from Lemma 2.11 and our choice of n that there exist strings $y_1, y_2 \in \Sigma^n - B_0$ such that $N_k^{A_{i-1} \bigcup \{y_1, y_2\}}(r_t)$ is noncategorical. Add y_1 and y_2 to the oracle, then freeze in their current state all strings of length at most $(n^j + j)^k + k$ and proceed to stage $i + 1$; we have ensured that N_k is noncategorical. (At this point, we could skip all future stages of the form $\langle i', k \rangle$, if we liked.) ∎

Corollary (to the Proof) 2.12 $(\exists A)[\mathrm{UP}_{\leq 2}^A \not\subseteq \mathrm{coNP}^{\mathrm{UP}^A}]$, and thus, clearly, $(\exists A)[\mathrm{FewP}^A \not\subseteq \mathrm{coNP}^{\mathrm{UP}^A}]$.

Note that it follows from Theorem 2.9 and Proposition 2.6 that:

Corollary 2.13 $(\exists A)[\mathrm{R}_{2-T}^p(\mathcal{UP}^A) \not\subseteq \mathrm{R}_{tt}^p(\mathcal{UP}^A)]$.

This result stands in contrast to the situation for UP and NP, for which bounded Turing (b-T) and bounded truth-table reductions (btt) are equivalent:

Fact 2.14

1. (see [Wag90]) $\mathrm{R}_{b-T}^p(\mathrm{NP}) = \mathrm{R}_{btt}^p(\mathrm{NP})$.

2. $\mathrm{R}_{b-T}^p(\mathrm{UP}) = \mathrm{R}_{btt}^p(\mathrm{UP})$.

In the next section, we'll look more closely at the two parts of the above fact, and will see that fine structures of the equalities are quite different.

3 Adaptive Versus Nonadaptive Reductions to UP

Corollary 2.13 shows that, for $\mathcal{UP}$, even very weak adaptive access is not robustly subsumed by nonadaptive access. It is not hard to see that the analog of Corollary 2.13 fails for UP. What *can* be proven of UP? In this section, we prove optimal (with respect to relativizable results) bounds on the relationship between adaptive and nonadaptive access to UP.

In fact, our results are sufficiently strong as to show that adaptive access to even $\mathcal{UP}$ does not (with respect to relativizable results) save even one query over nonadaptive access to UP, and that nonadaptive access to even NP does not (with respect to relativizable results) save even one query over the brute force nonadaptive simulation of adaptive access.

3.1 Adaptive Simulation of Nonadaptive Access

Consider the equalities of Fact 2.14 (both of which relativize). For the NP cases, there is a relatively crisp exponential tradeoff between adaptive and nonadaptive reductions: $R^p_{(2^k-1)-tt}(NP) \subseteq R^p_{k-T}(NP)$ [Bei91]. That is, adaptive reductions exponentially decrease the number of queries needed. The following result shows that such an improvement will be hard to obtain for unambiguous computation.

Theorem 3.1 For any constant k, there exists a recursive oracle A, such that $R^p_{k-tt}(UP^A) \not\subseteq R^{p,A}_{(k-1)-T}(\mathcal{UP}^A)$.

Some remarks are in order. We have chosen to compare the ordinary version of UP versus the promise version of $\mathcal{UP}$. This is a significant improvement of our earlier result [CHV92a,CHV92b] that $R^p_{k-tt}(UP^A) \not\subseteq R^{p,A}_{(k-1)-T}(UP^A)$. The reason is that in order to diagonalize away from $R^{p,A}_{(k-1)-T}(UP^A)$, it is sufficient to make a "potential" query to a noncategorical computation, even if by earlier queries and answers, this query would never be reached. In the more subtle case of $R^{p,A}_{(k-1)-T}(\mathcal{UP}^A)$, such a noncategorical computation accomplishes nothing. It is necessary to actually reach a query to a noncategorical computation to diagonalize. In other words, the construction has to be consistent with all previous queries and answers, if any.

In our earlier version dealing with UP^A, the constant k can be replaced by an arbitrary polynomial-time computable function $f(n)$ with polynomially bounded value. It remains open whether the claim of the current stronger version of Theorem 3.1 can be similarly generalized to non-constant access. We refer the interested reader to [CHV92a, CHV92b], which proves:

> Let us say that a function $f(\cdot)$ is nice if f is a total, polynomial-time computable function from $\mathcal{N} \longrightarrow \mathcal{N}$ such that f is polynomially bounded.[5] For every nice function $f(\cdot)$, there is a recursive oracle A such that $\text{R}^p_{f(n)+1-tt}(\text{UP}^A) \not\subseteq \text{R}^{p,A}_{f(n)-T}(\text{UP}^A)$.

Note that the distinction between $\text{R}^p(\cdot)$ and $\text{R}^{p,A}(\cdot)$ (e.g., in the above-quoted result and in Theorem 3.1) is that $\text{R}^{p,A}(\cdot)$ indicates that the deterministic transducer is allowed as many calls as it likes, up to a polynomial number of course, directly to A. The particular choice of $\text{R}^p(\cdot)$ and $\text{R}^{p,A}(\cdot)$ in Theorem 3.1 the one that makes this theorem technically strongest. Second, the actual test language in the proof is not only in $\text{R}^p_{k-tt}(\text{UP}^A)$, but in fact uses a fixed truth-table for all inputs—the parity truth table (that is, the truth table that accepts if and only if there are an odd number of queries whose answer is "yes").

Before we give the proof for Theorem 3.1, we first take a detour and consider a combinatorial game. We will prove a lemma about this game, and then we will apply the lemma to the construction of the oracle A.

---

[5]That is, there is a polynomial $p(\cdot)$ such that $(\forall n)[f(n) \leq p(n)]$. Since the theorem speaks of polynomial-time reductions, it makes complete sense to apply such a bound.

A Combinatorial Game

Suppose we are given the set $[n] = \{1, 2, \ldots, n\}$. The goal of this game is to produce eventually all the singleton sets $\{i\}$, for $1 \leq i \leq n$. The game is played in steps. At each step, we can produce an arbitrary collection of nonempty subsets of $[n]$, subject to the following restriction: for each pair of distinct subsets of $[n]$ this collection has, say A and B, there must be a subset C of $[n]$ that (a) satisfies $C \subseteq A \cup B$, and (b) was produced in some earlier step. For any set S let $\mathcal{P}(S)$ denote the power set of S.

Our combinatorial lemma states that, in order to produce all the singleton sets, the game must take at least n steps.

Lemma 3.2 If $\mathcal{S}_i \subseteq \mathcal{P}([n]) - \{\emptyset\}$, for $1 \leq i \leq m$, satisfy

1. $(\forall i : 1 \leq i \leq n)\,(\exists k_i)\,[\{i\} \in \mathcal{S}_{k_i}]$, and
2. $(\forall i : 1 \leq i \leq m)$ [if $A, B \in \mathcal{S}_i$ and $A \neq B$, then $(\exists j : j < i)\,(\exists C \in \mathcal{S}_j)\,[C \subseteq A \cup B]$],

then $m \geq n$.

We first make some easy remarks. We may without loss of generality assume that no $\mathcal{S}_i$ is empty; as deleting such an $\mathcal{S}_i$ will still satisfy our two conditions. We also explicitly required that no member of each $\mathcal{S}_i$ may be the empty set (otherwise this would trivialize condition 2).

It follows from condition 2 and the fact that no $\mathcal{S}_i$ is empty, that $||\mathcal{S}_1|| = 1$. We also note that without loss of generality we may assume that no subset $A \subseteq [n]$ need to appear more than once among the $\mathcal{S}_i$'s. By deleting all appearances of A except the first, we obtain a sequence $\mathcal{S}_i''$ still satisfying both conditions. More specifically, define, for $1 \leq i \leq m$,

$$\mathcal{S}_i'' = \mathcal{S}_i - \{\,A \in \mathcal{S}_i \mid (\exists j : j < i)[A \in \mathcal{S}_j]\,\}.$$

Then the sequence $\mathcal{S}_i''$ satisfies both condition 1 and 2. Note that after this "without loss of generality" transformation, it holds that for each i, there is a unique k_i (in the notation of condition 1).

We now turn to the proof of the Lemma.

Proof of Lemma 3.2 We prove by induction on n. The base case $n = 1$ is trivial. Assume the lemma is true for $n - 1$ and consider a sequence $\mathcal{S}_1, \ldots, \mathcal{S}_m$ satisfying both conditions, and with the above "without loss of generality" transformation already made.

We define a "relativized sequence with respect to n" as follows: $\mathcal{S}'_i$, where $1 \leq i \leq m$ and $i \neq k_n$ (in the notation of condition 1), will consist of nonempty subsets of $[n-1]$.

$$\mathcal{S}'_i = \begin{cases} \{ A - \{n\} \mid A \in \mathcal{S}_i \} & \text{if } i < k_n \\ \{ A \mid A \in \mathcal{S}_i \text{ and } n \notin A \} & \text{if } i > k_n \end{cases}$$

Note that since k_n is unique, this sequence is well-defined. Moreover, for $i \neq k_n$, if $A \in \mathcal{S}_i$, then $A \neq \{n\}$ and thus all sets collected in various $\mathcal{S}'_i$'s are nonempty, and by construction, are subsets of $[n-1]$.

We now prove that the "relativized sequence" $\mathcal{S}'_i$ satisfies our two conditions. This will imply that $m - 1 \geq n - 1$ by inductive hypothesis, and hence the induction will be completed.

For condition 1, let $i < n$, we show that there is an ℓ_i such that $\ell_i \neq k_n$ and $\{i\} \in \mathcal{S}'_{\ell_i}$. By hypothesis, there is a k_i such that $\{i\} \in \mathcal{S}_{k_i}$. If $k_i \neq k_n$, then we are done since $\{i\} = \{i\} - \{n\}$. Suppose $k_i = k_n$, then we apply condition 2 to the distinct sets $\{i\}$ and $\{n\}$ and we obtain a $j < k_n$ and a set $C \in \mathcal{S}_j$, such that $C \subseteq \{i\} \cup \{n\}$. Since $C \neq \emptyset$, and $C \neq \{i\}, \{n\}$ by the uniqueness of k_i and k_n, it must be that $C = \{i, n\}$. So $C - \{n\} = \{i\} \in \mathcal{S}'_j$.

To see that condition 2 is satisfied, let's take $A', B' \in \mathcal{S}'_i$ for some $i \neq k_n$, and $A' \neq B'$. If $i < k_n$ then $A' = A - \{n\}$ and $B' = B - \{n\}$ for some A and B in $\mathcal{S}_i$. Clearly $A \neq B$, thus there is a j, $j < i$, and a C, $C \in \mathcal{S}_j$, such that $C \subseteq A \cup B$. Then $C - \{n\} \in \mathcal{S}'_j$, and $C - \{n\} \subseteq A' \cup B'$.

If $i > k_n$ then $A', B' \in \mathcal{S}_i$ and $n \notin A', B'$. Thus there is a j, $j < i$, and a C, $C \in \mathcal{S}_j$, such that $C \subseteq A' \cup B'$. Since $n \notin A', B', n \notin C$, and $C - \{n\} = C$. Thus in case either $j < k_n$ or $j > k_n$, we are done as $C = C - \{n\} \in \mathcal{S}'_j$ and $C \subseteq A' \cup B'$. Finally suppose $j = k_n$. Since $C \neq \{n\}$, by condition 2, there is an ℓ, $\ell < j$, and a C^*, $C^* \in \mathcal{S}_\ell$, such that $C^* \subseteq C \cup \{n\}$. So we have $C^* - \{n\} \subseteq C \subseteq A' \cup B'$ and $C^* - \{n\} \in \mathcal{S}'_\ell$. ∎

An Intuitive Description of the Construction

We now give an outline of the proof of Theorem 3.1.

For an oracle set A, we will define a language L_A in $\mathrm{R}^p_{k-tt}(\mathrm{UP}^A)$. We will reserve a segment in A, which consists of k regions, for the membership in L_A for each input length. There will be at most one string in each of the k regions of the segment reserved for length n. The language L_A will be defined as a tally set of exactly those n such that the k regions that belong to the segment corresponding to length n contain an odd number of strings. Clearly, this language will be in $\mathrm{R}^p_{k-tt}(\mathrm{UP}^A)$.

To diagonalize against all deterministic polynomial-time Turing machines M that make at most $k-1$ sequential queries to $\mathcal{UP}^A$ oracles, we proceed as follows. We can expect to "use" one and only one region for each query and still be left with enough room to diagonalize. Initially, we will have no strings added to the segment at length n. When a query to a $\mathcal{UP}^A$ computation is made, it is expected that the answer to the query is "no," since a "yes" answer can be preserved via freezing an accepting path (which only excludes at most a polynomial number of strings from the oracle—a very small price to pay). If we have not added any strings at the segment corresponding to length n, we would expect all $k-1$ queries to be answered with a "no," and the machine M to halt with a result. If it rejects, which is correct for even parity, we would like to add one string in one of the k regions corresponding to length n to defeat this machine, and should preserve all the answers to the $k-1$ queries. The key issue is whether we can always do that, and indeed we can.

The first query, if the answer is "no" currently, might change later with the addition of some strings. So suppose it can be answered with a "yes," if we add a string in one particular region. If this happens in only one region, then this is not too bad, as we can exclude this one region from future consideration. We say that the first query "claims" this region. In the end, we would like to add one string, if necessary, so that the parity differs from the final answer of the machine M. The difficulty arises when a query may "claim" more than one region.

This is not a concern for the first query. In fact, if it "claims" more than one region, then by simultaneously adding two strings in two distinct "claimed" regions, we will have forced a noncategorical computation. The situation for the later queries is more subtle. For instance, the second query may "claim" both regions 1 and 2, but if we add one string in both regions, the first query may change its answer from "no" to "yes." In other words, the first query may have made a "claim on the pair of regions" 1 and 2, even though it may not have any "claim" on a single region. In this case, we may not add both strings, since this would "use up" two regions for just one query—the first query, and since it now answers with a "yes," the second query in general changes. Further complicating the matter is the fact that such claims of pairs or even larger tuples of regions are possible for all $k-1$ sequential queries, and we must prove that in the end we still have at least one region to add strings to without changing the answers to previous queries.

This is where we are going to use our combinatorial game and Lemma 3.2. We will consider the sequence of $k-1$ queries as $k-1$ moves in our combinatorial game, where a collection of non-empty subsets of $[k]$ is "claimed" at each step. If a subset $C = \{i_1, \ldots, i_s\} \subseteq [k]$ is "claimed" at step j, it intuitively means that there are "many" tuples of the form $(x_{i_1}, \ldots, x_{i_s})$, where x_{i_ℓ} belongs to the i_ℓth region, such that if we add them simultaneously to the oracle set A, we will have an accepting computation for the jth query. Since a "claim" implies many such tuples, when two claims on two distinct subsets C_1 and C_2 are made for the same query, we should be able to paste together a tuple corresponding to $C_1 \cup C_2$, which forces a noncategorical computation. This will finish off the current Turing machine M, *unless* it would produce some accepting computation for a previous query that was answered with a "no." Now here is the crucial observation: *This could happen only if there were a set D, claimed in a previous stage, such that $D \subseteq C_1 \cup C_2$.* In other words, the constraint for our game applies. Now Lemma 3.2 says that there must be one "unclaimed" region at the end, since we have only $k-1$ steps. This is the region with which we will diagonalize.

Proof of Theorem 3.1

We now give the details of the proof. For length n we will reserve the following segment of k regions, $S_{n,i} = \{x \mid (\exists y \in \{0,1\}^n)[x = 1^n 01^i 0y]\}$, $1 \le i \le k$. For $n \ge 1$, define $S_n = \bigcup_{i=1}^{k} S_{n,i}$. For all $n \ge 1$ and $1 \le i \le k$, we stipulate that $||A \cap S_{n,i}|| \le 1$. Let:

$$L_A = \{1^n \mid ||A \cap S_n|| \text{ is odd}\}.$$

Clearly as long as the oracle set A maintains the stipulation that $||A \cap S_{n,i}|| \le 1$, it holds that $L_A \in \text{R}^p_{k-tt}(\text{UP}^A)$. We show that, for some such oracle A, $L_A \notin \text{R}^p_{(k-1)-T}(\mathcal{UP}^A)$.

Let M_i be a standard enumeration of all deterministic polynomial-time Turing machines making (for any oracle and any input) at most $k-1$ queries to the oracle, and let N_i be a standard enumeration of nondeterministic polynomial-time Turing machines. Moreover, assume that machines M_i and N_i on inputs of size n run in time at most $n^i + i$. Let $\langle \cdot, \cdot \rangle$ be a standard pairing (i.e., one-one onto) function between $\mathcal{N} \times \mathcal{N}$ and $\mathcal{N}$ that is polynomial-time computable, polynomial-time invertible, and satisfies $(\forall i, j, k)[i = \langle j, k \rangle \implies \max(j, k) \le i]$. Using the standard correspondence between integers and strings, we will use this pairing function in both realms.

There is a countable set of requirements:[6]

$R_{\langle i, j \rangle}$:
There exists an n such that M_i on 1^n either makes some query x where $N_j^A(x)$ has more than one accepting path, or $M_i^{A, L(N_j^A)}(1^n) \ne L_A(1^n)$.

Suppose we have satisfied all the requirements R_ℓ, $\ell < s$. Let $s = \langle i, j \rangle$, and let A_s be the oracle set constructed so far, and let A'_s the set of strings forbidden membership in A. Initially both A_s and A'_s are empty. Choose a sufficiently large n_s so that no strings of length at least n_s are either in A_s or A'_s. Now consider the computation of M_i on 1^{n_s}. If M_i asks questions about A directly, we will answer them truthfully

---

[6] For any machine M and sets A and B, we'll use $M^{A,B}$ to denote $M^{A \oplus B}$, where $\oplus$ denotes disjoint union—$A \oplus B = \{0x \mid x \in A\} \cup \{1y \mid y \in B\}$.

Promise Problems and Guarded Access 121

according to A_s, and freeze all queries not in A_s out of A, that is, add them to A'_s. Since M_i is a deterministic polynomial-time machine, this can add at most a polynomial number $n_s^i + i$ of strings to A'_s. (We assume that n_s was chosen sufficiently large that this polynomially bounded quantity is much smaller than 2^{n_s}.)

The main concern is, of course, when M_i asks queries about a string x, where the computation of N_j on x is with respect to the oracle A and is supposedly an unambiguous computation (that is, one that has at most one accepting path). Let x be the first such query. If $N_j^{A_s}(x)$ accepts, we freeze one such accepting path, and add to A'_s the at most polynomial number of strings queried by N_j along this path that are not in A_s. We will move along the computation of M_i with the answer to its first query as "yes." This cannot change as long as no string from A'_s is ever added to A. (If there is more than one accepting path, then by freezing two such paths we are done, i.e., the requirement R_s is satisfied. Without loss of generality we assume this is not the case and we similarly assume this for later queries.)

Thus, we consider the queries of M_i along its computation, freezing accepting computations of N_j (if any), without adding any strings to the oracle. Since a "yes" answer from a query without adding any string to A only effectively decreases the number $k - 1$ of allowed queries, with only a small number (i.e., a polynomial number) of strings possibly excluded from future consideration, we may without loss of generality assume that all $k - 1$ queries are answered with "no" for the current A_s.

Let, initially,

$$X = ((S_{n_s,1} - A'_s) \cup \{\epsilon_1\}) \times ((S_{n_s,2} - A'_s) \cup \{\epsilon_2\}) \times \cdots \times ((S_{n_s,k} - A'_s) \cup \{\epsilon_k\})$$

be the *sample space* corresponding to an arbitrary configuration of the oracle for the length n_s, satisfying the global constraint $||A \cap S_{n_s,i}|| \leq 1$. Here, ϵ_i denotes the absence of all strings from region i. We will argue that there must be two configurations with different parity that the computation $M_i^{L(N_j^A)}$ cannot distinguish, unless one of the queries $N_j^A(x)$ has two accepting paths—an ambiguous computation.

For each query x_u, $1 \leq u \leq k-1$, consider all configurations of the oracle set A for which $N_j^A(x_u)$ has an accepting computation. We say that a query x_u makes a "C-claim," for some $C \subseteq [k]$, if there is a tuple $(z_1, \ldots, z_k) \in X$, where z_t is some $y_t \in S_{n_s,t} - A'_s$ if $t \in C$ and $z_t = \epsilon_t$ otherwise, such that the addition of exactly these strings y_t to A_s gives an accepting computation in $N_j^{A_s \cup \{y_t | t \in C\}}(x_u)$, but no proper subset of $\{y_t | t \in C\}$ will. Call such a tuple $y_t \in S_{n_s,t} - A'_s, t \in C$, a "triggering tuple." Say $C = \{1, 2, \ldots, c\}$. We call such a claim a "minor C-claim," if

$$\frac{\|B_C \cap X\|}{\|X\|} \leq \frac{2}{3},$$

where

$B_C = \{(y_1, y_2, \ldots, y_c, x_{c+1}, \ldots, x_k) \in X \mid (y_1, y_2, \ldots, y_c)$ is a triggering tuple, and $x_\ell \in S_{n_s,\ell}\}$.

A C-claim that is not a minor C-claim is called a "major C-claim."

We will "deal" with any minor C-claim by setting $X := X - B_C$. We note that any query that had a minor C-claim dealt with as above cannot make any C-claims in the future. There is one subtlety that should be mentioned. We defined the notion of a minor C-claim with respect to the current X, which can change whenever we deal with such a claim. Thus a minor claim may become a major claim, or cease to be a claim at all, when some other minor claim is dealt with. The important thing is that once a minor C-claim by a query is dealt with, this query can never make another C-claim. Therefore, we will just go up and down the query sequence of $k-1$ queries of M_i, dealing with any outstanding minor claims in any order, until there are no more minor claims left (with respect to the current X). There could be at most $(2^k - 1) \cdot (k - 1)$ such claims dealt with, and X will have cardinality at least $2^{nk}/3^{2^k k}$ through out.

Now suppose there are no more minor claims left. Let $\mathcal{S}_\ell$ consist of those C such that the ℓth query has a (major) C-claim left. We claim that the condition 2 of Lemma 3.2 is satisfied by the sequence $\mathcal{S}_\ell$.

In fact, let C and D be two distinct major claims by the ℓth query. For notational convenience, let $C = \{1, \ldots, c, c+1, \ldots, d\}$,

$D = \{1, \ldots, c, d+1, \ldots, e\}$, and $C \cap D = \{1, \ldots, c\}$, where $c \geq 0$, and $1 \leq d = ||C|| \leq k$, $1 \leq c + e - d = ||D|| \leq k$.

Since these claims are major claims,

$$||(B_C \cap X) \cap (B_D \cap X)|| > ||X||/3.$$

Thus, there are many strings in X of the form

$$(y_1, \ldots, y_c, y_{c+1}, \ldots, y_d, y_{d+1}, \ldots, y_e, x_{e+1}, \ldots, x_k),$$

where $T_C = (y_1, \ldots, y_c, y_{c+1}, \ldots, y_d)$ and $T_D = (y_1, \ldots, y_c, y_{d+1}, \ldots, y_e)$ are both triggering tuples to the computation of N_j for the ℓth query. Since T_C corresponds to an accepting path, along this path there can be at most a polynomial number of strings queried and not in A_s. More specifically, the set

$I_C = \{(y_1, \ldots, y_d, x_{d+1}, \ldots, x_k) \in X \mid (y_1, \ldots, y_d)$ is a triggering tuple, and along its accepting path some x_v, $d+1 \leq v \leq k$, was queried$\}$,

has cardinality $\leq 2^{n_s d} \cdot (k-d) \cdot (n_s^j + j) \cdot (2^{n_s(k-d-1)}) = k(n_s^j + j)2^{n_s(k-1)}$, which is much smaller than $||X||/3$. Similarly for I_D corresponding to T_D. Thus there exists $(y_1, \ldots, y_e, x_{e+1}, \ldots, x_k) \in X - (I_C \cup I_D)$, giving rise to triggering tuples T_C and T_D that are non-interfering. Hence, N_j will have two accepting computations if we add to the oracle the strings $y_1, \ldots, y_c, y_{c+1}, \ldots, y_d, y_{d+1}, \ldots, y_e$.

We claim that the sample point $(y_1, \ldots, y_c, y_{c+1}, \ldots, y_d, y_{d+1}, \ldots, y_e, \epsilon_{e+1}, \ldots, \epsilon_k)$ is still in X, i.e., it had not been removed due to any prior minor claims having been dealt with. Otherwise, such a minor claim must be a B-claim, for some $B \subseteq C \cup D$. But then the sample point $(y_1, \ldots, y_c, y_{c+1}, \ldots, y_d, y_{d+1}, \ldots, y_e, x_{e+1}, \ldots, x_k)$, which is in X, would have been taken out of X as well.

Thus it is consistent with any previous restrictions due to A_s' to add strings

$$y_1, \ldots, y_c, y_{c+1}, \ldots, y_d, y_{d+1}, \ldots, y_e.$$

Now either the computation of N_j is ambiguous, in which case we are done for $R_{\langle i, j \rangle}$, or else some earlier query $\ell' < \ell$ must have changed to "yes" with the addition of $y_1, \ldots, y_c, y_{c+1}, \ldots, y_d, y_{d+1}, \ldots, y_e$. This means that the ℓ'th query had made a (major) claim for some subset $B \subseteq C \cup D$.

Now our Lemma 3.2 applies. Since we have at most $k-1$ steps in the game, we will have some $\{w\}$, $1 \leq w \leq k$, such that there are no major $\{w\}$-claims. We claim that some $(\epsilon_1, \ldots, \epsilon_{w-1}, s_w, \epsilon_{w+1}, \ldots, \epsilon_k) \in X$, where $s_w \neq \epsilon_w$, must remain in X. Such a sample point can be only excluded by a minor $\{w\}$-claim. As there are at most k such claims (note: there are at most $k2^k$ minor claims altogether), there exists some $s_w \neq \epsilon_w$ so that $(\epsilon_1, \ldots, \epsilon_{w-1}, s_w, \epsilon_{w+1}, \ldots, \epsilon_k) \in X$. And of course $(\epsilon_1, \ldots, \epsilon_w, \ldots, \epsilon_k) \in X$ as well. These two configurations have different parity, and yet the computation of M_i will remain the same with or without s_w, while all the queries are answered "no." ∎

3.2 Nonadaptive Simulation of Adaptive Access

Just as Theorem 3.1 showed the limitations of adaptive simulations of nonadaptive access to UP, so does Theorem 3.3 study the limitations of nonadaptive simulations of adaptive access to UP. Intuitively, Theorem 3.3 and Corollary 3.4 say that the trivial $2^k - 1$-query brute-force parallel simulation of sequential access to UP is in fact optimal with respect to robust results. As in the case of Theorem 3.1, the choice of $R^p(\cdot)$ and $R^{p,A}(\cdot)$ below is that which makes the theorem the strongest.

Theorem 3.3 There is a recursive oracle A such that, for all $k \geq 1$, it holds that $R^p_{k\text{-}T}(\text{UP}^A) \not\subseteq R^{p,A}_{2^k-2\text{-}tt}(\text{NP}^A)$.

The proof is based on a "force your way through the tree" technique of Buhrman, Spaan, and Torenvliet [BST], who use their technique to prove that $\bigcup_{k>0} \text{NTIME}[2^{n^k}]$ has a set that is complete with respect to k-Turing reductions but not with respect to $2^k - 2$-truth-table reductions.

Proof of Theorem 3.3 Generally speaking, we'll define a language that k-Turing reduces to a UP^A set. Then we'll prune the $2^k - 1$ node tree implicit in the reduction in such a way that each time we remove a node from this tree we also fix the answer to one of the $2^k - 2$ truth-table queries being made. At the end of this process, we'll still have both accepting and rejecting possibilities in the Turing reduction tree, yet the behavior of the truth-table reduction will be settled. We now turn to the detailed proof.

Though the theorem states that $(\exists \text{ recursive } A)(\forall k \geq 1)$ $[R^p_{k\text{-}T}(\text{UP}^A) \not\subseteq R^{p,A}_{2^k-2\text{-}tt}(\text{NP}^A)]$, for clarity we'll instead show that $(\forall k \geq 1)(\exists \text{ recursive } A)[R^p_{k\text{-}T}(\text{UP}^A) \not\subseteq R^{p,A}_{2^k-2\text{-}tt}(\text{NP}^A)]$. It is not hard to see that by interleaving diagonalizations in the standard way one can easily transform our proof of the latter into a proof of the former.

For each $m \geq 1$, and each $j \in \{0,1\}^{\leq k-1}$, let $Q_{m,j} = \{\langle m, j, y\rangle_3 \mid |y| = m\}$, where $\langle \cdot, \cdot, \cdot \rangle_3$ is a 3-ary pairing function with the standard nice properties (a polynomial-time computable bijection that is polynomial-time invertible and honest). We'll later use $\langle \cdot, \cdot \rangle_2$ for a 2-ary pairing function that similarly has the standard nice properties.

Let
$$b_{m,1} = \begin{cases} 1 & \text{if } Q_{m,\epsilon} \cap A \neq \emptyset \\ 0 & \text{otherwise.} \end{cases}$$

We'll view $b_{m,1}$ as a bit string of length 1. Inductively, for $\ell \leq k$, let
$$b_{m,\ell} = \begin{cases} 1 & \text{if } Q_{m, b_{m,1} b_{m,2} \cdots b_{m,\ell-1}} \cap A \neq \emptyset \\ 0 & \text{otherwise.} \end{cases}$$

We can now state our test language: $L_A = \{0^m \mid b_{m,k} = 0\}$. Note that if we maintain, as we will, the global requirement that
$$(\forall m \geq 1)(\forall j \in \{0,1\}^{\leq k-1})[||Q_{m,j} \cap A|| \leq 1],$$
then L_A will be in $R^p_{k\text{-}T}(\text{UP}^A)$,

Initially, $A = \emptyset$.

Stage $\ell = \langle a, b \rangle_2$ We will ensure that $L_A \neq L(M_a^{A, L(N_b^A)})$, where $\{N_1, N_2, \cdots\}$ is a standard enumeration of nondeterministic polynomial-time Turing machines and $\{M_1, M_2, \cdots\}$ is a standard enumeration of deterministic polynomial-time Turing machines modified so that, though they may ask as many queries as they like to the first component of their oracle (A), they ask at most $2^k - 2$

queries, all in a truth-table fashion, to the second component of their oracle ($L(N_b^A)$). Without loss of generality, we may assume that M_i (N_i) runs in deterministic (nondeterministic) time at most $n^i + i$.

Choose m so large that (1) no string of length at least m has been in the reach of any previous stage (that is, so that $m > \max\{(q_s^{a_s} + a_s)^{b_s} + b_s \mid s < \ell, s = \langle a_s, b_s \rangle_2$, and q_s was the "m" of Stage $s\}$) (2) $(m^2+a)^b+b$ is less than $2^m/2^k$ (this gives us "room to diagonalize").

Consider the tree implicit in the definition of $b_{m,(\cdot)}$. That is, consider the language, which will be in UP^A due to the global constraint stated earlier, $L'_A = \{\langle 0^m, b_{m,1}b_{m,2}\cdots b_{m,r}\rangle_2 \mid r < k$ and $Q_{m, b_{m,1} b_{m,2} \cdots b_{m,r}} \cap A \neq \emptyset\}$. The intuitive machine for L_A will initially ask "$\langle 0^m, \epsilon \rangle_2 \in L'_A?$," and then depending on the answer will ask either "$\langle 0^m, 0\rangle_2 \in L'_A?$" or "$\langle 0^m, 1\rangle_2 \in L'_A?$," and so on. This gives an implicit "query tree" whose root has the label ϵ, and the left and right children of the root are respectively labelled "0" and "1," and so on. For each $1 \leq i \leq 2^k - 1$, let α_i be the ith node (viewed as a binary string from $\{0, 1\}^{\leq k-1}$) visited in an *inorder* traversal [AHU74]—a traversal in which we inorder traverse a node's left subtree, then visit the node, and then inorder traverse a node's right subtree.

If any of the at most $2^k - 2$ truth-table queries made to $L(N_b^A)$ by $M_a^{A, L(N_b^A)}(0^m)$ currently accepts, freeze (appropriately in or out of A) all strings queried on one accepting path. Consider each such truth-table query to be "neutralized."

We will go through at most $2^k - 1$ substages. The pth such substage is as follows.

Substage p, $1 \leq p \leq 2^k - 1$, of Stage ℓ If with the current oracle $0^m \notin L(M_a^{A, L(N_b^A)}) \iff 0^m \in L_A$, then we are entirely done with this stage. Freeze the status of all queried strings and go to Stage $\ell+1$. Let $b_{m,1}b_{m,2}...b_{m,k}$ be the bit-string defined early in this proof (computed with respect to the current oracle). Let z' be the largest i such that $b_{m,i}$ is 0. Note that in each of the $2^k - 1$ potential stages that we reach, such an i will exist, as to transform from the initial "b" vector of zeros to a final "b" vector of ones via changing the largest i with a zero to 1 will take $2^k - 1$ such moves, as (due to our construction and the choices of added strings) each time we change a zero to a one, all larger bits become zeros (e.g., for the case of $k = 2$, the sequence would go, assuming Case 2 is hit each time except the last, $00 \Rightarrow 01 \Rightarrow 10 \Rightarrow 11$). We'll say that $Q_{m, b_{m,1} b_{m,2} \cdots b_{m,z'-1}}$ is the current interesting region, and will denote it Q_{INT} (if $z' = 1$, then Q_{INT} is $Q_{m,\epsilon}$). Note that it is impossible, by our construction, that Q_{INT} already has one element, as the same "Q" is never interesting twice (since the interesting region's "b" value merely walks through the binary numbers, adding one each time). Let us say an extension of A is *legal* if it changes A by adding exactly one string w, and $w \in Q_{INT}$, and w was not previously frozen out of A, and adding w does not cause $||Q_{INT} \cap A||$ to have two elements, and "$w \in A$?" is not currently directly asked by M_a during the run of $M_a^{A, L(N_b^A)}(0^m)$. There are two cases.

Case 1 of Substage p, $1 \leq p \leq 2^k - 1$, of Stage ℓ: *There is no legal extension of the current oracle such that some non-neutralized truth-table query switches from rejecting to accepting.* In this case, we add to A any one string $w \in Q_{INT}$ forming a legal extension. Note that some such string will exist by our choice of m. Since adding w changed none of the truth-table queries, and since (by the definition of legal) w was not directly queried by M_a, extending in this way leaves the acceptance or rejection of $M_a^{A, L(N_b^A)}(0^m)$ unchanged. However, from the definition of L_A, the ordering of the substages via inorder traversal, and the fact that this case of this substage has been reached, it follows that extending this way will put 0^m in L_A if it was not in before the extension, and will remove 0^m from L_A if it was in before the extension. This is so as (a) a "b" string corresponds to acceptance if and only if its largest bit is a 0, and (b) the operation "change the largest 0 in a string to a 1 and set all bits larger than that bit to zero" always changes the last bit. (Note: (b) merely says—taking "larger" bits to correspond to lower order bits—that adding one to a binary number toggles it between odd and even.) Thus, we now have: $0^m \notin L(M_a^{A, L(N_b^A)}) \iff 0^m \in L_A$, and are done with Stage ℓ. Freeze everything touched and move to Stage $\ell + 1$.
End of Case 1
Case 2 of Substage p, $1 \leq p \leq 2^k - 1$, of Stage ℓ: *There is a legal extension of the current oracle such that some non-neutralized*

truth-table query switches from rejecting to accepting. Adopt one such extension, thus adding a string to A. Also, freeze an accepting path of each non-neutralized truth-table query that the extension causes to accept, and mark such queries as neutralized. Go to Substage $p + 1$.
End of Case 2
End of Substage p

Note that Case 2 reduces by at least one the number of non-neutralized queries (which initially was at most $2^k - 2$). In particular, if we reach Substage $2^k - 1$, then there are *no* non-neutralized truth-table queries left at the start of that case, and thus Case 2 cannot occur during Substage $2^k - 1$, and thus we indeed correctly diagonalize against $L(M_a^{A, L(N_b^A)})$ using at most $2^k - 1$ substages.
End of Stage m ∎

Corollary 3.4 *There is a recursive oracle A such that, for all $k \geq 1$, it holds that $R_{k\text{-}T}^p(\mathrm{UP}^A) \not\subseteq R_{2^k-2\text{-}tt}^{p,A}(\mathrm{UP}^A)$.*

3.3 Semi-Guarded Access

Let us define a notion we will, for the nonce, call "semi-guarded access": a P machine M is given access to a promise problem and, for any query z that M on input x might ask after some (possibly incorrect) sequence of oracle answers, it must hold that z obeys the promise. (The set of such queries is essentially[7] the set $Q(M, x)$ from Book, Long, and Selman's seminal work on quantitative relativization [BLS84, Definition 4.1]. Their paper presents various consequences that follow when this set is of limited size.) Note that the difference between guarded and semi-guarded access is that in guarded access, the promise must hold along each query asked in an *actual run* of the machine. In contrast, semi-guarded access means that every query that the machine

---

[7]We say "essentially," as their definition requires that a string be reachable via some oracle, and our definition requires that the string be reachable via some sequence of oracle answers. The distinction is whether one requires queries to the same string to be answered consistently if it is asked repeatedly. In the particular case here, that of Turing reductions, this issue is not of particular interest, as without loss of generality one may prevent (in one's standard enumeration of deterministic machines) machines from repeating any query.

could be fooled into asking (by giving it *incorrect* oracle answers for some of its questions) must also obey the promise. Note that every language in P^{UP} is trivially accepted by a P machine given semi-guarded access to $\mathcal{UP}$, and that every language accepted by a P machine given semi-guarded access to $\mathcal{UP}$ is trivially accepted by a P machine given guarded access to $\mathcal{UP}$ (that is, is in $P^{\mathcal{UP}}$).

In fact, it is easy to note that "semi-guarded access" to a promise problem gives no power beyond that of access to a machine that *globally* maintains the promise. The reason is that the fact that a question might be asked in a semi-guarded computation is itself a certificate that the promise holds. Taking unambiguous computation as an example, suppose L is accepted by P machine M given semi-guarded access to a member of $\mathcal{UP}$, namely (Q_i, R_i), where $Q_i = \{x \mid ||N_i(x)|| \leq 1\}$, and $R_i = \{x \mid ||N_i(x)|| \geq 1\}$, and N_i is a nondeterministic polynomial-time Turing machine (recall Definition 2.2). Then L is also accepted by deterministic polynomial-time machine M' given access to a UP set U, where $U = \{\langle x, b_1 b_2 b_3 \cdots b_\ell \rangle \mid \ell \in \mathcal{N}$ and if $M^{(\cdot)}(x)$ is run with the first ℓ oracle answers being, in turn, $b_1, b_2, \cdots$, then $M(x)$ asks an $(\ell+1)$st query that is in $R_i\}$ and M' is the machine that simulates M, except M' also keeps a history, h, of the oracle replies so far, and each time $M(x)$ is to make a query to (Q_i, R_i), instead M' asks U the query $\langle x, h\rangle$. Thus, for the case of Turing reductions, semi-guarded access to $\mathcal{UP}$ is equivalent to access to UP. It is easy to see that this also is true for all standard reducibilities (e.g., strong nondeterministic reducibility).

4 Fault-Tolerant Access to UP

This section asks whether fault-tolerant access to unambiguous computation is more powerful than UP itself, a question posed by Ko [Ko87]. In particular, we are interested in the class Ko dubbed $P_{1-help}(UP)$. Informally, a set A is in $P_{1-help}(UP)$ if there are a deterministic oracle Turing machine M and a UP set B such that: (1) if, on a given input $x \in A$, all the answers B—which is used as an oracle by M—gives are correctly conveyed to M, then M *quickly* accepts x, and (2) if, on a given input x (which may be in A or $\overline{A}$), regardless of what oracle is used, M will correctly (though perhaps not quickly) determine whether $x \in A$ or $x \notin A$. $\text{P}_{1-help}(\text{UP})$ may be viewed as a form of access to UP that is tolerant of faulty answers.

Definition 4.1 [Sch85,Ko87]

1. A deterministic oracle machine M is *robust* if for every oracle A, it holds that M^A halts on all inputs and $L(M^A) = L(M^\emptyset)$.

2. A set H *helps* a robust machine M on set S if there is a polynomial $p(\cdot)$ such that for all $x \in S$, it holds that $M^H(x)$ halts in $p(|x|)$ steps.

3. $\text{P}_{1-help}(H)$ denotes the class of all languages $L(M^\emptyset)$ that are computed by robust machines M such that H helps M on set $L(M^\emptyset)$. For each class $\mathcal{H}$ of sets, let $\text{P}_{1-help}(\mathcal{H}) = \{L \mid (\exists H \in \mathcal{H})[L \in \text{P}_{1-help}(H)]\}$.

Much is known about "helping," and the related notion of robust machines (see [Sch85,Ko87,Bal88,GB91,HH90]). For example, Ko proved that for NP fault-tolerance is so burdensome that no additional power is gained: $\text{NP} = \text{P}_{1-help}(\text{NP})$ [Ko87]. On the other hand, Ko also proved that $\text{UP} \subseteq \text{P}_{1-help}(\text{UP})$, and left as an open question whether $\text{UP} = \text{P}_{1-help}(\text{UP})$ [Ko87]. Theorems 4.2 and 4.4 provide a relativized world in which this equality fails, and prove that, though probably not equal to UP, the class $\text{P}_{1-help}(\text{UP})$ is equal to the class of sets that $\leq^p_{lpos}$ reduce to UP.

Theorem 4.2 There is an oracle A such that $\text{UP}^A \subsetneq \text{P}_{1-help}(\text{UP}^A)$.

Proof of Theorem 4.2 Let $L_A = \{\langle 0^i, 0^j \rangle \mid \|A \cap \Sigma^i\| \neq 0 \text{ or } \|A \cap \Sigma^j\| \neq 0\}$. We will construct A in such a way that $(\forall i)[\|A \cap \Sigma^i\| \leq 1]$. Thus $L_A \in \text{R}^p_d(\text{UP}^A)$, and so $L_A \in \text{P}_{1-help}(\text{UP}^A)$ (as Ko's proof that $\text{R}^p_d(\text{UP}) \subseteq \text{P}_{1-help}(\text{UP})$ [Ko87] in fact also shows that $\text{R}^p_d(\text{UP}^A) \subseteq \text{P}_{1-help}(\text{UP}^A)$). We need to construct A so that no UP^A machine can accept L_A. Let $N_1, N_2, \cdots$ be some standard enumeration of nondeterministic Turing machines such that N_i runs in time $n^i + i$. In the construction below we will ensure that for each i either:

1. N_i^A is not categorical, or

2. N_i^A does not accept L_A.

We will build $A = \bigcup_{i \geq 1} A_i$ in stages. Let $A_0 = \emptyset$ and let A_i, $i \geq 1$, denote the strings placed in the oracle by the end of stage i.

Stage i:

Pick an n sufficiently large so that $\binom{2^n}{2} \gg 2^n(n^i + i)$, and for $i > 1$ we also require that $n \gg m^i + i$, where m is the n of stage $i - 1$.

1. If $N_i^{A_{i-1}}(\langle 0^n, 0^{n+1}\rangle)$ accepts, go to stage $i+1$.

2. If there is a y, $|y| = n$ or $|y| = n+1$, such that $N_i^{A_i \cup \{y\}}(\langle 0^n, 0^{n+1}\rangle)$ rejects or is noncategorical, then add y to A and go to stage $i+1$.

3. Otherwise we may force $N_i^A(\langle 0^n, 0^{n+1}\rangle)$ to be noncategorical by adding two strings, one of length n and one of length $n + 1$, to A. This can be done by picking two strings u, v, $|u| = n$ and $|v| = n + 1$, such that u is not queried by $N_i^{A_{i-1} \cup \{v\}}$ on $\langle 0^n, 0^{n+1}\rangle$ and v is not queried by $N_i^{A_{i-1} \cup \{u\}}$ on $\langle 0^n, 0^{n+1}\rangle$. By our choice of n, such strings always exist (by Lemma 2.11). If we add both u and v to A_{i-1}, we will get at least two accepting paths on input $\langle 0^n, 0^{n+1}\rangle$, hence $N_i^{A_i}$ will not be categorical. Go to stage $i + 1$.

∎

Note that, though the notion of 1-helping UP has two potential relativizations, $P_{1-help}(UP^A)$ and $P^A_{1-help}(UP^A)$, Theorem 4.2 proves its result for the technically stronger of the two cases.

The original notion of polynomial-time positive Turing reductions, due to Selman [Sel82], required that the reductions be *globally positive*—positive not only with respect to additions and deletions to the set that was being reduced to, but also with respect to every set (even those incomparable to the set being reduced to). Hemachandra and Jain [HJ91] introduced "locally" positive reductions—reductions that need be positive only with respect to additions (or deletions or both) to the set being reduced to. Most, though not all, of Selman's results can be extended to the more flexible case of local positivity [HJ91]. Here, we need consider only their "lpos" reduction, a one-sided notion of local positivity.

Definition 4.3 [HJ91]

1. A query machine M is locally left positive with respect to B if $(\forall x)(\forall A)[x \in L(M^{B-A}) \Rightarrow x \in L(M^B)]$.

2. $A \leq^p_{lpos} B$ if $A \leq^p_T B$ by some machine that is locally left positive with respect to B.

The following can be viewed as an extension of Ko's: $\mathrm{R}^p_d(\mathrm{UP}) \subseteq \mathrm{P}_{1-help}(\mathrm{UP})$ [Ko87].

Theorem 4.4 $\mathrm{R}^p_{lpos}(\mathrm{UP}) = \mathrm{P}_{1-help}(\mathrm{UP})$.

Proof of Theorem 4.4 $\mathrm{P}_{1-help}(\mathrm{UP}) \subseteq \mathrm{R}^p_{lpos}(\mathrm{UP})$ follows easily from the definitions. For the other direction, let W be a set in $\mathrm{R}^p_{lpos}(\mathrm{UP})$. Let Y be the UP set to which W lpos reduces. Since $\mathrm{UP} \subseteq \mathrm{P}_{1-help}(\mathrm{UP})$, let Z be the UP set that 1-helps Y. Then W is 1-helped by Z as follows. Simulate the lpos reduction to Y, replacing each query to Y by running (only for the polynomial number of steps within which it will never incorrectly certify inclusion and will always, when given correct oracle answers, correctly certify actual inclusion) the 1-helping reduction (to Z). (We do not simulate the brute-force potentially exponential-time end of the 1-helping reductions from Y to Z.) At the end of the simulation of the lpos reduction, we immediately accept if the reduction dictates acceptance (by the definitions of lpos and 1-help, we are safe in doing so, and will always do so—given true answers from Z—on elements of W), and if the reduction dictates rejection we compute the right answer by brute force (in exponential time) and then give that answer. ∎

5 Fault-Tolerance Characterizes Probabilistic Computation

This paper has studied access to unambiguous computation, and has provided evidence that guarded access, nonadaptive access, and fault-tolerant access are each relatively powerful. We propose as an interesting research direction the analogous questions for such probabilistic classes (see [Gil77]) as BPP, R, coR, and ZPP. As a start in that direction, we now address both the issue of "promise" access and that of Schöning's and Ko's open questions (Open Question 5.2) for probabilistic classes.

Definition 5.1 ([Sch85] see also [Ko87]) $P_{help}(H)$ denotes the class of all languages $L(M^\emptyset)$ that are computed by robust machines M such that H helps M (see Part 2 of Definition 4.1) on set Σ^*. For each class $\mathcal{H}$ of sets, let $P_{help}(\mathcal{H}) = \{L \mid (\exists H \in \mathcal{H})[L \in P_{help}(H)]\}$.

Schöning [Sch85] proved that $P_{help}(\text{BPP}) \subseteq \text{ZPP}$ and Ko [Ko87] proved that $P_{1-help}(\text{BPP}) \subseteq \text{R}$. Schöning and Ko asked whether the containments were tight:

Open Question 5.2 [Sch85,Ko87]

1. $\text{R} = P_{1-help}(\text{BPP})$?
2. $\text{ZPP} = P_{help}(\text{BPP})$?

Though we do not know whether equality holds, we can prove that these statements hold in a natural model of weakened promises. We first present the currently standard approach to access to promise problems—that of "solutions," *globally* consistent assignments of values to strings for which the promise fails. Then we suggest an approach, the "arbitrary failure" model, that models the case when violating the promise yields a truly unreliable answer, and we note that in some sense this extremely strong protection can be obtained for "free." Thus, we suggest the arbitrary failure model as a particularly natural (and happily cost-free) framework in which to discuss fault-tolerance.

Definition 5.3 [ESY84]

1. A deterministic Turing machine M *solves* the promise problem (Q, R) if $(\forall x)[Q(x) \Rightarrow [M(x) \text{ halts} \land (M(x) = \text{"yes"} \iff R(x))]]$.

2. If a Turing machine M solves promise problem (Q, R), then the language $L(M)$ accepted by M is said to be a *solution* to (Q, R).

3. For each promise problem $S = (Q, R)$, we use $solns(S)$ to denote $\{L \mid L \text{ is a solution to } S\}$.

Definition 5.4

1. (see, e.g., [Val76,HH88]) 1SAT = $\{f \mid$ boolean formula f has at most one satisfying assignment$\}$.

2. [Sim75] MAJ = $\{f \mid$ boolean formula f has at least $\frac{1}{2}2^v$ distinct satisfying assignments, where v is the number of distinct variables in $f\}$.

3. BND = $\{f \mid$ boolean formula f has either less than $\frac{1}{4}2^v$ distinct satisfying assignments or more than $\frac{3}{4}2^v$ distinct satisfying assignments, where v is the number of distinct variables in $f\}$.

Definition 5.5 $\mathcal{BPP}$ ("promise BPP") is the following class of promise problems:
$$\{(Q_i, R_i) \mid i \geq 1\},$$
where $N_1, N_2, \cdots$ is a standard enumeration of probabilistic polynomial-time Turing machines, $Q_i = \{x \mid |\text{Prob}(N_i(x) \text{ accepts}) - \frac{1}{2}| \geq \frac{1}{4}\}$, and $R_i = \{x \mid \text{Prob}(N_i(x) \text{ accepts}) \geq \frac{1}{2}\}$.

Just as (1SAT,SAT), or equivalently $\mathcal{UP}$, is the "promise" version of unambiguous computation, so also is (BND,MAJ), or equivalently $\mathcal{BPP}$, the "promise" version of bounded probabilistic computation. (1SAT,SAT) has recently been used in a paper by Köbler, Schöning, and Torán.

Theorem 5.6 [KST92] Graph automorphism is polynomial-time disjunctive reducible to every solution L of the promise problem (1SAT,SAT).

Turning to Ko's open problems, we can prove the following.

Theorem 5.7 If L is any solution to the promise problem (BND,MAJ), then:

1. R $\subseteq$ P$_{1-help}(L)$, and
2. ZPP $\subseteq$ P$_{help}(L)$.

The proof of Theorem 5.7 is analogous to that of Theorem 5.10 below.

If what we want to capture is the fact that reductions exist to each fixed set that is correct when the promise holds, using solutions to promise problems (Definition 5.3) is a fine formulation. However, often, this is not the intuition one wants to capture. In the model of Definition 5.3, it is very hard to make any claims as to the simplicity of the class of sets that reduce to a given solution of a promise problem, as the places at which the promise fails may be filled in, for example, with the characteristic function of the halting problem. That is, the values taken on in the promise-breaking locations of a solution of a promise problem themselves form a consistent database available to a querying machine. This stands at odds with any intuition of "computation that is reliable only when the promise is fulfilled." Furthermore, it would seem natural to require a query not in the promise region to fail in some unreliable, arbitrary way, and to require the base machine to compute correctly in light of any answer given—that is, the same query might well be given a different answer if it is asked different times (perhaps on different inputs, or on different nondeterministic paths on the same input if our reducibility is nondeterministic, or ...).

We propose the following alternate model, which we'll call the *arbitrary failure promise model* (arbitrary failure, for short), and suggest that it captures the natural notion of "computation that is reliable only when the promise is fulfilled":

> Each time a promise-breaking query is made, an arbitrary answer is returned. The querying machine must "do the right thing" regardless of the answers it is given (i.e., it must be robust).

We provide a formal definition for the case of deterministic Turing access; other reducibilities can be defined in the obvious analogous ways. For the case of deterministic Turing access, we will prove that the new notion is just as flexible as the notion of solutions to promise problems: the two notions are equivalent.[8]

Definition 5.8 (Arbitrary Failure Access)

1. Let $A = (Q, R)$ be a promise problem. We say that $L \in \mathrm{P}^{\widehat{A}}$ if there is a deterministic Turing machine M such that:

   (a) The run time of M is bounded by some polynomial $q(\cdot)$, and

   (b) $L = L(M^R)$, and

   (c) For every input x and every $\alpha, \alpha' \in (0+1)^{\sigma(|x|)}$, it holds that $x \in simulate(M, Q, R, \alpha) \iff x \in simulate(M, Q, R, \alpha')$. Here, $x \in simulate(M, Q, R, \beta)$ means that if one simulates the run of $M^{(\cdot)}(x)$, with the oracle query (if any) z asked on the ith step given the answer:

   $$\begin{array}{l}\text{``yes'' if } z \in Q \cap R \\ \text{``no'' if } z \in Q \cap \overline{R} \\ \text{``yes'' if } z \in \overline{Q} \text{ and the } i\text{th bit of } \beta \text{ is } 1 \\ \text{``no'' if } z \in \overline{Q} \text{ and the } i\text{th bit of } \beta \text{ is } 0,\end{array}$$

   then M accepts at the end of its simulated run.

2. Let $\mathcal{C}$ be any class of promise problems, and let r be any reducibility for which the notion of arbitrary failure reductions to promise problems has been defined. Then $\mathrm{R}_r^p(\widehat{\mathcal{C}})$ denotes $\bigcup_{C \in \mathcal{C}} \mathrm{R}_r^p(\widehat{C})$.

The proof of Theorem 5.6 ([KST92], proof in [KST91]) is strong enough to hold perfectly well in this model, so graph automorphism is in $\mathrm{R}_d^p(\widehat{\mathcal{UP}})$.

---

[8]We leave as an open problem the issue of whether the notions are the same for nondeterministic reductions. However, we note that, for the case of promise classes that—like $\mathcal{UP}$ and $\mathcal{BPP}$—are built upon a set that is not just paddable but indeed is parsimoniously (solution-preservingly) paddable [Sim75], it is not hard to see that one also gets the nondeterministic reduction analog of the equivalence we note below between solutions to promise problems and arbitrary failure. This is simply because we can use the padding to ensure that if a question is asked on some path of the computation tree on a certain input, then it is asked on no other path of that tree, and no other path of any computation tree on any input.

As already mentioned, this isn't a stronger claim, but rather reflects a more general collapsing of models. We formalize this below, by claiming that reducing to every solution of a promise problem (which perhaps has different machines doing the reduction for different solutions, but with consistent answers for the non-promise regions of each solution) is equivalent to arbitrary failure access (which has a single machine that works robustly in the face of any failures). The proposition below should be thought of as a strengthened version of the uniformity theorem of Grollmann and Selman and Regan [GS88, Reg86].

Proposition 5.9 For every promise problem $S = (Q, R)$,

$$\{B \mid (\forall A \in solns(S))[B \leq_T^p A]\} = P^{\widehat{S}}.$$

Proof of Proposition 5.9 The $\supseteq$ direction is immediate. For the $\subseteq$ direction, let S be a promise problem, and let B be a set such that $(\forall A \in solns(S))[B \leq_T^p A]$. (Note: in the Grollmann-Selman [GS88] paper's notation, this could be written as $(\Sigma^*, B) \leq_T^{PromiseProblem} S$.) Via the uniformity theorem of Grollmann and Selman [GS88, p. 318] and Regan [Reg86], there is a single machine M' that uniformly reduces to the solutions of S. That is, each Turing reduction of the above displayed equation can be performed by M'. Now consider a machine M that on any input exactly mimics the behavior of M', with the following alteration. M remembers all queries and all answers that it sees, and if the simulation of M' ever attempts to ask a question that has been asked before (on an earlier query on the same input), M ignores the answer and instead uses the answer that was obtained previously. Noting Definition 5.8, this machine M proves that $B \in P^{\widehat{S}}$. The reader may wonder why—since in Definition 5.8, Part 1c, we don't worry about coherency of answers to queries not satisfying the promise—we just went to some effort to achieve this coherency. The reason is that the Grollmann-Selman uniformity theorem's "uniform" machine is merely guaranteed to act correctly *if there are consistent oracle responses*, and thus we expend the effort to ensure that the M being simulated by M' must obey the Grollmann-Selman results and perform as we wish (which makes M' a good machine for the purposes of Definition 5.8). ∎

Promise classes inherit this result. For example,

$$\mathrm{P}^{\widehat{\mathcal{BPP}}} = \mathrm{P}^{(\widehat{\mathrm{BND},\mathrm{MAJ}})} = \{B \mid (\forall A : A \in solns((\mathrm{BND},\mathrm{MAJ})))[B \leq_T^p A]\}.$$

For the case of helping, the arbitrary failure promise model also applies naturally: the machine being helped must be robust with respect to oracle answers, and must be "quick" (on all strings in the case of helping, and on strings in the set in the case of 1-helping) as long as its oracle gives correct answers to all queries for which the promise holds. Note that it will be considered fair play for an oracle to lie as it likes if it is asked a question for which the promise does not hold—and this in and of itself does not release the base machine from its obligation to compute quickly. The formal definition should be immediately clear from these comments and is omitted. We may now state the following theorem, which shows that Ko's open question regarding probabilistic helping is neatly resolved within the arbitrary failure model.

Theorem 5.10

1. $\mathrm{R} = \mathrm{P}_{1-help}(\widehat{\mathcal{BPP}})$.
2. $\mathrm{ZPP} = \mathrm{P}_{help}(\widehat{\mathcal{BPP}})$.

Proof of Theorem 5.10 We will prove the first equality, $\mathrm{R} = \mathrm{P}_{1-help}(\widehat{\mathcal{BPP}})$. Since $\mathrm{R} \cap \text{co-R} = \mathrm{ZPP}$ and $\mathrm{P}_{1-help}(\widehat{\mathcal{BPP}}) \cap \text{co-}\mathrm{P}_{1-help}(\widehat{\mathcal{BPP}}) = \mathrm{P}_{help}(\widehat{\mathcal{BPP}})$, the second equality is a corollary of the first.

$\mathrm{R} \supseteq \mathrm{P}_{1-help}(\widehat{\mathcal{BPP}})$: Let $L \in \mathrm{P}_{1-help}(\widehat{\mathcal{BPP}})$, and let M be a robust deterministic Turing machine accepting the language L using a promise problem $A = (Q, R)$, where for some probabilistic polynomial-time Turing machine N, $Q = \{x \mid |\mathrm{Prob}(N(x) \text{ accepts}) - \frac{1}{2}| \geq \frac{1}{4}\}$, and $R = \{x \mid \mathrm{Prob}(N(x) \text{ accepts}) \geq \frac{1}{2}\}$. Let the polynomial p bound the running time of N, and M when M is polynomial-time bounded.

We design a one-sided error probabilistic polynomial-time Turing machine M' accepting L as follows: Given input x; let $n = |x|$. M' on x will simulate M on x; when a new query (for queries already asked on the present path, use the answer already obtained) y is made by M, M' will run the probabilistic machine N on y, independently repeated for $2n+1$ times, and then take the majority of accepting or rejecting computations as the answer M receives from its oracle. Finally, M' accepts x *iff* the simulation concludes within time $p(n) \cdot p(p(n)) \cdot (2n+1) + p(n)$ with M accepting x. Otherwise M' rejects x on this path.

Clearly M' runs in probabilistic polynomial-time, and if M' accepts x, then $x \in L$, since x is accepted by M using some oracle.

On the other hand, suppose $x \in L$. Due to our amplification, any query M asked in its computation on x was answered correctly with high probability, *if it mattered at all*. More precisely, let $\alpha = \alpha_1 \alpha_2 \ldots \alpha_s \in \{0,1\}^s$ be the bit sequence corresponding to the answers M gets in the simulation. Surely $s \leq p(n)$. The probability that there exists some i, $1 \leq i \leq s$, such that the ith queried string $y_i \in Q$, and α_i is different from membership of y_i in R (i.e. $\alpha_i = 0 \iff y_i \in R$) is bounded by $s \cdot \sum_{k=0}^{n} \binom{2n+1}{k} \left(\frac{3}{4}\right)^k \left(\frac{1}{4}\right)^{2n+1-k}$, which can be shown to be less than $p(n) \cdot \left(\frac{3}{4}\right)^n$. Thus with probability close to 1, M' will accept x if $x \in L$, in polynomial-time.

$R \subseteq P_{1-help}(\widehat{\mathcal{BPP}})$: Let $L \in R$ and let N be a probabilistic polynomial-time Turing machine that accepts L with one-sided error, namely, if N accepts x on some path then $x \in L$, and if $x \in L$ then N accepts x with probability $\geq 7/8$. Let p be a polynomial bound on the running time of N. (Without loss of generality, we may assume that on any input of length n, any computation path of N takes exactly $p(n)$ independent random bits to reach its final configuration.)

We design our robust Turing machine M to accept L, using a promise problem $A = (Q, R)$ defined via a probabilistic polynomial-time Turing machine N^* as follows:

Let any input $\langle x, b_1 \ldots b_s, c \rangle$ be given, where $0 \leq s \leq p(|x|)$, $c, b_i \in \{0,1\}, 1 \leq i \leq s$. If $c = 0$, then N^* accepts it *iff* $N^{b_1 \ldots b_s}(x)$ accepts (in the BPP sense), where $N^{b_1 \ldots b_s}(x)$ denotes the computation of N as a probabilistic polynomial-time Turing machine, with input x, using $b_1 \ldots b_s$ as its initial s bits of probabilistic moves. If $c = 1$, then N^* independently repeats the following process $257p(|x|)^2$ times, and accepts $\langle x, b_1 \ldots b_s, 1 \rangle$ *iff* $s < p(|x|)$ and the majority of them are accepting computations.

The process is the following: Simulate $N^{b_1\ldots b_s}(x)$ with a fair coin for any additional random bits N may need, until a final state of N is reached. Let the first bit chosen by N after $b_1\ldots b_s$ be b. Then the process accepts on two paths if N accepts and $b = 0$, and the process rejects on two paths if N accepts and $b = 1$, and any non-accepting final state of N becomes one accepting and one rejecting path. (That is, in any of these three cases, we add an extra level to our tree—whose branchings so far corresponded to the uses of the fair coin—and accept/reject at the two leaves based on the rules just given.)

Clearly the difference of the number of accepting and rejecting paths in the process is equal to twice the difference of the number of accepting computations of $N^{b_1\ldots b_s 0}(x)$ and $N^{b_1\ldots b_s 1}(x)$. If g is the difference of accepting probabilities of the left and right subtrees of $N^{b_1\ldots b_s}(x)$, then the process accepts with probability $1/2 + g/4$; and if $|g| > 1/(4p(|x|))$, then the probability of N^* accepting $\langle x, b_1\ldots b_s, 1\rangle$ is either $\geq 3/4$ or $\leq 1/4$, (by our choice of repeating $257p(n)^2$ times and a simple estimate that $\sum_{k=0}^{128p^2(n)} \binom{257p^2(n)}{k} \left(\frac{1}{2} + \frac{1}{16p(n)}\right)^k \left(\frac{1}{2} - \frac{1}{16p(n)}\right)^{257p^2(n)-k} < 1/(2e) < 1/4$).

Now M, the 1-helper, works as follows: On input x, it queries $\langle x, \varepsilon, 0\rangle$. If the answer is no, run N on x by brute force (exhaustive search). Inductively, assume $b_1\ldots b_{s-1}$, $s \geq 1$, have been found, and the oracle answer to $\langle x, b_1\ldots b_{s-1}, 0\rangle$ is yes. M now queries $\langle x, b_1\ldots b_{s-1} 0, 0\rangle$ and $\langle x, b_1\ldots b_{s-1} 1, 0\rangle$, and let the answers be ℓr, where ℓ and $r = 1$ or 0 denoting yes or no. If $\ell r = 00$, then run N on x by brute force again. If $\ell r = 10$, set $b_s = 0$; if $\ell r = 01$, set $b_s = 1$. If $\ell r = 11$, we query $\langle x, b_1\ldots b_{s-1}, 1\rangle$, and set $b_s = 0$ if the answer is yes, and $b_s = 1$ if the answer is no.

If we found all $b_1\ldots b_{p(|x|)}$ without going brute force, then we accept x if N on this path accepts x, and if N on this path does not accept x we run brute force again. This completes the description of the machine M.

We show that this certifies that $L \in P_{1-help}(\widehat{\mathcal{BPP}})$. First, for any x to be accepted by M, using any oracle, an accepting path of N on x is found, either by the sequence $b_1\ldots b_{p(|x|)}$ or by brute force, thus

$x \in L$; for any x to be rejected by M, it can only come as a result of an exhaustive search, thus $x \notin L$; hence M is a robust machine accepting L.

Now suppose $x \in L$ and the promise problem A defined above is used as the oracle. We show that x is accepted in polynomial time.

The query $\langle x, \varepsilon, 0 \rangle$ by M must be answered with a yes, since $N^*(\langle x, \varepsilon, 0 \rangle)$ has acceptance probability $\geq 7/8 > 3/4$. Inductively, suppose $b_1 \ldots b_{s-1}$, $s \geq 1$, have been found, the query $\langle x, b_1 \ldots b_{s-1}, 0 \rangle$ by M has been answered with a yes, and the acceptance probability of $N^*(\langle x, b_1 \ldots b_{s-1}, 0 \rangle) \geq 7/8 - (s-1)/(8p(|x|))$.

Suppose one of the acceptance probability of $N^*(\langle x, b_1 \ldots b_{s-1}0, 0 \rangle)$ and $N^*(\langle x, b_1 \ldots b_{s-1}1, 0 \rangle)$ is $< 7/8 - s/(8p(|x|))$; let's say the right branch. Then the left branch has acceptance probability $> 7/8 - (s-2)/(8p(|x|)) \geq 7/8 - s/(8p(|x|)) \geq 3/4$, since the average of acceptance probability of the two is $\geq 7/8 - (s-1)/(8p(|x|))$. To the query $\langle x, b_1 \ldots b_{s-1}0, 0 \rangle$ the answer must be yes. On the other hand, the answer to $\langle x, b_1 \ldots b_{s-1}1, 0 \rangle$ could be either way. If it is no, we set b_s correctly (in this case $b_s = 0$). If it is also yes, then since the gap of the acceptance probabilities of the two branches is at least $1/4p(|x|)$, our query to $\langle x, b_1 \ldots b_{s-1}, 1 \rangle$ would set b_s correctly (in this case $b_s = 0$). Finally suppose both the acceptance probabilities of $N^*(\langle x, b_1 \ldots b_{s-1}0, 0 \rangle)$ and $N^*(\langle x, b_1 \ldots b_{s-1}1, 0 \rangle)$ are $\geq 7/8 - s/(8p(|x|)) \geq 3/4$, then both queries must have been answered with a yes, and the query $\langle x, b_1 \ldots b_{s-1}, 1 \rangle$ was made and some answer obtained, which led us to set either $b_s = 0$ or 1. Either way we are fine, since both acceptance probabilities are $\geq 7/8 - s/(8p(|x|))$. Finally the computation of N on x using $b_1 \ldots b_{p(|x|)}$ is accepting and we accepted x in polynomial time.

This completes the proof. ∎

Acknowledgments

We are grateful to Yenjo Han, Rajesh Rao, and Marius Zimand for proofreading earlier drafts of this paper, to an anonymous referee for helpful comments and for suggesting the current more general statement of Proposition 5.9, to Kenneth Regan for pointers to and helpful discussions about the Grollmann-Selman/Regan uniformity theorem, to Edith Spaan for helpful conversations, and to Andrew Yao and Uwe Schöning for arranging visits during which this work was done in part.

References

[AHU74] A. Aho, J. Hopcroft, and J. Ullman. *The Design and Analysis of Computer Algorithms*. Addison-Wesley, 1974.

[AR88] E. Allender and R. Rubinstein. P-printable sets. *SIAM Journal on Computing*, 17(6):1193–1202, 1988.

[Bal88] J. Balcázar. Self-reducibility structures and solutions of NP problems. Technical Report LSI-88-19, Universitat Politècnica de Catalunya, Facultat d'Informatica, 1988. Also appears as pages 175–184 of *Revista Matematica de la UCM 2 (1989)*.

[Bei89] R. Beigel. On the relativized power of additional accepting paths. In *Proceedings of the 4th Structure in Complexity Theory Conference*, pages 216–224. IEEE Computer Society Press, June 1989.

[Bei91] R. Beigel. Bounded queries to SAT and the boolean hierarchy. *Theoretical Computer Science*, 84(2):199–223, 1991.

[BI87] M. Blum and R. Impagliazzo. Generic oracles and oracle classes. In *Proceedings of the 28th IEEE Symposium on Foundations of Computer Science*, pages 118–126, October 1987.

[BLS84] R. Book, T. Long, and A. Selman. Quantitative relativizations of complexity classes. *SIAM Journal on Computing*, 13(3):461–487, 1984.

[BST] H. Buhrman, E. Spaan, and L. Torenvliet. Bounded reductions. In this volume.

[CGH+89] J. Cai, T. Gundermann, J. Hartmanis, L. Hemachandra, V. Sewelson, K. Wagner, and G. Wechsung. The boolean hierarchy II: Applications. *SIAM Journal on Computing*, 18(1):95–111, 1989.

[CH90] J. Cai and L. Hemachandra. On the power of parity polynomial time. *Mathematical Systems Theory*, 23(2):95–106, 1990.

[CHV92a] J. Cai, L. Hemachandra, and J. Vyskoč. Promise problems and access to unambiguous computation. In *Proceedings of the 17th Symposium on Mathematical Foundations of Computer Science*, pages 162–171. Springer-Verlag *Lecture Notes in Computer Science #629*, August 1992.

[CHV92b] J. Cai, L. Hemachandra, and J. Vyskoč. Promise problems and access to unambiguous computation. Technical Report TR-419, University of Rochester, Department of Computer Science, Rochester, NY, April 1992.

[ESY84] S. Even, A. Selman, and Y. Yacobi. The complexity of promise problems with applications to public-key cryptography. *Information and Control*, 61(2):159–173, 1984.

[EY80] S. Even and Y. Yacobi. Cryptocomplexity and NP-completeness. In *Proceedings of the 7th International Colloquium on Automata, Languages, and Programming*, pages 195–207. Springer-Verlag *Lecture Notes in Computer Science*, 1980.

[GB91] R. Gavaldà and J. Balcázar. Strong and robustly strong polynomial time reducibilities to sparse sets. *Theoretical Computer Science*, 88(1):1–14, 1991.

[Gil77] J. Gill. Computational complexity of probabilistic Turing machines. *SIAM Journal on Computing*, 6(4):675–695, 1977.

[GP86] L. Goldschlager and I. Parberry. On the construction of parallel computers from various bases of boolean functions. *Theoretical Computer Science*, 43:43–58, 1986.

[GS88]　　J. Grollmann and A. Selman. Complexity measures for public-key cryptosystems. *SIAM Journal on Computing*, 17(2):309–335, 1988.

[Hem89]　L. Hemachandra. The strong exponential hierarchy collapses. *Journal of Computer and System Sciences*, 39(3):299–322, 1989.

[HH88]　　J. Hartmanis and L. Hemachandra. Complexity classes without machines: On complete languages for UP. *Theoretical Computer Science*, 58:129–142, 1988.

[HH90]　　J. Hartmanis and L. Hemachandra. Robust machines accept easy sets. *Theoretical Computer Science*, 74(2):217–226, 1990.

[HH91]　　J. Hartmanis and L. Hemachandra. One-way functions and the non-isomorphism of NP-complete sets. *Theoretical Computer Science*, 81(1):155–163, 1991.

[HJ91]　　L. Hemachandra and S. Jain. On the limitations of locally robust positive reductions. *International Journal of Foundations of Computer Science*, 2(3):237–255, 1991.

[HR92]　　L. Hemachandra and R. Rubinstein. Separating complexity classes with tally oracles. *Theoretical Computer Science*, 92(2):309–318, 1992.

[Joh90]　　D. Johnson. A catalog of complexity classes. In J. Van Leeuwen, editor, *Handbook of Theoretical Computer Science*, chapter 2, pages 67–161. MIT Press/Elsevier, 1990.

[JY85]　　D. Joseph and P. Young. Some remarks on witness functions for non-polynomial and non-complete sets in NP. *Theoretical Computer Science*, 39:225–237, 1985.

[Ko85]　　K. Ko. On some natural complete operators. *Theoretical Computer Science*, 37:1–30, 1985.

[Ko87]　　K. Ko. On helping by robust oracle machines. *Theoretical Computer Science*, 52:15–36, 1987.

[KST91]　J. Köbler, U. Schöning, and J. Torán. Graph isomorphism is low for PP. Technical Report 91-05, Institut für Informatik, Universität Ulm, Ulm, Germany, July 1991.

[KST92] J. Köbler, U. Schöning, and J. Torán. Graph isomorphism is low for PP. In *Proceedings of the 9th Annual Symposium on Theoretical Aspects of Computer Science*, pages 401–411. Springer-Verlag *Lecture Notes in Computer Science #577*, February 1992.

[LLS75] R. Ladner, N. Lynch, and A. Selman. A comparison of polynomial time reducibilities. *Theoretical Computer Science*, 1(2):103–124, 1975.

[Lon82] T. Long. Strong nondeterministic polynomial-time reducibilities. *Theoretical Computer Science*, 21:1–25, 1982.

[OH91] M. Ogiwara and L. Hemachandra. A complexity theory for feasible closure properties. In *Proceedings of the 6th Structure in Complexity Theory Conference*, pages 16–29. IEEE Computer Society Press, June/July 1991. To appear in *Journal of Computer and System Sciences*.

[PZ83] C. Papadimitriou and S. Zachos. Two remarks on the power of counting. In *Proceedings 6th GI Conference on Theoretical Computer Science*, pages 269–276. Springer-Verlag *Lecture Notes in Computer Science #145*, 1983.

[Reg86] K. Regan. A uniform reduction theorem. In *Proceedings of the 13th International Colloquium on Automata, Languages, and Programming*, pages 324–333. Springer-Verlag *Lecture Notes in Computer Science #226*, July 1986.

[Sch85] U. Schöning. Robust algorithms: A different approach to oracles. *Theoretical Computer Science*, 40:57–66, 1985.

[Sel78] A. Selman. Polynomial time enumeration reducibility. *SIAM Journal on Computing*, 7(4):440–457, 1978.

[Sel82] A. Selman. Reductions on NP and P-selective sets. *Theoretical Computer Science*, 19:287–304, 1982.

[Sim75] J. Simon. *On Some Central Problems in Computational Complexity*. PhD thesis, Cornell University, Ithaca, N.Y., January 1975. Available as Cornell Department of Computer Science Technical Report TR75-224.

[Tar89] G. Tardos. Query complexity, or why is it difficult to separate $NP^A \cap coNP^A$ from P^A by random oracles A. *Combinatorica*, 9:385–392, 1989.

[Val76] L. Valiant. The relative complexity of checking and evaluating. *Information Processing Letters*, 5:20–23, 1976.

[Wag90] K. Wagner. Bounded query classes. *SIAM Journal on Computing*, 19(5):833–846, 1990.

[Wat88] O. Watanabe. On hardness of one-way functions. *Information Processing Letters*, 27:151–157, 1988.

The Complexity of Space Bounded Interactive Proof Systems

Anne Condon[*]
Computer Science Department
University of Wisconsin-Madison

1 Introduction

Some of the most exciting developments in complexity theory in recent years concern the complexity of interactive proof systems, defined in 1985 by Goldwasser, Micali and Rackoff [31] and independently by Babai [3]. In this paper, we survey results on the complexity of space bounded interactive proof systems and their applications.

An early motivation for the study of interactive proof systems was to extend the notion of NP as the class of problems with efficient "proofs of membership". Informally, a prover can convince a verifier in polynomial time that a string is in an NP language, by presenting a witness of that fact to the verifier. Suppose that the power of the verifier is extended so that it can flip coins and can interact with the prover during the course of a proof. In this way, a verifier can gather statistical evidence that an input is in a language.

As we will see, the interactive proof system model precisely captures this interaction between a prover P and a verifier V. In the model, the computation of V is probabilistic, but is typically restricted in time or space. A language is accepted by the interactive proof system if, for all inputs in

---
[*]Work supported by NSF grant number CCR-9100886

the language, V accepts with high probability, based on the communication with the "honest" prover P. However, on inputs not in the language, V rejects with high probability, even when communicating with a "dishonest" prover. In the general model, V can keep its coin flips secret from the prover. An important restriction is obtained by requiring that the verifier communicate all its coin flips to the prover as it flips them. Such interactive proof systems were first studied by Babai [3], who labeled them Arthur-Merlin games. They are also known as interactive proof systems with public coins, as opposed to the more general interactive proof systems with private coins.

There have been major breakthroughs in understanding the complexity of interactive proof systems. These breakthroughs have also had profound applications in diverse areas of computer science, including cryptography (zero-knowledge interactive proofs), program checking, formal language theory, group theory, stochastic processes, and in proving non-approximability results for NP-complete problems.

Our goal in this paper is to provide a survey of results for space bounded interactive proof systems - where space, rather than time, is the primary restricted resource of the verifier. We present bounds on the resulting complexity classes, and describe applications of these results to computational problems in areas such as formal language theory, stochastic processes and non-approximability of NP-complete problems. Unlike time bounded interactive proof systems, many fundamental problems on space bounded interactive proof systems still remain unsolved. Our goal is to describe these problems in a unified context, and to give the reader some insight into techniques that may be applicable to solving them.

Our decision to focus on space bounded interactive proof systems reflects our own biases and experience, and keeps our task within reasonable bounds. As a result, we do not describe in depth the remarkable results on time bounded interactive proof systems, which led to a complete characterization of both single and multi-prover, polynomial time bounded interactive proof systems [39], [45] [4], [26]. For completeness, we do compare these results with what is known about space bounded interactive proof systems. We also omit discussion of results on other related models, such as the games against nature of Papadimitriou [40] (which can be thought of as interactive proof systems with unbounded error) and the private alternating Turing machines of Peterson and Reif [42].

The interactive proof system model is defined in detail in Section 2. Results on space bounded interactive proof systems and selected proofs are described in the remaining sections. For convenience, all results are summarized in Figures 1 and 2 at the end of the paper. In the rest of this section, we highlight some of these results and their applications.

We begin by describing informally a very simple interactive proof system (P, V), due to Dwork and Stockmeyer [19], in order to make more concrete the model of an interactive proof system when the verifier uses limited space. This interactive proof system accepts the language Pal = $\{x \in \{a,b\}^* \mid x = x^R\}$, where x^R denotes the string x written backwards.

On input x, the prover P repeatedly sends x to the verifier V. V performs the following computation each time it receives a string, say w, from a prover. First, V flips a coin. If the outcome is heads, V checks that the string w matches the input x, by scanning the input from left to right while receiving w. If the outcome of the coin flip is tails, V checks that w matches x^R, by scanning the input from right to left. If the check succeeds on all iterations, the verifier accepts the input.

Note that the verifier uses $O(1)$ space in this interactive proof system. We call such a verifier a 2pfa, since it is essentially a probabilistic finite state automaton with a 2-way input head. If the input $x \in$ Pal, then (P, V) accepts with probability 1, whereas if $x \notin$ Pal, then on each iteration the verifier finds a mismatch with probability at least $1/2$, no matter what string the prover sends. This is true because the verifier keeps its coin flips hidden from the prover. Thus, (P, V) accepts all strings in Pal with probability 1, whereas the probability that (P^*, V) accepts a string not in Pal is at most $1/2^k$, if there are k iterations of the above protocol.

This example does not illustrate the full power of $O(1)$ space bounded interactive proof systems. We will see that in fact, any language in DTIME($2^{O(n)}$) has an interactive proof which is $O(1)$ space bounded. Furthermore, any language in DTIME($2^{poly(n)}$) has an interactive proof system which is log space bounded. We denote the classes of languages accepted by interactive proof systems which are $O(1)$ and log space bounded by IP(2pfa) and IP(log-space), respectively. The best known upper and lower bounds on these classes are as follows.

$$\mathrm{DTIME}(2^{poly(n)}) \subseteq \mathrm{IP}(\text{log-space}) \subseteq \mathrm{ATIME}(2^{2^{poly(n)}}) \text{ and}$$

$$\mathrm{DTIME}(2^{O(n)}) \subseteq \mathrm{IP}(\text{2pfa}) \subseteq \mathrm{ATIME}(2^{2^{O(n)}}),$$

where DTIME and ATIME refer to deterministic and alternating time bounded classes, respectively, and $poly(n)$ denotes $n^{O(1)}$. Note that there is a large gap between the upper and lower bounds here.

The ability of the verifier to keep its coin flips hidden from the prover is crucial in both the Pal example and in the above bounds. If the interactive proof system is an Arthur-Merlin game, its power is considerably weaker. We denote by AM(2pfa) and AM(log-space) the classes of languages accepted by public coin interactive proof systems with $O(1)$ and log space, respectively. Also, we denote by 2PFA the class of languages accepted by 2-way probabilistic finite state automata with bounded error. Then,

$$\mathrm{2PFA} \subset \mathrm{AM}(\text{2pfa}) \subset \mathrm{AM}(\text{log-space}) = \mathrm{P}.$$

An example of a language separating 2PFA from AM(2pfa) is Center, the set of strings over the alphabet $\{a, b\}$ which have a b in the center. An example of a language separating AM(2pfa) from P is Pal.

The expected time needed by a $O(1)$ space bounded interactive proof system to recognize a language in $\mathrm{DTIME}(2^{O(n)})$ may be double exponential in the size of the input. It is therefore useful to consider complexity classes where the time, as well as the space, used by the interactive proof system is limited. In the statement of the following results, the restriction poly-time is added to denote complexity classes where, in addition to a space bound, the number of steps taken by the verifier is bounded by a polynomial.

$$\mathrm{IP}(\text{log-space, poly-time}) = \mathrm{IP}(\text{poly-time}) = \mathrm{PSPACE} \text{ and}$$

$$\mathrm{NC} \subseteq \mathrm{AM}(\text{log-space,poly-time}) \subseteq \mathrm{P} \subseteq \mathrm{AM}(o(\log^2 n)\text{-space, poly-time}).$$

NC denotes the class $\mathrm{ASPACE,TIME}(\log n, \log^{O(1)} n)$.

In the case of $O(1)$ space bounded interactive proof systems, the class IP(2pfa,poly-time) contains an NP-complete language, and properly contains the class AM(2pfa). Again, Pal separates the two classes.

Yet another possible restriction on the verifier is that it uses few random bits. We denote the complexity classes of interactive proof systems in which the verifier uses log random bits, by adding the notation log-random-bits. Note that the complexity classes IP(log-space,log-random-bits) and AM(log-space,log-random-bits) are contained in IP(log-space,poly-time) and AM(log-space,poly-time), respectively. This is because if only $O(\log n)$ random bits are used, the verifier can flip them all at the start and behave deterministically thereafter; and a $O(\log n)$ space bounded computation that halts must run in polynomial time. The following additional results indicate that the containments may be strict.

$$\text{IP(log-space,log-random-bits)} = \text{NP and}$$

$$\text{NLOG} \subseteq \text{AM(log-space,log-random-bits)} \subseteq \text{LOGCFL}.$$

Here, NLOG is nondeterministic log space and LOGCFL is the class of languages that are log-space reducible to context free language recognition (see Sudborough [46]).

So far, we have described results on space bounded interactive proof systems with additional restrictions on the time or amount of randomness used by the verifier. We next consider more fundamental variations of the model. The first is the multiple-prover model, where the verifier can interact with two or more provers. The provers cannot communicate with each other during the proof. Intuitively, it is potentially more powerful than the single prover model because the verifier can ask overlapping sets of questions of each prover, and use the consistency of the provers on the common questions to verify that both are honest. We denote the class of languages accepted by interactive proof systems with two provers and a verifier which is a 1-way probabilistic finite state automaton, or pfa, by 2IP(pfa). This model is extremely powerful.

$$\text{2IP(pfa)} = \text{Recursive languages and}$$

$$\text{2IP(pfa, poly-time)} = \text{NTIME}(2^{poly(n)}).$$

The other variation of the model that we consider is zero knowledge interactive proof systems. Roughly, an interactive proof system is zero knowledge if on all accepted inputs, the verifier can learn nothing from the proof other than the fact that the input is in the language. For a reasonable formalization of this notion for space bounded interactive proof systems, the following results are known. We denote by ZKIP(2pfa) and ZKIP(log-space,poly-time) the classes of languages which have zero knowledge interactive proof systems when the verifier is a 2pfa or is simultaneously log space and polynomial time bounded, respectively.

$$\text{ZKIP(2pfa)} \subset \text{IP(2pfa)}.$$

In fact, Pal is an example of a language in IP(2pfa) but not in ZKIP(2pfa).

$$\text{IP(log-space,poly-time)} = \text{ZKIP(log-space,poly-time)}.$$

To conclude this section, we give some examples of applications of results on space bounded interactive proof systems to other computational problems. Most of these applications are discussed in more detail in future sections.

We first cite two examples of undecidability results that follow from results on $O(1)$ space bounded interactive proof systems. Lipton [16] applied a result on "weak" interactive proof systems, which are $O(1)$ space bounded, to show that the emptiness problem for 1-way probabilistic finite state automata with unbounded error probability is undecidable, a problem that had been open since the late 60's (see Theorem 3.3). Feige and Shamir [23] applied results on space bounded multiple prover interactive proof systems to prove that a game-theoretic problem proposed by Reif [43] in 1979 is undecidable. Roughly the problem is to decide if, in a game from a certain class of "reasonable" 2-player games of incomplete information, a given player has a strategy which is expected to win.

The first application of interactive proof systems to obtain a non-approximability result for an NP-complete problem arose from the study of space bounded interactive proof systems. Condon [11] showed that a variation of the word problem for matrices, called the "max word problem for matrices" is NP-complete and furthermore, that the corresponding optimization problem cannot be approximated within any constant factor, unless P = NP. The result also has applications to the emptiness problem for 1-way probabilistic finite state automata with unbounded error.

A nice application to problems in automata theory is due to Dwork and Stockmeyer [20]. They showed that 2pfa's that are restricted to run in expected polynomial time and have bounded error accept exactly the regular languages. The techniques used to prove this were derived from their work on $O(1)$ space bounded verifiers.

A final application is in the area of bounding the rate of convergence of stochastic processes. In studying space bounded interactive proof systems, Condon and Lipton [16] obtained tight bounds on the rate of convergence of certain classes of discrete time-varying Markov chains to their absorbing states. A time-varying Markov chain is a sequence of random variables over a finite state space, with the following property. For all positive integers i, a transition matrix P_i determines the value of the $(i+1)st$ random variable, given the value of the ith random variable. Let $\mathcal{M}$ be the family of n-state time-varying Markov chains such that the matrices P_k are all from some finite set of stochastic matrices, say $\{A, B\}$. We assume that all the entries of A and B are rational, of the form p/q where p and q are integers, $p \leq q \leq 2^n$. A special case of the results of Condon and Lipton on time-varying Markov chains can be stated simply as follows. Suppose that for all chains M in $\mathcal{M}$, n is a halting state which is eventually reached from the initial state with probability 1. Then the expected time to reach the halting state n is $2^{2^{\Theta(n)}}$. A well known result for stationary Markov chains under similar conditions is that the expected time to reach a halting state n is $2^{\Theta(n)}$.

The rest of the paper is organized as follows. We first define precisely the model of an interactive proof system and related complexity classes. In Sections 3 and 4, we present results on log space bounded interactive proof systems with private and public coins, respectively. We consider $O(1)$ space bounded interactive proof systems in Section 5. We give an overview of some of the proofs of these results. We note that the results of Sections 3 and 4 can be extended to other space bounds $s(n) = \Omega(\log n)$ and the results of section 5 can be extended to sublogarithmic space bounds. Finally, two variations on the model – multiple prover interactive proof systems and zero knowledge interactive proof systems are considered in Section 6, and known results are stated without proof. Some open problems are discussed in the concluding section.

2 Definitions

In this section, we describe the interactive proof system model. The definitions we use here are probably closest to those of Dwork and Stockmeyer [19], although many alternative, equivalent definitions can be found in the literature ([6], [10] [31]).

An interactive proof system consists of a prover P and a verifier V. The verifier is a probabilistic Turing machine with a 2-way, read-only input tape, a read-write work tape and a source of random bits. The states of the verifier are partitioned into reading and communication states. In addition, the Turing machine has a special communication cell that allows the verifier and prover to communicate.

A transition function describes the one-step transitions of the verifier. Whenever the verifier is in a reading state, the transition function of the verifier determines the next configuration of the verifier, based on the symbol under the tape heads, the state and possibly the outcome of an unbiased coin toss. Whenever the verifier is in a communication state, the next configuration is determined as follows. Associated with each communication state is a symbol; without loss of generality we assume that the set of such symbols is $\{0,1\}$. When in communication state c, the verifier writes the symbol associated with c in the communication cell and in response, the prover writes a symbol in the cell. Based on the state and the symbol written by the prover, the verifier's transition function defines the next state of the verifier.

The prover P is specified by a prover transition function. This function determines what communication symbol is written by the prover in response to a symbol of the verifier, based on the input and the sequence of all past communication symbols written by the verifier. Without loss of generality we assume that all symbols written by the prover in the communication cell are from the set $\{a,b\}$ and that the input alphabet is Σ. Thus the prover's transition function is a mapping from $\Sigma^* \times \{0,1\}^*$ to $\{a,b\}$.

The probability that (P,V) accepts (rejects) x is the limit as $k \to \infty$ of the probability, (taken over all coin tosses of the verifier), that (P,V) reaches the accepting (rejecting) state on x in k steps. The probability that (P,V) halts is defined to be the probability that (P,V) accepts or rejects.

The prover-verifier pair (P, V) is an interactive proof system for L with error probability $\epsilon < 1/2$ if

1. for all $x \in L$, the probability that (P, V) accepts x is $> 1 - \epsilon$,
2. for all $x \notin L$, and all provers P^*, the probability that (P^*, V) rejects x is $> 1 - \epsilon$.

In the paper, we assume that $\epsilon = 1/4$, unless otherwise specified. In most of the results of this paper (except for those in which the number of random bits is limited), the constant ϵ can be replaced by any function of the form $1/2^{O(n)}$.

A different, weaker, definition of language acceptance for space bounded interactive proof systems obtained by replacing condition 2. above by the following.

2'. for all $x \notin L$ and all provers P^*, the probability that (P^*, V) accepts x is $\leq \epsilon$.

For most complexity classes, the definitions are equivalent. A notable exception is the class of languages accepted by interactive proof systems which are $O(1)$ space bounded. In this case, we use the notation weak-IP(2pfa) to refer to the class of languages accepted by interactive proof systems with respect to the weaker definition, that is, with condition 2' instead of condition 2. Lipton [16] showed that the class weak-IP(2pfa) contains all the recursively enumerable languages. We discuss this class further in Section 3.

In many of the interactive proof systems that we describe, the roles of the prover and verifier are typically to send strings to each other, and informally we say that "the verifier sends a string w to the prover", or "the verifier receives a string w from the prover". This can be made precise in our model of a single communication cell, as follows. Suppose the verifier wishes to send a string $w = w_1 w_2 \ldots w_k$ to the prover, where for all i, $w_i \in \{0, 1\}$. To do this, the verifier can write $w_1 1 w_2 \ldots 1 w_k 0$ in the communication cell. The prover can recognize the end of the string by the appearance of a 0 at an even numbered position. In a similar fashion, the prover can send a string over $\{a, b\}^*$ to the verifier.

Just as for Turing machines, a *configuration* of the verifier of an interactive proof system for a fixed input is a tuple containing an encoding of the work tapes, the positions of the tape heads on the input and work tapes of the verifier, and the state and the contents of the communication cell. A configuration that contains a communication state is called a communication configuration, and one which contains a reading state is called a reading configuration.

Two well-studied special cases of the general definition of interactive proof systems represent two extremes; one in which the verifier sends the prover complete information about its current configuration and the other in which the verifier sends the prover no information. In an *Arthur-Merlin game*, whenever the verifier flips a coin, the outcome is written in the communication cell. ¿From this, a prover has complete information about the computation of the verifier. In this case, we can make certain simplifying assumptions about the prover P, namely that the response of P depends only on the input x and current configuration of V, and not on the complete sequence of symbols written by V in the communication cell. (Condon [10] shows why such an assumption can be made without loss of generality). We say in this case that the prover P uses a *Markov strategy*. At the other extreme is a *oneway* interactive proof system, where the verifier sends no information to the prover. In this model, the verifier simply writes the same symbol in the communication cell whenever it needs another symbol from the prover. In this case, the prover can be simply represented as an infinite string over the alphabet $\{a,b\}$, where the ith symbol of the string is the symbol written by the prover in the communication cell the ith time that the verifier enters a communication state.

The notion of a single-prover interactive proof system was generalized to two and more provers by Ben-Or, Goldwasser, Kilian and Wigderson [8]. In a 2-prover interactive proof system, two provers interact with the verifier, but the provers cannot communicate with each other during the proof. We discuss this model in Section 6.

We next describe the resource bounds considered in the paper. An interactive proof system (P,V) is $s(n)$ *space bounded* if on any input of length n, for all P^*, the number of work tape cells read by V is at most $s(n)$, during the computation of (P^*,V). If the number of work tape cells used by the verifier is $O(1)$, the verifier is a probabilistic 2-way finite state automaton, or 2pfa. We say that (P,V) is $t(n)$ *time bounded* if on any input of length n, for all P^*, the *expected* number of steps taken by the verifier during the computation of (P^*, V) is at most $t(n)$. We consider expected time, rather than worst case time, because we will consider interactive proof systems which are $O(1)$ space bounded, and expected time is a more natural definition in this case. A third bound we consider is a limit on the number of random bits used by the verifier. We say that (P, V) uses $r(n)$ *random bits* if on any input of length n, for all provers P^*, the number of random bits used by V during the computation of (P^*, V) is at most $r(n)$.

2.1 Notation

As we stated earlier, we use IP and AM to refer to interactive proof systems with private or public coins, respectively. We use oneway-IP to denote oneway interactive proof systems. Thus, IP(<restrictions>) is the class of languages which have interactive proof systems with restrictions denoted by <restrictions>. The most common restrictions we consider are: (i) log-space, (ii) 2pfa, (iii) poly-time and (iv) log-random-bits, which mean that the interactive proof system (i) is $O(\log n)$ space bounded (ii) is $O(1)$ space bounded, (iii) is polynomial time bounded and (iv) uses $O(\log n)$ random bits, respectively. Thus, IP(log-space,poly-time) is the class of languages accepted by interactive proof systems with private coins, which are simultaneously $O(\log n)$ space bounded and polynomial time bounded. Also, oneway-IP(log-space, log-random-bits) is the class of languages accepted by oneway interactive proof systems which are simultaneously $O(\log n)$ space bounded, and use $O(\log n)$ random bits. In Section 6, we also consider interactive proof systems which are $O(1)$ space bounded, and in addition, the verifier can read its input only in one direction. In this case, we say that the verifier is a pfa. Then, AM(pfa) is the class of languages accepted by Arthur-Merlin games, or public coin interactive proof systems, where the verifier is a pfa.

2.2 Other Complexity Classes and Related Work

We refer to the alternating Turing machines of Chandra, Kozen and Stockmeyer [9] widely in the paper, and so we give a brief description here. An alternating Turing machine is a generalization of a nondeterministic Turing machine, with both existential and universal states. The roles of these states with respect to language acceptance is specified using a computation tree as follows. The nodes of a computation tree of an alternating Turing machine on an input w are labeled by configurations. The root is labeled by the initial configuration of the machine on x, and leaves are labeled by halting configurations. Each internal node labeled by a universal configuration has one child labeled by each possible configuration that is reachable in one step. Each internal node labeled by an existential configuration has exactly one child, labeled by some configuration reachable in one step. The tree is accepting if the tree is finite and all leaves are labeled with accepting configurations. The input w is accepted by the machine if there is an accepting computation tree corresponding to w.

An alternating Turing machine is $t(n)$ time bounded, or $s(n)$ space bounded, if on all accepted inputs of length n, there is an accepting computation tree of height $\leq t(n)$, or whose nodes are labeled by configurations of length $\leq s(n)$ respectively. We assume that the input can be accessed by writing an address on a special index tape, so that sublinear time bounds give rise to meaningful complexity classes.

We let ATIME($t(n)$) and ASPACE($s(n)$) denote the class of problems which are accepted by $O(t(n))$ time bounded and $O(s(n))$ space bounded alternating Turing machines, respectively. Also, the class of languages accepted by alternating Turing machines which are simultaneously $O(s(n))$ space bounded and $O(t(n))$ time bounded is denoted by
ASPACE,TIME($s(n), t(n)$). Chandra, Kozen and Stockmeyer [9] showed that ATIME(poly(n)) = PSPACE and ASPACE($\log n$) = P. We will also consider the complexity class NC [17] of problems which have $\log^{O(1)} n$ time algorithms on a PRAM (parallel random access machine) with a polynomial number of processors. Ruzzo [44] showed that NC = ASPACE,TIME($\log n$, $\log^{O(1)} n$).

For completeness, we next discuss results on polynomial time bounded, single prover interactive proof systems. Although a detailed treatment of these results are beyond the scope of this paper, we will compare them with the results on space bounded interactive proof systems presented in this paper (see Figure 1). We denote by IP(poly-time) and AM(poly-time) the classes of languages accepted by polynomial time bounded interactive proof systems with private and public coins, respectively.

It is not too hard to see from the definitions that NP ⊆ AM(poly-time) ⊆ IP(poly-time) ⊆ PSPACE (see [10]). In 1986, Goldwasser and Sipser [32] showed that IP(poly-time) = AM(poly-time). In 1990, Lund, Fortnow, Karloff and Nisan [39] found an interactive proof system for the permanent function, which is hard for the class #P of Valiant [48] and thus hard for the polynomial time hierarchy, PH, by a result of Toda [47]. Thus they showed that any language in PH is in IP(poly-time). Their proof uses a result of Lipton [38] that the permanent of square matrices over a finite field is random self-reducible. Lipton's proof is based on Beaver and Feigenbaum's [7] construction of "instance hiding schemes" for arbitrary Boolean functions. Building on this work, Shamir [45] showed that all languages in PSPACE have interactive proof systems. Similar techniques have been used to obtain results on the power of multiple prover interactive proof systems, which we discuss in Section 6, most notably the result of Babai, Fortnow and Lund [4] that any language in nondeterministic exponential time is accepted by a polynomial time bounded, two-prover, interactive proof system.

There has been much other work on polynomial time bounded interactive proof systems with various restrictions on the prover and verifier, other than space. We mention two examples here, which are closely related to results in this paper. In [15], Condon and Ladner introduce a variation of the model of interactive proof system, in which the power of the prover is restricted and study resulting complexity classes, including a model in which the verifier is log space bounded. Arora and Safra [1] and Arora, Lund, Motwani, Sudan and Szegedy [2] considered a model in which the verifier can only use a limited number of symbols received from the prover in its computation. Their techniques show that any language in NP can be accepted by a polynomial time bounded interactive proof system, which uses $O(\log n)$ random bits and in which the verifier uses only $O(1)$ (randomly chosen) symbols received from the prover in its computation. This result was used to show that, unless P=NP, the problem of approximating the size of the largest clique in a graph is NP-complete. Earlier results on the hardness of approximating the clique number were proved by Feige, Goldwasser, Lovasz, Safra and Szegedy [22]. The proofs of Arora et al. also build on previous work of Babai, Fortnow, Levin and Szegedy [5] on checking computations and of Babai, Fortnow and Lund [4] on multiple prover interactive proof systems.

3 Log Space; Private Coins

In this section, we consider the power of interactive proof systems with no time bounds, in which the verifier is $O(\log n)$ space bounded. The main results of this section show that

$$\text{DTIME}(2^{poly(n)}) \subseteq \text{IP(log-space)} \subseteq \text{ATIME}(2^{2^{poly(n)}}).$$

The lower bound on IP(log-space) was proved by Condon [12] and independently by Dwork and Stockmeyer [19]. The proof is based on the following idea: a computation of a polynomial space bounded alternating machine is repeatedly sent by the prover to the verifier and the verifier probabilistically checks for errors in the computation. If none are found, and all computations end in an accepting state, the verifier accepts the input. It is interesting to note in the proof that the interactive proof system requires double exponential expected time on accepted inputs, so it appears that even with $O(\log n)$ space bounded verifiers, interactive proof systems can run "usefully" for double exponential time.

The upper bound was proved by Condon and Lipton [16], and requires a proof that interactive proof systems must halt with high probability in double exponential time. The proof of this exploits a close relationship between space bounded interactive proof systems and families of time varying Markov chains, and in fact has applications in the theory of time-varying Markov chains.

The lower and upper bounds on IP(log-space) are presented in Theorems 3.1 and 3.2. Following this, we briefly discuss the complexity of log space bounded interactive proof systems, when there are additional resource bounds on the verifier, such as time and randomness. Applications of these results to the
non-approximability of NP-complete problems are discussed at the end of this section.

Theorem 3.1 *DTIME($2^{poly(n)}$) $\subseteq$ IP(log-space).*

The Complexity of Space Bounded Interactive Proof Systems 161

Proof: To build up to the proof, we first describe why NP is contained in IP(log-space). Suppose that L is accepted by a nondeterministic Turing machine M in polynomial time. On input x of length n, the prover sends a computation, or sequence of configurations of M on x, to the verifier. The verifier checks that the computation ends in an accepting computation. The verifier also checks that the computation is *valid*, namely that it starts in the initial configuration and that the $(i+1)$st configuration follows from the ith configuration according to the rules of M. However, this check cannot be done deterministically in $O(\log n)$ space. Instead, the verifier chooses one symbol, say the jth symbol, from each configuration, uniformly at random. (Without loss of generality, we can assume that the length of a configuration is a power of 2, to make this possible.) Then V checks that symbol j is correct. Using standard encodings of configurations, this can be done using $O(\log n)$ space, by storing only the index j and a constant number of symbols from the ith configuration. If the prover sends an invalid computation, the verifier detects that it is invalid with probability at least $1/poly(n)$. To reduce the error, the verifier repeats the above protocol polynomially many times and accepts if and only if no computation is found to be invalid. Note that the above argument actually shows that NP $\subseteq$ oneway-IP(log-space,poly-time), since the verifier never sends information to the prover.

This idea can be extended to show that DTIME($2^{2^{poly(n)}}$) is contained in IP(log-space). The following argument is essentially due to Dwork and Stockmeyer [19]. Recall that ASPACE(poly(n)) = DTIME($2^{poly(n)}$); thus it is sufficient to show that ASPACE(poly(n)) is contained in IP(log-space). In this case, the verifier must determine that on input x, there is an accepting subtree of some alternating Turing machine M on input x. We assume that every leaf of the tree has depth exponential in the input length, and thus the tree has $2^{2^{poly(n)}}$ leaves. The prover sends the verifier computations, corresponding to paths in the computation tree, and the verifier tests just as above that each computation is valid and that the final configuration is accepting. The verifier sends the prover random coins, to direct the choice of the path taken by the prover at universal nodes of the tree. The verifier accepts if and only if all computations are valid and end in an accepting configuration. $2^{2^{p(n)}}$ computations must be sent by the prover to the verifier, for some polynomial $p(n)$, in order to ensure that with high probability, all paths of the computation subtree are checked.

However, V cannot count to $2^{2^{p(n)}}$ in log space. Thus, we extend the description of V, to ensures that the expected number of computations V receives from P is $2^{2^{p(n)}}$, for some polynomial $p(n)$. To do this, V performs a "halt test" each time it receives a computation from P. V flips a polynomial number of coins for each configuration received from P, and halts in an accepting state at the end of the computation if all coin flips are heads. Since there are exponentially many configurations in a computation, this ensures that the probability that V halts at the end of a given configuration is $1/2^{2^{poly(n)}}$.

This completes our informal description of (P, V). To prove that (P, V) accepts L, it must be shown that given an input $x \notin L$, (P^*, V), rejects x with high probability, where P^* is any prover. To see why this is true, note that if the prover P^* sends V an infinite computation, there must be infinitely many i such that the $(i+1)$st configuration does not follow from the ith configuration. In this case, V will reject with probability 1. Also, if P^* sends V an invalid computation, V is more likely to find an error than to halt as a result of the halt test. □

Theorem 3.2 (i) $IP(log\text{-}space) \subseteq ATIME(2^{2^{poly(n)}})$, and
(ii) $oneway\text{-}IP(log\text{-}space) \subseteq NTIME(2^{2^{poly(n)}})$.

Proof: We describe only the proof that oneway-IP(log-space) is contained in $NTIME(2^{2^{poly(n)}})$. The proof that $IP(log\text{-}space) \subseteq ATIME(2^{2^{poly(n)}})$ is a generalization using similar techniques. Suppose that L is a language in the class oneway-IP(log-space), where L is accepted by (P, V). To further simplify the presentation here, we assume that on inputs not in L, (P^*, V) halts with probability 1 for all provers P^*. (We do not know if this assumption can be made without loss of generality, but the proof is similar in the case that (P^*, V) halts with probability $> 1/2$.)

The key to the proof is to show that if an input x of length n is not in L, then for all P^*, (P^*, V) reaches a rejecting state with high probability in time $2^{2^{poly(n)}}$. To prove this, we need the following notation. Let m be the number of communication configurations of V on x, that is, the number of configurations in which the verifier is in a communication state. Number them $1, \ldots, m$, and without loss of generality assume that 1 is the initial configuration and that m is a unique rejecting configuration. For communication configurations i and j, let $p(i,j,a)$ be the probability that from configuration i, if the prover's response is a, the next communication configuration reached (eventually) is j. Define $p(i,j,b)$ similarly, replacing a by b. Note that these probabilities are completely determined by x, i, j, a, b and the transition function of V, and hence can be computed in polynomial time. In fact, these probabilities are rational numbers of the form p/q where $p \leq q \leq 2^{m+1}$. The proof of this is very similar to a proof of Gill [29] on the transition probabilities of $\log n$ space bounded probabilistic Turing machines. Finally let A and B be the $m \times m$ matrices whose ijth entries are $p(i,j,a)$ and $p(i,j,b)$, respectively.

Since the interactive proof system is oneway, each prover P^* corresponds to an infinite string $\sigma_1 \sigma_2 \ldots, \sigma_i \ldots$ where each $\sigma_i \in \{a,b\}$. Suppose that at some time t, the computation of (P^*, V) has reached communication configuration I. We claim that with probability > 0, a halting (accepting or rejecting) configuration is reached within the next 2^m communication configurations. Suppose not. For $0 \leq l \leq 2^m$, let S_l be the set of communication configurations reachable from I after receiving l further symbols from the prover. By the pigeon-hole principle, since each $S_i \subseteq \{1, \ldots, m\}$, $S_j = S_k$ for some $0 \leq j < k \leq 2^m$. Then, if P^{**} is the prover corresponding to the sequence $\sigma_1 \ldots \sigma_{t+j} (\sigma_{t+j+1} \ldots \sigma_{t+k})^*$, the computation of (P^{**}, V) does not halt with probability 1, contradicting our assumption that on all provers, the computation halts with probability 1.

Thus, with probability > 0, a halting configuration is reached within $2^{poly(n)}$ steps, if it has not been reached already. In fact, since the probabilities $p(i,j,a)$ and $p(i,j,b)$ are bounded below by $1/2^{poly(n)}$, the probability of halting is at least $1/2^{2^{poly(n)}}$. It is straightforward to conclude from this that with probability $> 1/2$, a halting configuration is reached in $t(n) = 2^{2^{poly(n)}}$ steps. Since $x \notin L$, then with probability $> 1/4$, the rejecting configuration has been reached.

We can now describe a simple nondeterministic algorithm for L, that runs in $2^{2^{poly(n)}}$ time. On an input of length n, nondeterministically guess a string $\sigma_1 \ldots \sigma_{t(n)}$. This string represents the first $t(n)$ symbols of the strategy of a prover of the interactive proof system. Next, compute the product $M_1 \ldots M_{t(n)}$, where M_i is A or B if σ_i is a or b, respectively. Let p be the $(1,m)$th entry of this product. Recall that the mth configuration is the rejecting configuration. Hence, p is the probability that (P^*, V) rejects after V has received $t(n)$ symbols from P^*, where P^* is the prover corresponding to the string $\sigma_1 \ldots \sigma_{t(n)}$. If $p > 1/4$, reject, else accept. □

The previous theorem should be contrasted with the following result of Lipton, which shows that with respect to the weak definition, interactive proof systems are extremely powerful.

Theorem 3.3 *Any recursively enumerable language L is in weak-IP(2pfa).*

The proof of this theorem generalizes a result of Frievalds [28], who showed that the emptiness problem for 2-way probabilistic finite state machines is undecidable. As an application of this result, Lipton showed also that the emptiness problem for 1-way probabilistic finite state machines is undecidable.

We now turn to complexity classes where the time as well as the space is limited. If a log space bounded interactive proof system has the additional restriction that the time is polynomially bounded, the following results are known.

Theorem 3.4 *(i) PSPACE = IP(log-space,poly-time), and*
(ii) NP = oneway-IP(log-space, poly-time).

Proof: The proof of (i) follows from the fact that PSPACE = IP(poly-time) [45] and that IP(poly-time) = IP(log-space,poly-time). The latter can be proved using essentially the same techniques that were developed in Theorem 3.1; the proof can be found in [12] and was independently proved by Rompel.

We give a brief description of the proof of (ii). Note that we actually showed in Theorem 3.1 that NP ⊆ oneway-IP(log-space, poly-time). To prove the other direction of (ii), we reduce the problem of deciding if a string is in L, where L is accepted by a oneway interactive proof system (P, V), which is log space bounded, to the following problem, which is easily seen to be in NP.

The *max word problem for matrices* is: given a tuple (S, v, w, k, c), where S is a set of $m \times m$ matrices, v and w are m-vectors, k is an integer and c is a constant, is there a way to select a sequence of k matrices $M_1, \ldots, M_k$ (not necessarily distinct) from S in such a way that the product $vM_1 \ldots M_k w^T$ is greater than c? All entries of the matrices and the vectors, as well as the bound c, are rational numbers expressed in binary and k is an integer, expressed in unary notation.

Given x, we reduce the problem of deciding if $x \in L$ to the max word problem for matrices as follows. In the reduction, k is polynomially bounded in $|x|$ and S consists of two matrices A and B are the matrices defined in Theorem 3.2. Thus, the entries of A and B are the transition probabilities of the V between communication configurations, with respect to the two possible symbols a and b that V can receive from the prover. (Recall that these matrices can be constructed in polynomial time). The constant $c = 1/2$, and the vectors v and w are such that all entries are 0, except that entry 1 of v is 1, where again 1 is the number of the initial configuration, and entry $m-1$ of w is 1, where $m-1$ is the number of the unique accepting configuration. Then, the value $vM_1 \ldots M_k w^T$ is the probability that V accepts when the prover sends the string $\sigma_1 \ldots \sigma_k$, where $\sigma_i = a$ if $M_i = A$ and $\sigma_i = b$ if $M_i = B$. □

As a consequence of this proof that the max word problem for matrices is NP-complete, we can conclude that in fact, the problem cannot be approximated by any constant factor, unless NP=P. To our knowledge, this is the first example of the use of interactive proof systems to prove a non-approximability result for an NP-complete problem. Also, this result on the complexity of the max word problem has applications in the theory of probabilistic finite state automata, rational series and k-regular sequences. We describe one of these applications briefly. We consider probabilistic finite state automata (*pfa*'s) with rational transition probabilities, as defined in Paz [41]. Suppose we define the *k-emptiness problem* for pfa's as follows. Given a pfa and a number k, expressed in unary notation, does the pfa reject every string of length $\leq k$? By a simple reduction from the max word problem, we prove that the k-emptiness problem for pfa's is complete for co-NP. Moreover, unless NP=P, it is not possible to approximate the maximum probability that a string of length n is accepted by a pfa.

In the last result of this section, we consider the complexity classes resulting when the verifier is restricted to log space and log random bits.

Theorem 3.5 *NP = oneway-IP(log-space, log-random-bits) = IP(log-space, log-random-bits).*

The proof that NP ⊆ oneway-IP(log-space, log-random-bits) can be found in Condon and Ladner [14]. The main technique of the proof is due to Lipton [38]. Briefly, it is possible to efficiently reduce the problem of testing if a Boolean formula is satisfiable to that of testing if two multisets are equal. Lipton described a test for equality of two multisets using very few random bits and limited space, by comparing a short *fingerprint* of each multiset. The proof that IP(log-space,log-random-bits) ⊆ NP is fairly straightforward.

4 Log Space; Public Coins

The results of this section show that if space is restricted, interactive proof systems with public coins are significantly less powerful than those with private coins.

The first main theorem of this section, Theorem 4.1, due to Condon [10], shows that in fact, AM(log-space) = P. One direction of the proof is based on the fact that when coins are public, a space bounded interactive proof system can be modeled as a Markov decision process, and the probability of reaching an absorbing state of such a process can be computed using linear programming.

We then consider the class AM(log-space,poly-time). From Theorem 4.1, it immediately follows that NLOG ⊆ AM(log-space,poly-time) ⊆ P. By cleverly adapting the techniques of Lund et al. [39] and Shamir [45] to space bounded interactive proof systems, Fortnow and Lund [26] improved the lower bound of NLOG to show that NC ⊆ AM(log-space,poly-time). Even more, they showed that P ⊆ AM($\frac{\log^2 n}{\log \log n}$-space, poly-time). We present their techniques in Theorem 4.2. Thus, "slightly more" than log space is sufficient for a public coin, polynomial time bounded interactive proof system to recognize all languages in P. It is an intriguing open question whether in fact AM(log-space,poly-time) = P.

Theorem 4.1 *AM(log-space) = P.*

Proof: We show that AM(log-space) $\subseteq$ P. Let $L \in$ AM(log-space), and suppose that L is accepted by (P, V). We can assume without loss of generality that for all provers P^*, (P^*, V) halts with probability 1 on all inputs. (Roughly, this is because the verifier V can be modified to flip $p(n)$ coins after each step for some polynomial p and reject if all are heads. In this way, with exponentially small probability at each original step, the computation halts, and so eventually halts with probability 1. If the polynomial p is sufficiently large, the probability of halting because of this coin-flipping test is so small that it does not significantly affect the error probability).

Fix an input x, and number the configurations of V on x so that the initial configuration is numbered 1. For each i, define $p(i)$ to be the maximum probability of reaching an accepting configuration from configuration numbered i, maximized over all provers which use a Markov strategy. (Recall from Section 2 that it is sufficient to consider Markov strategies in a public coins interactive proof system). To determine whether x is accepted by (P, V) for some P, it is sufficient to determine whether $p(1) > 1/2$. If $p(1) < 1/2$, then on all Markov provers P^*, the probability that (P^*, V) accepts x is $< 1/2$, and so x is not in L. If $p(1) > 1/2$, then x must be in L.

Thus, to prove that AM(log-space) is in P, it is sufficient to show that the values $p(i)$ for all i can be computed in polynomial time. The proof relies on the fact that the probabilities $p(i)$ satisfy the following equations. If i is an accepting or rejecting configuration, then $p(i)$ is 1 or 0, respectively. Otherwise, suppose that j and k (j may equal k) are the configurations reachable from i in one step of the verifier. Then if i is a reading configuration, $p(i) = 1/2(p(j) + p(k))$ and if i is a communication configuration, $p(i) = \max\{p(j), p(k)\}$. Intuitively, this is because, at a reading configuration, where a random move is made, the probability of eventually accepting is the average of the probabilities of accepting from the configurations reachable in one step, whereas at a communication configuration, the probability of eventually accepting is the maximum of the probabilities of accepting, taken over the two possible responses of the prover. A precise justification can be found in [10], and is based on results of Howard [34] on Markov decision processes.

Derman [18] showed that the values $p(i)$ which satisfy the above equations are the unique solution to the following linear programming problem.

Let m be the number of configurations of v on x. Minimize $\sum_{l=1}^{m} p(l)$, subject to the constraints

$p(i) \geq p(j)$,   if i is a communication configuration from which j and k are reachable in one step

$p(i) \geq 1/2(p(j) + p(k))$,   if i is a reading configuration from which j and k are reachable in one step

$p(i) = 0$,   if i is a rejecting configuration

$p(i) = 1$,   if i is an accepting configuration

$p(i) \geq 0$, $1 \leq i \leq m$.

Since the linear programming problem is in P (Khachiyan [35]), the values $p(i)$ can be computed in polynomial time, completing the proof.

The other direction, that P $\subseteq$ AM(log-space), is proved by simulating an alternating machine which is log space bounded by an Arthur-Merlin game which is log space bounded. The Arthur-Merlin game has exponential expected running time. □

4.1 Log Space and Polynomial Time

From Theorem 4.1, it follows that AM(log-space,poly-time) $\subseteq$ P. This was also shown by Fortnow [25], who also showed that NLOG $\subseteq$ AM(log-space,poly-time). However, it is open whether AM(log-space, poly-time) = P. In Theorem 4.3, we describe the techniques of Fortnow and Lund [26], which shed light on this question. They show how an alternating machine can be simulated by an Arthur Merlin game, with only a modest increase in the time and space used.

$$\text{ASPACE,TIME}(s(n), t(n)) \subseteq$$

$\bigcap_{\epsilon > 0} \text{AM}(\frac{s(n) \log t(n)}{\log s(n)}\text{-space}, (s^2(n)t(n) + n \log n)s^\epsilon(n)\log^2 t(n)\text{-time})$.

The proof illustrates how the techniques of "arithmetizing" Boolean formulas can be applied to gain insight to space bounded interactive proof systems. From this and well known relationships between alternating machine classes and P and NC, the following results are obtained.

Theorem 4.2 *(i) $NC \subseteq AM(\text{log-space, poly-time})$, and*
(ii) $P \subseteq AM(\frac{\log^2 n}{\log \log n}\text{-space, poly-time})$.

The Complexity of Space Bounded Interactive Proof Systems   169

We will describe the proof of a slightly different result, which is simpler to present than the above result of Fortnow and Lund, but which illustrates all of the important techniques.

Theorem 4.3

$$ASPACE, TIME(s(n), t(n))$$

$$\subseteq AM(s(n)\log t(n)\text{-space}, (s^2(n)t(n) + n\log n)\log^2 t(n)\text{-time}).$$

Proof: Lund and Fortnow show that if $L \in ASPACE, TIME(s(n), t(n))$, then L is accepted by an alternating Turing machine M that uses $O(s(n))$ space, $O(t(n))$ time and has the following additional properties. M uses one tape, and alternates between existential and universal states at each step, with two possible transitions at each step.

Let $\phi_i(I, x)$ be a Boolean predicate which is true if and only if on input x, there is an accepting subtree of M on x which is rooted at I and has depth $2i$. Input x is accepted by M if and only if $\phi_N(I_0, x)$ is true, where I_0 is the initial configuration and $2N$ is the running time of M on x. Because of the properties of M, $\phi_i(I, x)$ has a nice inductive definition.

$$\phi_i(I, x) =$$

$$\begin{cases} g(I), & \text{if } i = 0, \\ \exists z_1 \in \{0,1\} \forall z_2 \in \{0,1\} \exists I' \in \{0,1\}^{k-2} : f(I, z_1, z_2, I') \wedge \phi_{i-1}(I', x), & \text{otherwise.} \end{cases}$$

Here, $k - 2$ is the length of a binary encoding of a configuration of M on x, and $f(I, z_1, z_2, I')$, is a predicate over $\{\wedge, \vee, ^-\}$ which is true if and only if M on input x moves from existential configuration I to existential configuration I', when the existential and universal choices in the next two moves are z_1 and z_2, respectively. Similarly, $g(I)$ is a predicate over $\{\wedge, \vee, ^-\}$ which is true if and only if I is an accepting configuration.

The next step is to define an arithmetic formula $A_i(I, x)$ such that for all i, $\phi_i(I, x)$ is true if and only if $A_i(I, x) = 1$ and $\phi_i(I, x)$ is false if and only if $A_i(I, x) = 0$. $A_i(I, x)$ has the form

$$A_i(I, x) =$$

$$\begin{cases} G(I), & \text{if } i = 0 \\ \amalg_{z_1 \in \{0,1\}} \prod_{z_2 \in \{0,1\}} \sum_{I' \in \{0,1\}^{k-2}} F(I, z_1, z_2, I') \cdot A_{i-1}(I', x), & \text{otherwise.} \end{cases}$$

Here, $\coprod_{a \in \{0,1\}} \sigma(a) = \sigma(0) + \sigma(1) - \sigma(0)\sigma(1)$.

The functions F and G are arithmetic formulas obtained from f and g with the following properties. $F(I, z_1, z_2, I')$ is either 0 or 1 and is 1 if and only if $f(I, z_1, z_2, I')$ is true. Similarly, $G(I)$ is either 0 or 1 and is 1 if and only if $g(I)$ is true. F and G are obtained from the Boolean functions f and g by the following inductive rules. If a is a variable, then the corresponding function A is a. If A and B are the arithmetic functions obtained from Boolean expressions a and b, then AB, $1 - (1 - A)(1 - B)$ and $1 - A$ are the arithmetic functions obtained from $a \wedge b$, $a \vee b$ and $\bar{a}$, respectively. If f and g are suitably expressed as Boolean functions (details omitted), then F has size $O(s^2(n))$, depth $O(\log s(n))$ and constant degree, and G has size $O(n \log n)$, depth $O(\log n)$ and has constant degree.

The verifier checks that $A_N(I_0, x) = 1$ inductively, by reducing the problem of verifying that $A_i(I, x) = \beta$ to the problem of verifying that $A_{i-1}(I', x) = \beta'$. When $i = 0$, $A_i(I, x)$ is evaluated by computing $G(I)$. All calculations are done over the field $\mathbf{Z}/p\mathbf{Z}$, for some prime p.

To describe this reduction, we introduce the following notation. Suppose that

$$A_i(I, x) = Q^{(1)}_{z_1 \in \{0,1\}} Q^{(2)}_{z_2 \in \{0,1\}} .. Q^{(k)}_{z_k \in \{0,1\}} F(I, z_1, .., z_k) A_{i-1}(z_3, .., z_k, x).$$

Here, the vector $(z_3, \ldots, z_k)$ corresponds to I' and each $Q^{(j)}$ is from the set $\{\prod, \coprod, \sum\}$. Given numbers $r_1, \ldots, r_k \in \mathbf{Z}/p\mathbf{Z}$, for $1 \leq j \leq k$ let $P_j(z_j)$ be the function

$$Q^{(j+1)}_{z_{j+1} \in \{0,1\}} .. Q^{(k)}_{z_k \in \{0,1\}} F(I, r_1, .., r_{j-1}, z_j, ..z_k) A_{i-1}(r_3, .., r_{j-1}, z_j, .., z_k, x).$$

Note that $P_j(z_j)$ is a univariate polynomial in z_j which has constant degree. Then the following protocol of the verifier V reduces the problem of verifying that $A_i(I, x) = \beta$ to the problem of verifying that $A_{i-1}(I', x) = \beta'$.

1. Choose a sequence $r_1, \ldots, r_k$ of values in $\mathbf{Z}/p\mathbf{Z}$ independently and uniformly at random. Let $j = 1$.

2. Receive from the prover a polynomial $f_j(z_j) \in \mathbf{Z}/p\mathbf{Z}[z_j]$. (The prover P is defined so that $f_j(z_j) = P_j(z_j) \bmod p$.)

3. If $j > 1$, check that $Q^{(j)}_{z_j \in \{0,1\}} f_j(z_j) = f_{j-1}(r_{j-1}) \bmod p$ and if $j = 1$, that $Q^{(1)}_{z_1 \in \{0,1\}} f_1(z_1) = \beta \bmod p$.

4. If the check fails, halt and reject. Otherwise if $j < k$, send r_j to the prover, set $j = j + 1$, and repeat the protocol from step 2. If $j = k$, let $\beta' = f_k(r_k)/F(I, z_1, \ldots, z_k) \bmod p$ and let $I' = r_3, \ldots, r_k$.

The protocol has the property that if $A_i(I, x) = \beta \bmod p$ then also $A_{i-1}(I', x) = \beta' \bmod p$. However, if $A_i(I, x) \neq \beta \bmod p$ then the probability that $A_{i-1}(I', x) \neq \beta'$ is high, regardless of what the prover sends to V.

The key to the proof of this second property is that for all $j \geq 1$, if the polynomial $f_j(z_j)$ sent by the prover to the verifier is not equal to $P_j(z_j)$, then with high probability, $f_j(r_j) \neq P_j(r_j)$. To see this, note that if two polynomials of constant degree are not equal, they agree at at most a constant number of points. Hence, the probability that they agree at a point r_j chosen randomly and uniformly from $\mathbf{Z}/p\mathbf{Z}$ is $O(1/p)$. There are $k = \Theta(s(n))$ iterations of the above protocol for a given i, and the protocol is repeated $\Theta(t(n))$ times, since $N = \Theta(t(n))$. Hence if $p = \Theta(s(n)t(n))$, the error probability of the protocol is small.

The space needed is dominated by the space to store $k = O(s(n))$ elements $r_1, \ldots, r_k$ of $\mathbf{Z}/p\mathbf{Z}$, each of which has length $O(\log p) = O(\log t(n))$. Hence the space is $O(s(n) \log t(n))$. We next consider the time needed by (P, V). The time to execute the above protocol is dominated by the time to execute step 4, where an arithmetic formula F of size $O(s^2(n))$ must be evaluated. This requires $O(s^2(n))$ additions and multiplications over the field $\mathbf{Z}/p\mathbf{Z}$, and each of these operations can be done in $O(\log^2 p)$ steps. The protocol is executed $N = O(t(n))$ times; hence the total time required for the protocol is $O(t(n)s^2(n)\log^2 p)$. Once the protocol has been executed N times, G is evaluated, which takes time $O(n \log n \log^2 p)$. Since $p = \Theta(s(n)t(n))$, it follows that $\log p = O(\log t(n))$, the total time is $O((t(n)s^2(n) + n \log n)\log^2 t(n))$. □

Finally, we consider public coin interactive proof systems with log space and log random bits. The following relationships were proved by Condon and Ladner [15].

Theorem 4.4 *NLOG $\subseteq$ AM(log-space, log-random-bits) $\subseteq$ LOGCFL.*

The left containment is immediate.

The proof that AM(log-space, log-random-bits) $\subseteq$ LOGCFL, uses the equivalence of LOGCFL and the class of languages accepted by nondeterministic pushdown automata which have $O(\log n)$ auxiliary storage and run in polynomial time [46], and describes how a *computation tree*, which represents all of the computations of an interactive proof system (P, V), can be traversed using the pushdown store. This is possible with only $O(\log n)$ auxiliary space since the number of branching points in the tree is $O(\log n)$, due to the limited number of random bits used by the verifier. When traversing a path, only the direction at the branching points taken so far, and the current configuration need be stored. For more details, see [15].

5 Constant Space

Dwork and Stockmeyer [19] have proved a number of strong results about the power of interactive proof systems with constant space bounded verifiers. In this restricted setting, they have obtained separation results that are not possible for polynomial time, or even log space bounded interactive proof systems. For example, we will see in Section 5.2 that AM(2pfa,poly-time) is properly contained in IP(2pfa, poly-time). The techniques used to obtain these results laid the foundations for new results on the power of probabilistic finite state automata. In [20], Dwork and Stockmeyer showed that 2-way probabilistic finite state automata with bounded error, that run in polynomial expected time, accept exactly the regular languages.

We describe the results on constant space bounded interactive proof systems in two sections, one for private coins and the other for public coins. In the constant space bounded model, we assume that the input is presented on a finite tape, with endmarkers # at both ends of the input.

5.1 Private Coins

In the introduction, we saw that the language Pal, of strings that read the same backwards as forwards, is in the class IP(2pfa,poly-time). The following theorem shows that languages much more complex than Pal have interactive proof systems which are $O(1)$ space bounded, if more than polynomial time is allowed.

Theorem 5.1 *DTIME($2^{O(n)}$) $\subseteq$ IP(2pfa) $\subseteq$ ATIME($2^{2^{O(n)}}$).*

The lower bound is due to Dwork and Stockmeyer [19] and the upper bound to Condon and Lipton [16]. The proof of the lower bound is similar to that of Theorem 3.1. In this case, on input x, the prover sends the verifier the computation of a linear space bounded alternating Turing machine, and the verifier must check that the computation is valid. One of the main differences between this proof and that of Theorem 3.1 is in the way that the verifier checks that the jth symbol of the $(i+1)$st configuration of the computation follows correctly from the ith configuration. In Theorem 3.1, V stores j, but this is not possible here since the verifier has only $O(1)$ space. Instead, when receiving the computation from the prover, the verifier uses the input as a "ruler", in order to locate the jth symbol of the $(i+1)$st configuration in the string obtained from the prover, starting from the j symbol of the ith configuration. This is possible since the length of a configuration is linear in the length of the input. Another difference is that it is no longer possible to choose j uniformly at random. It turns out that an alternative simple method, which favors symbols early in the computation, suffices for the correctness of the protocol.

The next theorem, proved using similar techniques, shows that the class IP(2pfa,poly-time) is also quite powerful.

Theorem 5.2 *AM($O(n)$-space,poly-time) $\subseteq$ IP(2pfa,poly-time).*

It follows immediately that IP(2pfa,poly-time) contains an NP-complete language. Moreover, AM(2pfa) $\subseteq$ AM($O(n)$-space,poly-time), and hence is contained in IP(2pfa,poly-time). (The proof that AM(2pfa) $\subseteq$ AM(O(n)-space,poly-time) is similar to the proof of Theorem 4.1, except instead of using linear programming to compute the values $p(i)$, they are sent by the prover to the verifier.)

5.2 Public Coins

We now consider Arthur-Merlin games which are $O(1)$ space bounded. In Theorem 5.3, we describe the result of Dwork and Stockmeyer that this class does not contain Pal. It follows that AM(2pfa) is properly contained in P and also that AM(2pfa) is properly contained in IP(2pfa, poly-time). In addition, we state other results of Dwork and Stockmeyer, which show that AM(2pfa) contains languages that are neither in AM(2pfa, poly-time) nor in 2PFA.

Theorem 5.3 *Pal $\notin$ AM(2pfa).*

Proof: Suppose to the contrary that (P, V) is a public coin interactive proof system which is $O(1)$ space bounded and accepts Pal. To simplify the proof, we assume that for all P^*, (P^*, V) halts with probability 1 and that the computation ends with the input head at the right end of the tape. Only slight modifications to the following proof are necessary when these conditions are not satisfied. There are three main steps to the proof: (i) we define a notion of "closeness" of two strings; (ii) we argue that for sufficiently large m, there are two distinct strings w_i and w_j of length m which are close and (iii) we show that for some P^*, (P^*, V) accepts $w_j w_i^R$ with probability greater than $1/2$, achieving a contradiction.

To define closeness, we need the following notation, which describes the possible conditions in which a sub-computation of (P, V) can start or end on the string w, when the input is ww^R. Define a *starting condition* to be a pair (q, η), where q is a state of M and $\eta \in \{\text{Left}, \text{Right}\}$; intuitively it means that the computation of (P, V) is started in state q at the end of w denoted by η. Similarly, a pair (q, η) denotes a *stopping condition*, intuitively that the head of V falls off the η end of w with V in state q. Let $p(w, a, b)$ be the probability that on the computation of (P, V) on input ww^R, the stopping condition is b, given that the starting condition is a. Note that $p(w, a, b)$ depends on P and hence indirectly on the fact that w is followed by w^R on the input tape.

We say two numbers x and y are β-*close* for $\beta > 1$ if (a) $x = 0$ if and only if $y = 0$ and (b) if $x > 0$ and $y > 0$, then $1/\beta \leq x/y \leq \beta$. Also, two strings w_i and w_j are β-close if for all a, b, $p(w_i, a, b)$ and $p(w_j, a, b)$ are β-close. This completes (i), the definition of closeness.

We next outline a proof of (ii), that given any constant $\beta > 1$, for sufficiently large m, there is a pair of strings w_i, w_j, both of length m, which are β-close. Let d be the number of pairs (a, b) where a is a starting

The Complexity of Space Bounded Interactive Proof Systems   175

condition and b is a stopping condition. Note that there is a set S of strings of length m such that $|S| \geq 2^{m-d}$, and for all pairs (a,b), either $p(w,a,b) > 0$ for all $w \in S$, or $p(w,a,b) = 0$ for all $w \in S$. Moreover, if w is of length m and $p(w,a,b) \neq 0$, then it can be shown that $p(w,a,b) \geq 1/c^m$ for some constant c. Hence the range in which $p(w,a,b)$ lies is $[1/c^m, 1]$. From this, a pigeon-hole argument can be used to show that for sufficiently large m, two of the 2^{m-d} strings of S are β-close. Let these be w_i and w_j.

We finally describe the proof of (iii), that there is a prover P^* such that (P^*, V) accepts $w_j w_i^R$ with probability $> 1/2$. P^* is the prover which, when V's head is in the string $\#w_j$, responds to V as if the input is $w_j w_j^R$, and when V's head is in the string $w_i^R \#$, responds to V as if the input is $w_i w_i^R$. It remains to show that the probability that (P^*, V) accepts $w_j w_i^R$ is at least $1/2$. To do this, we show that the probability that (P^*, V) accepts $w_j w_i^R$ is within some small constant of the probability that (P, V) accepts $w_i w_i^R$. The computations of (P^*, V) on $w_i w_i^R$ and on $w_j w_i^R$ are modeled as Markov chains H_i and H_j, respectively, which have the same number of states. Both chains have special initial, accept and reject states, and in addition, have two states $(q, l), l = 1, 2$ for each state q of V. State $(q, 1)$ of H_i means that the verifier is in state q with the head at the right end of w_i, and state $(q, 2)$ means that the verifier is in state q with the head at the left end of w_i^R. The initial state of H_i means that the verifier is in its initial state with the head under the left endmarker $\#$, and the accept and reject states mean that the verifier has halted in an accepting or rejecting state, respectively. The probability transitions between these states model the computation of (P, V) on $w_i w_i^R$. H_j is defined similarly, except that the state $(q, 1)$ of H_j means that the verifier is in state q with the head at the right end of w_j. Because w_i and w_j are β-close, and because P and P^* are the same on the right half of the strings $w_i w_i^R$ and $w_j w_i^R$, these Markov chains satisfy the property that the transition probabilities between any two states are β-close.

A result of Leighton and Rivest [37] shows that if two Markov chains satisfy this property, then the probabilities of reaching the accept states of both chains are β^{2s}-close, where s is the number of states in the Markov chains. Thus, the probability that $w_i w_j^R$ is accepted is at least β^{-2s} times the probability that $w_i w_i^R$ is accepted, which is at least $\beta^{-2s} 3/4$. For sufficiently large m, we can choose β so that this is greater than $1/2$, completing the proof of (iii). □

The argument in the above theorem was generalized by Dwork and Stockmeyer to prove the following theorem, which can be used to identify other languages that are not in AM(2pfa).

Theorem 5.4 *Let $L \subseteq \Sigma^*$. Suppose there is an infinite set I of positive integers and, for each $m \in I$, sets $W_m = \{w_1, w_2, \ldots, w_{N(m)}\}, U_m = \{u_1, u_2, \ldots, u_{N(m)}\}$, and $V_m = \{v_1, v_2, \ldots, v_{N(m)}\}$ of words such that*

1. *$|w| \leq m$ for all $w \in W_m$,*

2. *for every integer k there is an m_k such that $N(m) \geq m^k$ for all $m \in I$ with $m \geq m_k$, and*

3. *for all $1 \leq i, j \leq N(m)$, $u_j w_i v_j \in L$ if and only if $i = j$.*

Then $L \notin AM(2pfa)$.

Similar techniques can be used to separate the classes 2PFA and AM(2pfa,poly-time) from AM(2pfa). leading to the following results.

Theorem 5.5 *(i) $2PFA \subset AM(2pfa)$,*
(ii) $AM(2pfa,poly\text{-}time) \subset AM(2pfa)$ and
(iii) $2PFA \not\subset AM(2pfa,poly\text{-}time)$.

The language Upal = $\{a^n b^n \mid n \geq 0\}$ is an example of a language in 2PFA but not in AM(2pfa,poly-time). Frievalds [28] showed that Upal is in the class 2PFA. The language Center = $\{wbx \mid w, x \in \{a,b\}^* \text{ and } |w| = |x|\}$ is an example of a language in the class AM(2pfa) but is not in either of the classes AM(2pfa,poly-time) or 2PFA. The proof that Center is in AM(2pfa) is a simple generalization of Frievalds' proof that Upal is in the class 2PFA. (In this case, the prover is needed to direct the verifier to the center of the string.)

The proofs that Center and Upal are not in AM(2pfa,poly-time) and that Center is not in 2PFA are refinements of the techniques used to prove Theorems 5.3 and 5.4.

6 Variations on the Model

Applications in cryptography, distributed computation and other fields have prompted studies of variations of the model of interactive proof systems which we have considered so far. We describe two of these variations here, and state without proof known results for the models. We consider multiple provers in Section 6.1. Feige and Shamir [23] and independently Condon [16] showed that, even when the verifier is a probabilistic finite state automaton with a 1-way input head, there is a 2-prover interactive proof system which can accept any recursive language.

Zero knowledge interactive proof systems are described in Section 6.2, and both positive and negative results about the existence of zero knowledge interactive proof systems for certain languages are presented. Unlike many results on polynomial time bounded interactive proof systems, these results are not based on any unproven assumptions. Results on log space bounded zero knowledge interactive proof systems are due to Kilian [36] and on constant space bounded interactive proof systems are due to Dwork and Stockmeyer [19].

6.1 Multiple Provers

The multiple prover model, in which the verifier interacts with $k \geq 2$ provers, $P_1, \ldots, P_k$, was originally introduced by Ben-Or, Goldwasser, Kilian and Wigderson [8]. It is natural to ask whether multiple provers increases the power of the model. The answer appears to be "yes" when the interactive proof system is polynomial time bounded, since Babai, Fortnow and Lund [4] showed that polynomial time bounded, 2-prover interactive proof systems accept all languages in nondeterministic exponential time. We will see in this section that, also in the case of log space bounded verifiers, the answer is a resounding "yes".

In the multiple prover model, the verifier has k communication cells, and the communication states of the verifier are partitioned into k groups. Whenever the verifier's state is a communication state in the ith group, the next configuration is determined by communicating with the ith prover via the ith communication cell, as in the single-prover model. Each prover P_i is specified by a prover transition function from $\Sigma^* \times \{0,1\}^*$ to $\{a, b\}$.

$P_i(x, b_1 \ldots b_j)$ is the response of prover P_i on input x, when the verifier has just sent b_j to the prover, and $b_1, \ldots, b_{j-1}$ is the sequence of all past communication symbols written by the verifier in the ith communication cell. Language acceptance is defined in a similar manner to language acceptance for single prover interactive proof systems, with $(P_1, \ldots, P_k, V)$ replacing (P, V) and $(P_1^*, \ldots, P_k^*, V)$ replacing (P^*, V).

In what follows, we restrict our attention to 2-prover interactive proof systems. The results below are also true for k-prover interactive proof systems, for any constant k. We denote by 2IP(<restrictions>) the class of languages which have 2-prover interactive proof systems with the the restrictions denoted by <restrictions>.

The first result states that the class of languages accepted by multi-prover interactive proof systems is exactly the recursive languages, even when the verifier is a pfa, or a 1-way probabilistic finite state automaton.

Theorem 6.1 *2IP(pfa) is exactly the set of recursive languages.*

This result was proved by Feige and Shamir [23] and independently by Condon [16]. The proof of Feige and Shamir actually proves something stronger - that the result is true even if the verifier acts *synchronously*, which means that the verifier communicates with each prover at regular intervals. This can be formalized by requiring that the verifier communicates with prover P_i at step t if and only if $t = i \bmod k$. The main technique introduced in their proof is a method whereby the verifier can simulate a Turing machine computation, using the provers to "store" the contents of the tape. Roughly, this is done by giving each prover random information about the tape contents. This information is meaningless to a single prover, but by combining the information of both provers, the verifier can reconstruct the tape cells whenever necessary.

Feige and Shamir extended their technique to show that when the verifier is a pfa, and is simultaneously polynomially time bounded, then 2IP(pfa,poly-time) = 2IP(poly-time). Combining this with the result of Babai, Fortnow and Lund [4] that 2IP(poly-time) = NTIME($2^{poly(n)}$), the following theorem is obtained.

Theorem 6.2 *2IP(pfa,poly-time) = NTIME($2^{poly(n)}$).*

A variation of the multi-prover model, called the noisy oracle model, was proposed by Feige, Shamir and Tennenholtz [24]. The paper of Feige and Shamir [23] also contains results on the complexity of the noisy oracle model when space is restricted.

6.2 Zero Knowledge

Informally, a zero knowledge interactive proof system for a language L is one in which, on input $x \in L$, a verifier can learn nothing from the prover P, other than the fact that $x \in L$. The notion of a zero knowledge interactive proof system was first introduced by Goldwasser, Micali and Rackoff [31] for polynomial time bounded interactive proof systems. The notion has been central to much recent work in cryptography, and also has interesting applications in distributed computing (see for example Feige, Fiat and Shamir [21] and Goldreich, Micali and Wigderson [30]).

However, many of the results on polynomial time zero-knowledge interactive proof systems are based on unproven assumptions. Because of this, Dwork and Stockmeyer [19] and later Kilian [36] studied space bounded zero knowledge interactive proof systems, with the goal of proving, without using any unproven assumptions, both positive and negative results on the existence of zero knowledge interactive proof systems for certain languages.

Formalizing the intuitive notion of zero knowledge, especially that of "learning nothing" from the prover, is no easy task. The example of the language $\text{Pal} = \{x \mid x = x^R\}$, introduced in Section 1, gives some insight to the difficulty. We saw there an interactive proof system (P, V) for Pal in which the prover P simply sends the input repeatedly to the verifier. This seems to be a zero knowledge interactive proof system, since the verifier is obtaining nothing from the prover that it does not already "know". However, Dwork and Stockmeyer argue intuitively that it is not a zero knowledge interactive proof system, as follows. Let A be the set of all strings that are double palindromes, that is, a string x is in A if and only if $x = ww^R$ and w is also a palindrome. Let $B = \text{Pal} - A$. Techniques similar to those of Theorem 5.3 show that no 2pfa can separate A from B. However, there is a 2pfa V^* such that (P, V^*) can separate A from B. In order to check if the second half of the input is a palindrome, the verifier can perform the following computation, as it receives the input from the prover. Starting at the left end of the tape, V^* moves its head two steps to the right for every symbol it receives from P, until the right end of the tape is reached. At this point, if the input is ww^R, P has finished sending w and is ready to send w^R to V^*. Now, V^* can test if $w = w^R$.

Dwork and Stockmeyer proposed a definition of zero knowledge, based on the following notion of separating sets. We say (P, V) *separates* sets $A, B \subseteq \Sigma^*$ with $A \cap B = \emptyset$ if for all $x \in A$, the probability that (P, V) accepts x is at least $3/4$ and for all $x \in B$, the probability that (P, V) accepts x is at most $3/4$. Similarly, we can define what it means for a 2pfa, M to separate two sets. Let $\mathcal{V}$ be a class of verifier machines and let $(\emptyset, \mathcal{V})$ be the subset of machines in $\mathcal{V}$ that do not communicate with the prover. Let (P, V) be an interactive proof system for the language L where $V \in \mathcal{V}$. Then (P, V) is a recognition zero knowledge interactive proof system for L with $\mathcal{V}$ verifiers if, for any $V^* \in \mathcal{V}$ and any $A, B \subseteq L$ with $A \cap B = 0$ such that (P, V^*) separates A and B, there is an $M_{V^*} \in (\emptyset, \mathcal{V})$ such that M_{V^*} separates A and B. We denote by ZKIP(<restrictions>) the class of languages which have recognition zero knowledge interactive proof systems with the restrictions denoted by <restrictions>. With this definition, Dwork and Stockmeyer proved the following results.

Theorem 6.3 *Pal and the graph isomorphism problem are not in ZKIP(2pfa).*

We have already seen that interactive proof systems do exist for these languages. Dwork and Stockmeyer also prove a positive result for a restricted class of verifiers. A *sweeping-2pfa* is a 2pfa which, on any input, only switches the direction of head movement when at the left or right end of the input.

Theorem 6.4 *The language Upal $= \{a^n b^n \mid n \geq 0\}$ is in ZKIP(sweeping-2pfa, poly-time).*

The proof of this result is quite intricate, and uses deep results from the theory of Markov chains. The theorem is not vacuous, as Greenberg and Weiss [33] showed that Upal is not accepted by any automaton which runs in polynomial expected time.

Kilian [36] studied zero knowledge interactive proof systems with verifiers which are simultaneously log space bounded and polynomially time

The Complexity of Space Bounded Interactive Proof Systems

bounded. He proposed a stronger definition of zero knowledge than that of Dwork and Stockmeyer, and proved the following result.

Theorem 6.5 *IP(log-space, poly-time) = ZKIP(log-space,poly-time).*

The techniques of Kilian are quite different from those of Dwork and Stockmeyer, and involve tools from communication complexity and cryptography.

7 Open Problems

In the preceding sections, we have mentioned numerous unsolved problems on the complexity of space bounded interactive proof systems. We discuss some of these further here.

• There is a large gap between the best known upper and lower bounds for *IP(2pfa)*. The best known lower bound is deterministic exponential time and the best known upper bound is alternating double exponential time. Can either of these bounds be improved?

One way to improve these bounds is suggested by the structure of interactive proof systems for languages in DTIME($2^{poly(n)}$), as described in Theorem 3.1. All known interactive proof systems with a log space bounded verifier have the property that on any input, the verifier and prover iterate a double exponential number of times a protocol that takes only exponential time. We call such an interactive proof system *periodic*. It would be interesting to find a non-periodic interactive proof system that accepts languages not known to be in DTIME($2^{poly(n)}$). Alternatively, a proof that all languages accepted by log space interactive proof systems are also accepted by periodic interactive proof systems would imply better bounds for IP(log-space).

• Dwork and Stockmeyer [19] asked the following question. Is AM(2pfa,poly-time) equal to the class of regular languages?

Dwork and Stockmeyer show that there is a pair of sets A and B which can be separated by an Arthur-Merlin game which is polynomial time bounded and $O(1)$ space bounded, but which cannot be separated by 2pfa. The sets are defined as follows. Let U be the set of words of the form $x\#^k$ where $k = 2^{|x|}$. Then, A is the set of $x\#^k \in U$ such that $x \in$ Center and B

is the set of $x\#^k \in U$ such that $x \notin$ Center. However, the set U is not in AM(2pfa,poly-time), which makes it difficult to extend this separation problem to show that AM(2pfa,poly-time) accepts a non-regular language.

- The work of Fortnow and Lund [26] motivates the question: is AM(log-space, poly-time) = P? Their result that P is contained in AM($o(\log^2 n)$-space, poly-time) leads one to conjecture that perhaps the answer to the above open question is yes. However, it is not clear that their techniques can be extended to prove this. A positive answer to the question might imply that Markov decision processes can be evaluated by new polynomial time algorithms that do not involve linear programming techniques.

- ¿From Section 6, it appears that questions on the complexity of space bounded multiple prover interactive proof systems are completely resolved. However, one intriguing question remains. What languages are in the class oneway-2IP(2pfa)? The best bounds we know are (i) any language in nondeterministic linear space is in oneway-2IP(2pfa) and (ii) oneway-2IP(2pfa) is contained in the class of recursive languages. This is a rather large gap, and one of these bounds can surely be improved.

- Finally, can other nonapproximability results, such as the result on the max word problem presented in Section 3, be obtained from further study of space bounded interactive proof systems?

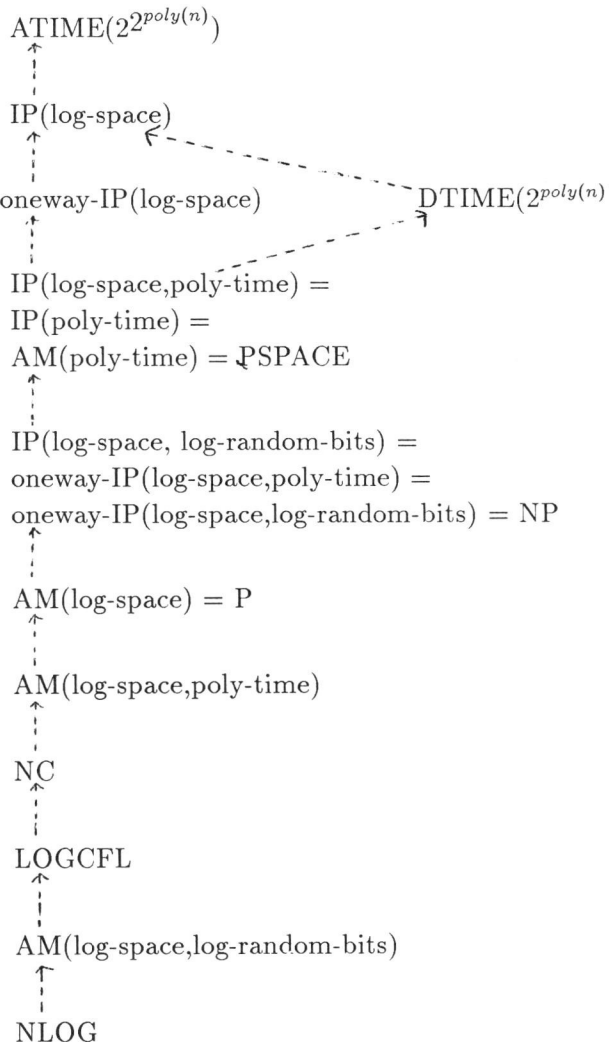

Figure 1: Results on log space bounded interactive proof systems. Each class is contained in the one above it. No containments are known to be proper.

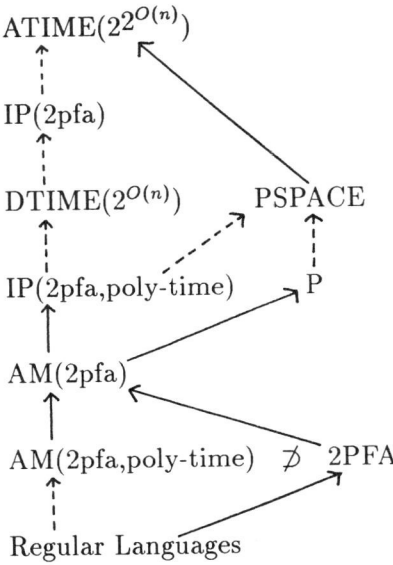

Figure 2: Results on constant space bounded interactive proof systems. Solid arrows denote proper containments, and dotted arrows denote containments which are not known to be proper.

References

[1] S. Arora and S. Safra. Probabilistic checking of proofs; a new characterization of NP, Manuscript, 1992.

[2] S. Arora, C. Lund, R. Motwani, M. Sudan and M. Szegedy. Proof verification and hardness of approximation problems, Manuscript, 1992.

[3] L. Babai. Trading group theory for randomness. Proceedings of 17th Annual ACM Symposium on the Theory of Computing, May 1985, 421-429.

[4] L. Babai, L. Fortnow and C. Lund. Non-deterministic exponential time has two-prover interactive protocols. *Computational Complexity 1* 1991, 3-40.

[5] L. Babai, L. Fortnow, L. Levin and M. Szegedy. Checking computations in polylogarithmic time, Proceedings of the 23rd Annual ACM Symposium on the Theory of Computing, May, 1991, 21-31.

[6] L. Babai and S. Moran. Arthur-Merlin games: A randomized proof system, and a hierarchy of complexity classes. *J. Comput. System Sci.*, 36 (1988), 254-276.

[7] D. Beaver and J. Feigenbaum. Hiding instances in multioracle queries. Proceedings of the 7th Annual Symposium on Theoretical Aspects of Computer Science, Springer Verlag Lecture Notes in Computer Science 415, February 1990, 37-48.

[8] M. Ben-Or, S. Goldwasser, J. Kilian and A. Wigderson. Multi-prover interactive proofs: how to remove intractability, Proceedings of the 20th Annual ACM Symposium on the Theory of Computing, May, 1988, 113-131.

[9] A. K. Chandra, D. C. Kozen and L. J. Stockmeyer. Alternation, *J. ACM*, 28(1), 1981, 114-133.

[10] A. Condon. *Computational Models of Games*, MIT Press, July 1989.

[11] A. Condon. The complexity of the max word problem and the power of one-way interactive proof systems, Proceedings of the 8th Annual Symposium on Theoretical Aspects of Computer Science, Springer Verlag Lecture Notes in Computer Science 480, February 1991, 456-465.

[12] A. Condon. Space bounded computational games, *J. ACM*, 38(2), April 1991, 472-494.

[13] A. Condon. The complexity of stochastic games, *Information and Computation*, 96:2, February 1992, 203-224.

[14] A. Condon and R. Ladner. Probabilistic game automata, *J. Comput. System Sci.* 36(3), June 1988, 452-489.

[15] A. Condon and R. Ladner. Interactive proof systems with polynomially bounded strategies, To appear in the Proceedings of the Seventh Annual Structure in Complexity Theory Conference, June 1992.

[16] A. Condon and R. J. Lipton. On the complexity of space bounded interactive proof systems, Proceedings of the 30th Annual IEEE Symposium on the Foundations of Computer Science, October 1989, 462-467. See also University of Wisconsin, Madison Technical Report Number 841, 1989.

[17] S. A. Cook. Deterministic CFL's are accepted simultaneously in polynomial time and log squared space, Proceedings of the 11th Annual ACM Symposium on Theory of Computing, 1979, 338-345.

[18] C. Derman. *Finite State Markov Decision Processes*, Academic Press, 1972.

[19] C. Dwork and L. Stockmeyer. Interactive proof systems with finite state verifiers, Tech. Report RJ 6262, IBM Research Division, Almaden Research Center, San Jose, CA, 1988. See also "Finite state verifiers I: the power of interaction", *J. ACM*, to appear, 1992, and "Finite state verifiers II: zero knowledge, *J. ACM*, to appear, 1992.

[20] C. Dwork and L. Stockmeyer. On the power of 2-way probabilistic finite state automata, Proceedings of the 30th Annual IEEE Symposium on

the Foundations of Computer Science, October 1989, 480-485. See also "A time complexity gap for two-way probabilistic finite state automata", SIAM J. Comput. 19, 1990, 1011-1023.

[21] U. Feige, A. Fiat and A. Shamir. Zero knowledge proofs of identity, Proceedings of the 19th Annual ACM Symposium on Theory of Computing, May 1987, 210-217.

[22] U. Feige, S. Goldwasser, L. Lovasz, S. Safra, and M. Szegedy. Approximating clique is almost NP-complete, Proceedings of the 32nd Annual IEEE Symposium on the Foundations of Computer Science, October 1991, 2-12.

[23] U. Feige and A. Shamir. Multi-oracle interactive protocols with space bounded verifiers, Proceedings of the Fourth Annual Structure in Complexity Theory Conference, June 1989, 158-164.

[24] U. Feige, A. Shamir and M. Tennenholtz. The noisy oracle problem, Proceedings of CRYPTO 1988.

[25] L. J. Fortnow. Complexity-theoretic aspects of interactive proof systems, Ph. D. Thesis, Technical Report Number TR-447, Laboratory for Computer Science, MIT.

[26] L. Fortnow and C. Lund. Interactive proof systems and alternating time-space complexity, Proceedings of 8th Annual Symposium on Theoretical Aspects of Computer Science, February, 1991, 263-274.

[27] L. Fortnow, J. Rompel and M. Sipser. On the power of multi-prover interactive protocols, Proceedings of the Third Annual Conference on Structure in Complexity Theory, 1988, 156-161.

[28] R. Frievalds. Probabilistic two-way machines, Proceedings of the International Symposium on Mathematical Foundations of Computer Science, Springer-Verlag Lecture Notes in Computer Science 118, 1981, 33-45.

[29] J. Gill. Computational complexity of probabilistic Turing machines, *SIAM J. Comput.* 6(4), 1977, 675-695.

[30] O. Goldreich, S. Micali and A. Wigderson. Proofs that yield nothing but their validity and a method of cryptographic protocol design, Proceedings of the 27th Annual IEEE Symposium on Foundations of Computer Science, October 1986, 218-229.

[31] S. Goldwasser, S. Micali and C. Rackoff. The knowledge complexity of interactive proof systems, *SIAM J. Comput.* 18, 1989, 186-208.

[32] S. Goldwasser and M. Sipser. Public coins vs. private coins in interactive proof systems, *Randomness and Computation*, Volume 5 of Advances in Computing Research, JAI Press, Greenwich, 1989 73-90.

[33] A. G. Greenberg and A. Weiss. A lower bound for probabilistic algorithms for finite state machines, *J. Comput. System Sci.* 33, 1986, 88-105.

[34] R. A. Howard. *Dynamic Programming and Markov Processes*, M.I.T. Press, 1960.

[35] L. G. Khachiyan. A polynomial algorithm for linear programming, Soviet Math Dokl. 20, 1979, 191-194.

[36] J. Kilian. Zero knowledge with log-space verifiers, Proceedings of the 29th Annual IEEE Symposium on Foundations of Computer Science, 1988, 25-35.

[37] F. T. Leighton and R. L. Rivest. Estimating a probability using finite memory, *IEEE Trans. Information Theory* IT-32, 1986, 733-742.

[38] R.J. Lipton. New directions in testing, *Distributed Computing and Cryptography, DIMACS Series on Discrete Mathematics and Theoretical Computer Science*, 2 1991, American Mathematical Society, 191-202.

[39] C. Lund, L. Fortnow, H. Karloff and N. Nisan. Algebraic methods for interactive proof systems, Proceedings of the 30th Annual IEEE Symposium on the Foundations of Computer Science, October 1990, 2-10.

[40] C. H. Papadimitriou. Games against nature, *J. Comput. System Sci.*, 31, 1985, 288-301.

[41] A. Paz. *Introduction to Probabilistic Automata*, Academic Press, 1971.

[42] G. L. Peterson and J. H. Reif. Multiple-person alternation, Proceedings of 20th Annual IEEE Symposium on Foundations of Computer Science, 1979, 348-363.

[43] J. H. Reif. The complexity of two-player games of incomplete information, *J. Comput. System Sci.* 29, 1984, 274-301.

[44] W. L. Ruzzo. On uniform circuit complexity, *J. Comput. System Sci.* 22, 1981, 365-383.

[45] A. Shamir. IP=PSPACE, Proceedings of the 30th Annual IEEE Symposium on the Foundations of Computer Science, October 1990, 11-15.

[46] I.H. Sudborough. On the tape complexity of deterministic context-free languages, *J. ACM*, 25(3), 1978, 405-114.

[47] S. Toda. PP is as hard as the polynomial time hierarchy, *SIAM J. Comput.*, 20(5), 1991, 865-877.

[48] L. Valiant. The complexity of computing the permanent, *Theoretical Computer Science*, 8, 1979, 189–201.

Fixed-Parameter Tractability and Completeness III: Some Structural Aspects of the W Hierarchy

Rod Downey[1]
Mathematics Department
Victoria University
Wellington, New Zealand

Michael Fellows[2]
Computer Science Department
University of Victoria
Victoria, B.C. Canada

---

[1] Research partially supported by a grant from Victoria University IGC, by the United States / New Zealand Cooperative Science Foundation, and by the University of Victoria and Simon Fraser University during a visit to British Columbia in August, 1991.

[2] Research supported by the United States Office of Naval Research, by the United States National Science Foundation, and by the National Science and Engineering Research Council of Canada.

Abstract

We analyze basic structural aspects of the reducibilities used to describe fixed parameter tractability and intractability, the model we introduced in earlier papers in this series. Results include separation and density, the latter for the strongest reducibility.

1. Introduction

A wide variety of natural computational problems have the property that their input consists of two or more parameters. Consider the following examples.

Example 1. The Vertex Cover problem takes as input a pair (G, k) consisting of a graph G and a positive integer k, and determines whether there is a set of k vertices in G having the property that every edge in G has at least one endpoint in this set.

Example 2. The Graph Genus problem takes as input a pair (G, k) as above, and determines whether the graph G embeds on the surface of genus k.

Example 3. The Planar Improvement problem takes as input a pair (G, k) as above, and determines whether G is a subgraph of a planar graph G' of diameter at most k.

Example 4. The Graph Linking Number problem takes as input a pair (G, k) as above, and determines whether G can be embedded in 3-space so that at most k disjoint cycles in G are topologically linked.

Example 5. The Dominating Set problem takes as input a pair (G, k) as above, and determines whether there is a set of k vertices in G having the property that every vertex of G either belongs to the set, or has a neighbor in the set.

Example 6. The Weighted CNF Satisfiability problem takes as input a pair (ϕ, k) where ϕ is a propositional (boolean) formula in conjunctive normal form, and k is a positive integer, and determines whether there is a weight k satisfying truth assignment to the variables of ϕ. (A truth assignment has *weight* k if it assigns exactly k variables the value *true* and all others the value *false*.)

With the exception of examples 3 and 4, the above problems are known to be NP-complete. We consider the question of what can be said about the complexity of these problems when the parameter k is held fixed. In many practical applications of computational problems having this form, efficient algorithms for a small range of parameter values may be quite useful.

For each of examples 1–4 above, there is a constant α such that for every fixed parameter value k the problem can be solved in time $O(n^\alpha)$. For example 1, we may take $\alpha = 1$. This means that for each fixed k there is an algorithm A_k that determines whether there is a vertex cover of size k in an input graph G in time $C_k n$ [BG]. For examples 2–4 we may take $\alpha = 3$ by the deep results of Robertson and Seymour [RS1,RS2].

Examples 5 and 6 illustrate the contrasting situation where for fixed values of k we seem to be able to do no better than a brute force examination of all possible solutions. In both cases the best known algorithm is $O(n^{k+1})$ for fixed k.

We are thus concerned with an issue in computational complexity that is very much akin to the central issue in P versus NP. In the previous papers of this series [DF1,DF2,DF3] we have established the framework of a completeness theory with which to address the apparent fixed-parameter intractability of problems such as examples 5 and 6. In particular, we defined a hierarchy of classes of parameterized problems and showed that a variety of natural problems are complete for various levels of this hierarchy. Dominating Set, for example, is complete at the second level.

In this paper, we study structural aspects of the fixed-parameter complexity hierarchy. For the remainder of this section we briefly recap the main points of this theory.

Definition. A *parameterized problem* is a set $L \subseteq \Sigma^* \times \Sigma^*$ where Σ is a fixed alphabet. In the interests of readability and with no effect on our theory, we consider in this paper that a parameterized problem L is a subset $L \subseteq \Sigma^* \times N$. Furthermore, in this context we consider N as bring represented as tally sets, that is $N = \{1^n : n = 0, 1, 2...\}$. We simply write n for 1^n in these circumstances. We will tend to use k, i, j for members of N and x, y, z for strings. For $k \in N$ we write $L_k = \{y | (y, k) \in L\}$. We refer to L_k as the kth *slice* of L.

Careful analysis of problem examples 1–4 above leads to three flavours of tractability.

Definition. We say that a parameterized problem L is
(1) *nonuniformly fixed-parameter tractable* if there is a constant α and a sequence of algorithms Φ_x such that, for each $x \in N$, Φ_x computes L_x in time $O(n^\alpha)$;
(2) *uniformly fixed-parameter tractable* if there is a constant α and an algorithm Φ such that Φ decides if $(x, k) \in L$ in time $f(k)|x|^\alpha$ where $f : N \to N$ is an arbitrary function;
(3) *strongly uniformly fixed-parameter tractable* if L is uniformly fixed-parameter tractable with the function f recursive.

The reader familiar with classical recursion theory will note that these notions might be considered as analogues of piecewise recursive recursively enumerable sets. Most reasonable variations of the above definitions can be seen to coincide with one of the three flavors offered. For example, if in (1) we require the sequence of algorithms Φ_x to be recursive, then equivalently we have (2). In Section 2 we will show that the three forms of fixed-parameter tractability defined above are distinct, even on the recursive sets.

Problem example 1 is strongly uniformly f.p.tractable (as are most examples of fixed-parameter tractability obtained without essential use of the Graph Minor Theorem). Example 2 can be shown to be strongly uniformly f.p. tractable by the methods of [FL2]. The reader should note that the graph minor theorem would only give nonuniform tractability and to get uniformity needs additional algebraic techniques. Example 3 can be shown to be uniformly f.p. tractable by the method of [FL1] (since the technique of [FL2] is not presently known to apply, we do not know a strongly uniform algorithm). Example 4 is at present only known to be nonuniformly f.p. tractable.

If $P = NP$ then examples 5 and 6 are also f.p. tractable. Thus aside from proving $P \neq NP$, a completeness program would seem to be the best we can do with respect to explaining the apparent fixed-parameter intractability of these problems.

We define three flavors of problem reducibility corresponding to the three flavors of f.p. tractability.

Definition. Let A, B be parameterized problems. We say that A is *uniformly P-reducible* to B if there is an oracle algorithm Φ, a constant α, and an arbitrary function $f : N \to N$ such that
(a) the running time of $\Phi(B; \langle x, k \rangle)$ is at most $f(k)|x|^\alpha$,
(b) on input $\langle x, k \rangle$, Φ only asks oracle questions of $B^{(f(k))}$ where

$$B^{(f(k))} = \bigcup_{j \leq f(k)} B_j = \{\langle x, j \rangle : j \leq f(k) \& \langle x, j \rangle \in B\}$$

(c) $\Phi(B) = A$.

If A is uniformly P-reducible to B we write $A \leq_T^u B$. Where appropriate we may say that $A \leq_T^u B$ via f. If the reduction is many:1 (an *m-reduction*), we will write $A \leq_m^u B$.

Definition. Let A, B be parameterized problems. We say that A is *strongly uniformly P-reducible* to B if $A \leq_T^u B$ via f where f is recursive. We write $A \leq_T^s B$ in this case.

Definition. Let A, B be parameterized problems. We say that A is *nonuniformly P-reducible* to B there is a constant α, a function $f : N \to N$, and a collection of procedures $\{\Phi_k : k \in N\}$ such that $\Phi_k(B^{(f(k))}) = A_k$ for each $k \in N$, and the running time of Φ_k is $f(k)|x|^\alpha$. Here we write $A \leq_T^n B$.

Note that the above are good definitions since whenever $A < B$ with $<$ any of the reducibilities, if B is f.p. tractable so too is A. Note also that the above definitions allow us to specify the notions of f.p. tractability we had before. Now nonuniformly f.p. tractability corresponds to being $\leq_T^n \emptyset$. We will henceforth write $FPT(\leq)$ as the f.p. tractable class corresponding to the reducibility $\leq$. We next turn to the complexity classes of parameterized problems introduced in [DF1,DF2]. These classes correspond, in a finely resolved way, to the complexity of checking a solution, as measured by circuit depth.

Fix attention on any of the above reducibilities. We consider circuits in which some gates have bounded fan-in and some have unrestricted fan-in. It is assumed that fan-out is never restricted.

Definition. A Boolean circuit is of *mixed type* if it consists of circuits having gates of the following kinds.

(1) *Small gates*: *not* gates, *and* gates and *or* gates with bounded fan-in. We will usually assume that the bound on fan-in is 2 for *and* gates and *or* gates, and 1 for *not* gates.

(3) *Large gates*: *And* gates and *Or* gates with unrestricted fan-in.

We will use lower case to denote small gates (*or* gates and *and* gates), and upper case to denote large gates (*Or* gates and *And* gates).

Definition. The *depth* of a circuit C is defined to be the maximum number of gates (small or large), not counting *not* gates, on an input-output path in C. The *weft* of a circuit C is the maximum number of large gates on an input-output path in C.

Definition. We say that a family of circuits F has *bounded depth* if there is a constant h such that every circuit in the family F has depth at most h. We say that F has *bounded weft* if there is constant t such that every circuit in the family F has weft at most t. F is a *decision circuit family* if each circuit has a single output. A decision circuit C *accepts* an input vector x if the single output gate has value 1 on input x. The *weight* of a boolean vector x is the number of 1's in the vector.

Definition. Let F be a family of boolean circuits. We allow that F may have many different circuits with a given number of inputs. To F we associate the parameterized circuit problem $L_F = \{(C,k) : C \in F \text{ and } C \text{ accepts an input vector of weight } k\}$.

Definition. A parameterized problem L belongs to $W[t]$ if L reduces to the parameterized circuit problem $L_{F(t,h)}$ for the family $F(t,h)$ of mixed type decision circuits of weft at most t, and depth at most h, for some constant h.

Definition. A parameterized problem L belongs to $W[P]$ if L reduces to the circuit problem L_F where F is the set of all circuits (no restrictions).

Definition. We designate the class of fixed-parameter tractable problems FPT.

In the papers [DF1,DF2,DF3,ADF] we have identified many natural complete problems for these classes. We mention the following variant of Satisfiability.

Definition. A boolean expression X is termed t-*normalized* if:
(1) $t = 2$ and X is in product-of-sums (P-o-S) form,
(2) $t = 3$ and X is in product-of-sums-of-products (P-o-S-o-P) form,
(3) $t = 4$ and X is in P-o-S-o-P-o-S form,
... etc.

WEIGHTED t-NORMALIZED SATISFIABILITY
Input: A t-normalized boolean expression X and a positive integer k.
Question: Does X have a satisfying truth assignment of weight k?

Our analogue of Cook's Theorem is the following.

Theorem. *For any fixed reducibility,*
(1) (Downey-Fellows [DF1,2]) For $t \geq 2$, Weighted t-Normalized Satisfiability is complete for $W[t]$.
(2) (Downey-Fellows [DF3]) Weighted Satisfiability for 2CNF formulas is complete for $W[1]$.

The above leads to an interesting hierarchy

$$FPT \subseteq W[1] \subseteq W[2] \subseteq ... \subseteq W[P]$$

Note that if $P = NP$ then the hierarchy collapses. We conjecture that each of the containments is proper. Many natural problems are complete for various levels. For example, Independent Set is complete for $W[1]$ and Dominating Set is complete for $W[2]$.

In the present paper we explore some structural aspects of the W hierarchy. In Section 2 we present some basic preliminary results and review the needed background material in recursion theory and computational complexity. In Section 3 we turn to results concerned with embeddings into the relevant degree structures. In Section 4 we prove some density results akin to the well-known Ladner theorem for NP-completeness. In Section 5 we present some related results on relativizations.

2. The Basics

We shall need a little basic recursion theory (see Soare[So]). The reader should recall that $K_0 = \{\langle x, y \rangle : \phi_x(y) \downarrow\}$ encodes the halting problem. (Here ϕ_x denotes the x-th partial recursive function.) There is a natural notion of reducibility $\leq_T$, called Turing reducibility, between languages and the equivalence classes are called *degrees (of unsolvability)*. The most complex (with respect Turing reducibility) recursively enumerable (r.e.) degree is the degree of K_0 above and this degree is denoted by $0'$. We shall need the following result.

(2.1) Lemma*(Shoenfield limit lemma)* $B \leq_T A$ *iff there is a recursive function $f(,)$ such that, for all x,*
(i) $\lim_s f(x, s) = f(x)$ *exists,(i.e. $f(x, s) \neq f(x, s+1)$ only finitely often) and*
(ii) $f(x) = B(x)$.

Here we identify sets with their characteristic functions. Similarly it follows that a function g is recursive in $0'$ (and we say g is Δ_2^0) iff $g(x) = lim_s G(x, s)$ for a recursive G with $G(x, s) \neq G(x, s+1)$ only finitely often. We shall additionally say that G is an r.e. function if $G(x, s+1) > G(x, s)$ whenever $G(x, s+1) \neq G(x, s)$. Such highly undecidable sets and functions are relevant to our studies as can be seen from the following theorem.

(2.2) Theorem (i) Suppose that $A \leq_T^u B$(or $A \leq_m^u B$) with A and B recursive. Then there exists an r.e. function f such that $A \leq_T^u B$ (resp. $A \leq_m^u B$) via f.

(ii) Suppose that $A \leq_T^n B$ (or $A \leq_m^n B$ with A and B recursive. Then there exists an r.e. function f such that $A \leq_T^n B$ (resp. $A \leq_m^n B$) via f.

Proof We do (i) for $\leq_T^u$, the others being essentially similar. So suppose that A and B are recursive and $A \leq_T^u B$. Then there is a procedure Φ, a constant α and a function g so that for all k
(2.3) $(\forall z)((\langle z, k\rangle \in A$ iff $\Phi(B^{(g(k))}; \langle z, k\rangle) = 1$ and runs in time $\leq g(k)|z|^\alpha)$.

We claim that $0'$ can compute a value that works in place of $g(k)$ in the above. That is for each k, $0'$ can compute $m = m(k)$ satisfying (2.3) with m in place of g. Call this $(2.3)'$. The reason is that the expression in the scope of the universal quantifier is recursive and hence the whole expression is $\leq_T K_0$. (For the reader who has forgotten this sort of thing, we briefly remind them that for each pair $\langle n, k\rangle$ we can enumerate a partial recursive function $\psi_{\langle n,k\rangle} = \phi_{h(n,k)}$ whose index $h(n, k)$ is given by the s-m-n theorem with $dom\psi_{\langle n,k\rangle}$ equal to N if there is some z with $\langle z, k\rangle \notin A$ but $\Phi(B^{(n)}; \langle z, k\rangle) = 1$, or $\langle z, k\rangle \in A$ and $\Phi(B^{(n)}; \langle z, k\rangle) = 0$, or $\Phi(B^{(n)}; \langle z, k\rangle)$ not running in time $n|z|^\alpha$; and we have $\psi_{\langle n,k\rangle}$ the empty function otherwise. Now K_0 can decide if $\langle h(n, k), h(n, k)\rangle \in K_0$ and hence can compute the least n such that $dom\psi_{\langle n,k\rangle} = \emptyset$. For such an n we have that
$$A_k = \Phi(B^{(n)}) \text{ in running time } n|z|^\alpha.$$
Now it is clear that we can define such an $n(k)$ via a function where values only increase and hence we can take an m to perform the role of g that is r.e.. □

As we mentioned earlier, one possible variation for the definition would be to consider a recursive collection $\Phi_{g(k)}$ of reductions with $\Phi_{g(k)}(B^{g(k)}) = A_k$ in running time $O(|z|^\alpha)$. This gives nothing new.

(2.4) Remark (i) Suppose A and B are recursive sets with $A \leq_T^n B$ via a recursive collection $\{\Phi_{g(k)} : k \in N\}$ of reductions all running in time $O(|z|^\alpha)$. Then $A \leq_T^u B$.
(ii) Furthermore, if the running time is recursively bounded (and hence can be taken to be $g(k)|z|^\alpha$), then $A \leq_T^s B$.

Proof (i). We shall define a single reduction Δ that takes the role of each of the $\Phi_{g(k)}$. On input $\langle z, k \rangle$, Δ first computes a stage $s = s(k)$ where $g(k) \downarrow$ in $|s|$ steps. Thereafter Δ simulates $\Phi_{g(k)}$. For (ii), use this and $g(k) + s(k)$ in place of $g(k)$ for the running time. □

We will now construct examples to show that the basic classes are indeed different for the various reducibilities.

(2.5) Theorem (i) $FPT(\leq_T^u) \subset FPT(\leq_T^n)$ **even for recursive sets.**
(ii) $FPT(\leq_T^s) \subset FPT(\leq_T^u)$

Proof (ii) We prove this by a simple diagonalization argument. Let $\{\langle \Phi_e, \phi_e \rangle : e \in N\}$ denote an enumeration of all pairs consisting of a procedure and a partial recursive function. We shall satisfy the requirements for $e \in N$:

$R_{\langle e,n \rangle}$: Either ϕ_e is not total, or
for some k, x, $\Phi_e(\emptyset; \langle x, k \rangle) \neq A(\langle x, k \rangle)$, or
$\Phi_e(\emptyset; \langle x, k \rangle)$ does not run in time $\phi_e(|k|)|x|^n$

Additionally, we must ensure that $A \in FPT(\leq_T^u)$. We devote $A_{\langle e,n \rangle}$ to meeting $R_{\langle e,n \rangle}$. We ensure that at most one element of row $\langle e, n \rangle$ enters A, and if z enters $A_{\langle e,n \rangle}$ then z is of the form $\langle 1^m, \langle e, n \rangle \rangle$ for some m.

We shall build A in stages. At stage s we decide the fate of $\langle 1^s, k \rangle$ for all $k \in N$. At stage s, the construction runs as follows:
For each $\langle e, n \rangle \leq s$, if $R_{\langle e,n \rangle}$ is not yet declared satisfied, compute s steps in the computation of $\phi_e(\langle e, n \rangle)$. (Call this $\phi_{e,s}(\langle e, n \rangle)$.) If $\phi_{e,s}(\langle e, n \rangle) \uparrow$ do nothing for $\langle e, n \rangle$ at this stage keeping $m(\langle e, n \rangle, s) = m(\langle e, n \rangle, s - 1)$. If $\phi_{e,s}(\langle e, n \rangle) \downarrow$ declare $R_{\langle e,n \rangle}$ as satisfied and perform the following diagonalization for $\langle e, n \rangle$. Run $\Phi_e(\emptyset; \langle 1^s, \langle e, n \rangle \rangle)$ for $\phi_e(\langle e, n \rangle)s$ many steps. If this does not halt in this many steps we need do nothing since the running time is wrong. If $\Phi_e(\emptyset; \langle 1^s, \langle e, n \rangle \rangle) \downarrow$ in $\phi_e(\langle e, n \rangle)s$ many steps set

$$A((\langle 1^s, \langle e, n \rangle \rangle)) = 1 - \Phi_e(\emptyset; \langle 1^s, \langle e, n \rangle \rangle).$$

In either case set $m(\langle e,n\rangle, 1^s) = 2\phi_e(\langle e,n\rangle)(s+1)$. It is clear that the diagonalization succeeds ensuring that $A \neq \Phi_e(\emptyset)$. Note that $A \in FPT(\leq_T^u)$ since for any k, $\langle z,k\rangle \in A$ iff z is of the form 1^t and $\langle 1^t, k\rangle$ is put into A at stage t. This can be decided in time $m(k,t)t$ and since $m(k,t) \neq m(k,t+1)$ at most once we see that $A \in FPT(\leq_T^u)$.

(i) Again we use a simple diagonalization argument. Now we need a family of reductions $\{\Delta_k : k \in N\}$ with $\Delta_k(\emptyset)$ computing A_k. By the limit lemma we need to meet the following requirements.

$R_{\langle e,n\rangle}$: Either $lim_s \phi(\langle e,n\rangle, s)$ fails to exist, or
$\Phi_e(\emptyset) \neq A_{\langle e,n\rangle}$, or
it does not run in time $\phi_e(\langle e,n\rangle)|z|^\alpha$.

Here we are working with pairs consisting of a procedure and a binary recursive function. We are denoting by $\phi(p)$ the value of $lim_s \phi(p,s)$ if it exists. We shall additionally, and without loss of generality assume that ϕ_e is nondecreasing in both variables where defined. In the construction to follow a value can be *used* for $\langle e,n\rangle$.

At stage s, if $R_{\langle e,n\rangle}$ is not as yet declared satisfied and $\langle e,n\rangle \leq s$ find the least unused $j \leq s$, if any, such that $j = \phi_{e,s}(\langle e,n\rangle, t) \downarrow$ for some $t \leq s$. If either $\phi_{e,s}(\langle e,n\rangle, t) \uparrow$ for all $t \leq s$ or there is no unused j do nothing. If j and hence t exists declare j as used. Now compute js steps in the computation of $\Phi_e(\emptyset; \langle 1^s, \langle e,n\rangle\rangle)$.

It is clear that A is recursive. Now Δ_k is one of the following two reductions:
Either $\Delta_k = \Psi$ which, on input $\langle x,k\rangle$ says that $\langle x,k\rangle \notin A$, or $\Delta_k = \Delta$ which, on input $\langle x,k\rangle$ computes s where R_k is satisfied, and then has $\langle y,k\rangle \in A$ iff y is of the form 1^s and $1 - \Phi_{e(k)}(\emptyset; \langle 1^s, k\rangle) = 1$ and $\langle y,k\rangle \notin A$ otherwise. Note that the algorithm runs in constant time, so that $A \in FPT(\leq_T^n)$. □

3. Embedding Type Results

In this section we shall analyze the general degree structures associated with the various reducibilities. That is, we look at $(REC, \leq_T^q)$, the recursive sets under $\leq_T^q$ for $q \in \{u, n, s\}$. We will concentrate on embedding type results such as Ladner[Ld], Ambos-Spies[AS], Melhorn[Me] etc, which eventually give enough definability to calculate the degree of the theory of $(REC, \leq_T^p)$ in Shinoda-Slaman[SS], and to get the undecidability result of Ambos-Spies and Nies[AN1,2]. We shall describe a basic technique that allows us to prove analogues of all these results. Local definability however presents special problems in our setting and we treat this in the next section, where we look, for instance, at density.

We begin with the easiest illustration of our technique.

(3.1) Theorem *If C is any complexity class generated by a superpolynomial function f, then there exist recursive sets A and B in C such that $A \not\leq_T^n B$ and $B \not\leq_T^n A$. (We write this as $A|_T^n B$.) Furthermore A and B can be chosen so that for all k, A_k and B_k are in P-time.*

Proof In C we build A and B to meet the requirements below.

$R_{2\langle e,n\rangle}$: For $k = 2\langle e, n\rangle$ either
$\phi_e(k, s)$ has no limit, or
there is an x such that $\Phi_{\phi_e(k)}(A; \langle x, k\rangle) \neq B(\langle x, k\rangle)$.

$R_{2\langle e,n\rangle+1}$: Same as $R_{2\langle e,n\rangle}$ but with A and B reversed.

Fix a recursive superpolynomial f for C. To meet the requirements we employ a priority argument. We first describe the basic module: that is the method whereby we meet a *single* requirement, $R_{2\langle e,n\rangle}$, say.

Again we shall have a notion of *used*. Again we employ row $2\langle e, n\rangle$ to meet $R_{2\langle e,n\rangle}$. Let $k = 2\langle e, n\rangle$. Initially we await a stage s_0 such that for some $t_0 \leq s_0$ and some corresponding least unused j_0 we have

$$j_0 = \phi_{e,s_0}(k, t_0)$$

We then declare j_0 as used. Note that j_0 is our current guess for the final value of $\phi_e(k)$. (Further note that this value may not exist.) Now await a stage $u_0 > s_0$ where

$$f(|\langle 1^{u_0}, k\rangle|) > j_0 |\langle 1^{u_0}, k\rangle|^n.$$

As f is superpoly such a stage u_0 must exist. Further note that we can wait and arrange matters so that we can see this in time $O(|x|^{n+1})$. See if $\Phi_{j_0}(A_{u_0-1}; \langle 1^{u_0}, k \rangle)$ halts in at most $j_0 u_0$ many steps.(Here A_t denotes the portion of A that we have decided by stage t.) We say that $R_{2\langle e,n\rangle}$ *receives attention via* $\langle j_0, u_0 \rangle$. If the computation does indeed halt then set

$$B_{u_0}(\langle 1^{u_0}, k \rangle) = 1 - \Phi_{j_0}(A_{u_0-1}; \langle 1^{u_0}, k \rangle)$$

and declare that $A[m_0] = A_{u_0-1}[m_0]$ where $m_0 = u(\Phi_{j_0}(A_{u_0-1}; \langle 1^{u_0}, k \rangle))$, the maximum length of an element queried in the computation.(This is called the *use* of the computation. We remind the reader that here we are using the notation that $Q[x] = \{z : z \in Q \text{ and } |z| \leq |x|\}$.) If the computation does not halt, do nothing.

By the above process, if $R_{2\langle e,n\rangle}$ receives attention via $\langle j_0, u_0 \rangle$ then j_0 is not a possible value of $lim_s \phi_e(k, s)$ if indeed $\Phi_{\phi_e(k)}(A) = B$. But it follows that either $\Phi_{\phi_e(k)}(A) \neq B$ or $R_{2\langle e,n\rangle}$ receives attention infinitely often. The latter case means that it has no limit.

The reader should realize that there is the usual priority conflict in the above. A R_{2d} requirement usually requires us to change B and preserve A, to preserve any disagreement made. An R_{2d+1} type requirement asks us to change A and preserve B. The idea with such arguments is to define a priority ordering that allows all of the requirements to be met. In this particular construction, it is easiest to break the R_j into infinitely many subrequirements of the form:

$R_{2\langle e,n\rangle,m}$: Either $\phi_e(k, s)$ changes value at least k times, or
$\Phi_{\phi_e(k)}(A)$ does not run in time $\phi_e(k)|x|^n$, or
there is some x with $\Phi_{\phi_e(k)}(A; \langle x, k \rangle) \neq B(\langle x, k \rangle)$.

We have similar requirements of type $R_{2\langle e,n\rangle+1,m}$. We then use the finite injury method to combine strategies. In the formal construction to follow we use the convention that all computations etc halting at stage s use elements of length below s.

(3.2) Definition We say that $R_{2\langle e,n\rangle,m}$ *requires attention* at stage $s+1$ if $\langle 2\langle e, n \rangle, m \rangle$ is least so that (i),(ii) and (iii) below all hold.

(i) No requirement is currently under attack.
(ii) $Count(2\langle e,n\rangle, s) = m - 1$.
(iii) There is some unused $j \leq s$ such that for some $t \leq s$, $\phi_{e,s}(2\langle e,n\rangle, t) = j$.
For the least such j declare that $R_{2\langle e,n\rangle, m}$ to require attention *via j*.

Construction
Stage 0. Do nothing
Stage $s+1$ If no requirement is currently under attack see if there is an R_q for $q \leq s$ which requires attention. If no such R_q exists do nothing. If R_q, for $q = \langle 2\langle e,n\rangle, m\rangle$, say, requires attention, declare j as used, and that R_q to be currently under attack with parameter j. Reset $Count(2\langle e,n\rangle, s) = m$.

If some $R_{2\langle e,n\rangle, m}$ is currently under attack and has parameter j (say), see if $f(s) > j|\langle 1^s, 2\langle e,n\rangle\rangle|^n$. If not do nothing. If so declare $R_{2\langle e,n\rangle, m}$ to be no longer under attack *at stage* $t = j|\langle 1^s, 2\langle e,n\rangle|^n + 1$. (This will protect any possible $\Phi_j(A_{s-1}; \langle 1^s, 2\langle e,n\rangle\rangle)$ computation that halts in at most $j|\langle 1^s 2\langle e,n\rangle\rangle|^n$ many steps.)

Now see if $\Phi_j(A_{s-1}; \langle 1^s, 2\langle e,n\rangle\rangle)$ halts in at most $j|\langle 1^s, 2\langle e,n\rangle\rangle|^n$ many steps. If not do nothing else. If so then set

$$B(\langle 1^s, 2\langle e,n\rangle\rangle) = 1 - \Phi_j(A_{s-1}; \langle 1^s, 2\langle e,n\rangle\rangle).$$

Proceed analogously for the $R_{2\langle e,n\rangle+1, m}$ type requirements.

End of Construction

To see that the construction succeeds, if ever we attack some R_k we eventually conclude this attack. We get to set $B(\langle 1^s, k'\rangle)$ (or $A(\langle 1^s, k'\rangle)$ as the case may be) and hence by the assumption that $|s| > u(\Phi_j(A_{s-1}; (\langle 1^s, k'\rangle))$ it must be that

$$\Phi_j(A_{s-1}; \langle 1^s, k'\rangle) = \Phi_j(A; \langle 1^s, k'\rangle),$$

and hence $\Phi_j(A) \neq B$. Note also that if we argue by priorities, we see that each R_k will be attacked if necessary and hence we either force $\phi_e(k', s)$ to change infinitely often, or the running time is wrong. □

Simple variations of the above technique can be used to improve (2.5).

(3.3) Theorem (i) $FPT(\leq_T^n)$ contains infinitely many problems pairwise incomparable with respect to $\leq_T^u$).
(ii) $FPT(\leq_T^u)$ contains infinitely many problems pairwise incomparable with respect to $\leq_T^s$.

Proof sketch of (e.g.) (ii) We build $\{A_i : i \in N\}$ in stages to meet the following requirements:

$R_{i,j,e,n}$: If $i \neq j$ then either ϕ_e is not total, or
the running time of $\Phi_e(A_i; \langle x, k \rangle)$ exceeds $\phi_e(k)|x|^n$, or
for some x, k, $\Phi_e(A_i; \langle x, k \rangle) \neq A_j(\langle x, k \rangle)$.

Additionally we must ensure that for all i, $A_i \in FPT(\leq_T^u)$. Again we use row $k = \langle i, j, e, n \rangle$ to meet $R_{i,j,e,n}$. assume $i \neq j$. We will describe the basic module. We wait for a stage s where $\phi_{e,s}(k) \downarrow$. At stage s, if no other R_q is under attack, $R_{i,j,e,n}$ asserts control (as with (3.1), assuming that $\langle i, j, e, n \rangle$ is least), and sees if $\Phi_e(A_{i,s-1}; \langle 1^s, k \rangle)$ halts in fewer than $\phi_e(k)s^n$ many steps. If so we set

$$A_j(\langle 1^s, k \rangle) = 1 - \Phi_e(A_{i,s-1}; \langle 1^s, k \rangle).$$

In either case it declares R_e as satisfied and declares it to be under attack until stage $\phi_e(k)s^n + 1$. (Again this is to protect the $\Phi_e(A_{i,s-1}; \langle 1^s, k \rangle)$ computation from A_i enumeration.)

It is clear that $A_i \in FPT(\leq_T^u)$ for all i giving (ii), and (i) is essentially similar. □

The idea of using the current guess as to the values of the constants to meet the requirements for constructions analyzing the structure of $(REC, \leq_T^q)$ for $q \in \{u, s, n\}$ is very flexible and allows us to show that a lot of classical $\leq_T^p$ have analogues in our setting.

For instance using this idea it is quite straightforward to show that each complexity class properly containing P contains minimal pairs, that is $A, B \notin FPT$ such that if $C \leq A, B$ then $C \in FPT$. We can also construct recursive $D \notin FPT$ such that if $E \oplus G \equiv D$ then E and G do not form a minimal pair. (Analogue of Downey[Do]). This would seem to indicate that the central strategies of Shinoda-Slaman[SS] ought to extend to be able to verify the following conjecture:

(3.4) Conjecture: (i) The degree of the (first order) theory of $(P(N), \leq_T^q)$ for $q \in \{s, u, n\}$ is that of second order arithmetic, and the degree of $(REC, \leq_T^q)$ is that of first order arithmetic. Thus they are as 'complicated as possible'.
(ii) All recursively presentable lattices can be embedded into $(REC, \leq_T^q)$.

Again similar comments apply for the structures $(REC, \leq_m^q)$. In this case we would be looking at the analogue of Ambos-Spies and Nies[AN]:

(3.5) Conjecture: The theory of $(REC, \leq_m^q)$ is undecidable for any $q \in \{s, u, n\}$.

In both cases we remark that there are some differences since the partial orderings generated by the reducibilities need one or two more quantifiers in the non-strongly uniform cases so an approach more like that used for the r.e. weak truth-table degrees ([ANS]) may be necessary. In the next section we examine local embeddings where we must work below, say, a given degree. Here analogues do not always work .

4. Density

From the last section, it would seem that most results from $(REC, \leq_T^p)$ ought to lift to our setting. However if we look at the structure $\{B : B \leq_T^u A\}$ for a *given* A, this is not in general true. First we shall examine the analogue of Ladner's result that the polynomial degrees of recursive sets are dense, which in particular show that if $P \neq NP$ then there are NP languages that are neither polynomial time nor NP- complete. The analogous fact for the weft hierarchy would be that *if $W[t] \neq W[t+1]$ then there exist infinitely many intermediate problems between $W[t]$ and $W[t+1]$*. This is indeed true for strong uniform reducibility as we now see.

(4.1) Theorem. *If A and B are recursive with $A <_q^s B$, then there exists a set C with $A <_q^s A \oplus C <_q^s B$, where $q \in \{m, T\}$.*

Fixed Parameter Tractability and Completeness

Proof We begin by briefly recalling the construction of Ladner[Ld]. Recall that this worked as follows. There were given recursive sets $A \not\leq B$ (working with $\leq_m^p$, say). Let $\{z_n : n \in N\}$ be a standard P-time length/lexicographic P-time ordering of Σ^*. We can assume that A and B are given as the range of p-time functions with domain N in unary notation. We write $A_s = \{f(1^0), ..., f(1^s)\}$ if $f(N) = A$ in this sense. We can also ask that if $|f(1^y)| > |f(1^{y-1})|$ then for all $z > y$, $|f(1^z)| \geq |f(1^y)|$. We call this a P-*standard enumeration*. So we will assume that we have such enumerations of A and B. Recall also for a reduction Δ on a set E, $u(\Delta(E; x))$ denotes the length of the longest element used in the computation. Let $\{\Phi'_e : e \in N\}$ denote a standard enumeration of all P-time m-procedures.

We must build C to satisfy the requirements:

$$R'_{2e} : \ \Phi'_e(A \oplus C) \neq B$$

$$R'_{2e+1} : \ \Phi'_e(A) \neq C$$

additionally ensuring that $C \leq_m^p B$. For the sake of the R'_j we define a polynomial time relation $R(n)$ on $N = \{1\}^*$. Then we declare that $x \in C$ iff $R(|x|) = 0$ and $x \in B$. Clearly this makes $C \leq_m^p B$.

Now we meet the R'_j *in order* by 'delayed' diagonalization. So we begin with R'_0. We set at each stage s, $R(s) = 1$ until a stage t is found where (i) - (iv) below hold. (Here we consider s, t etc as being in N.)
(i) $\Phi'_{0,t}(A_t \oplus \emptyset; z_n) \downarrow$ in less than t steps.
(ii) $A_t[q] = A[q]$ if $|q| < u(\Phi'_0(A_t \oplus \emptyset; z_n))$.
(iii) $B_t[z_n] = B[z_n]$.
(iv) $\Phi'_{0,t}(A_t \oplus \emptyset; z_n)(= \Phi'_0(A \oplus \emptyset; z_n)) \neq B(z_n) = B_t(z_n)$.

At stage t we say that we have diagonalized R'_0 at z_n, this being found by *looking back for an A- and a B- certified disagreement*.

The idea is then to move to R'_1 and then to R'_2 etc. For R'_1 we set $R(t+1) = 0$, causing C to look like B locally. So we keep $R(u)$ for $u > t$ equal to zero until a stage v is found with some $m \leq v$ and

$$\Phi'_{0,v}(A_v; z_m) \neq C_v(z_m),$$

via $A-$ and $B-$ certified computations. We then move to R_2' setting $R(v+1)$ to be 1 again. Thus the set C so constructed looks like B with 'holes' in it.

Keeping the above ideas in mind we turn to the result at hand. Now we are given $A \leq B$ with $\leq$ either $\leq_T^s$ or $\leq_m^s$. Again we must construct C, now to meet the following requirements

$R_{2\langle e,n\rangle}$: Either ϕ_e is not total,
or $(\exists k)(B_k \neq \Phi_e(A \oplus C^{(\phi_e(k))}))$
or $(\exists x, k)(\Phi_e(A \oplus C^{(\phi_e(k))}; \langle x, k\rangle))$ does not run in time $\phi_e(k)|x|^n$.

$R_{2\langle e,n+1\rangle}$: Either ϕ_e is not total,
or $(\exists k)(C_k \neq \Phi_e(A^{(\phi_e(k))}))$
or $(\exists x, k)(\Phi_e(A^{(\phi_e(k))}; \langle x, k\rangle))$ does not run in time $\phi_e(k)|x|^n$.

To aid the discussion we will use several conventions. First, if $\phi_{e,s}(k) \downarrow$, then the computation $\Phi_e(E^{(\phi_e(k))}; \langle x, k\rangle)$ cannot call any y of the form (k', z) for $k' > \phi_e(k)$. Also since we get a win for free if $\phi_{e,s}(k) \downarrow$ and the running time of $\Phi_e(E^{(\phi_e(k))}; \langle x, k\rangle)$ exceeds $\phi_e(k)|x|^n$, we shall assume that in the above the third option does not pertain to R_j and concentrate on the first two. This is because if the running time exceeds the bounds during the construction, we can *cancel* the relevant requirement. The argument to follow is a priority one with the Ladner strategy embedded.

Without loss of generality we can take ϕ_e to be strictly increasing. Again there will be long intervals with $C(\langle x, k\rangle)$ equal to $\emptyset$ and long intervals where it looks like B, for 'many' k. We have problems, since, for instance, we cannot decide if ϕ_e is total. We first focus on the satisfaction of a single $R_0 = R_{\alpha\langle e,n\rangle}$. We then describe the basic module for an odd type requirement, and finally describe the coherence mechanism whereby we combine strategies.

The Basic R_0-Module.

To meet R_0 above, we perform the following cycle. We have a parameter $k(0, s)$ that is nondecreasing in s and such that $\lim_s k(0, s) = k(0)$ exists. This is meant to be the number of "rows" devoted to R_0. It remains constant until we change it.

1. (Initialization.) Pick $k(0,0) = 1$.
2. Wait until a stage s occurs with one of the following holding:

2(a). (Win.) "Looking back" we see a disagreement. That is, as with the Ladner argument, we see an $n < s$ with $z_n \in \{\langle x,j \rangle : j < k(0,s)\}$.

$$\Phi_{e,s}(A \oplus C^{(\phi_e(k(0,s)-1))}; z_n) \neq B(z_n)$$

via A- and B-certified computations, or

2(b). Not (2a) and $\phi_{e,s}(k(0,s)) \downarrow$.

Comment If s does not occur then $\phi_e(k(0,s)) \uparrow$ and hence ϕ_e is not total. In this case we call $k(0,s)$ a *witness to the nontotality* of ϕ_e.

If 2(a) pertains, we declare R_0 to be *satisfied* (forever) and end its effect (forever). If 2(b) pertains, then we perform the following action.

3. R_0 *asserts control of* $C^{(\phi_e(k(0,s)))}$. That is, R_0 asks that for all $t \geq s$, until 2(b) pertains, we promise to set $C^{(\phi_e(k(0,s)))}(y) = 0$ for all y with $|y| = t$ and $y \in (\Sigma^*)^{(\phi_e(k(0,s)))}$. This can be achieved *via* a restraint $r(n,k)$.

4. Reset $k(0, s+1) = k(0,s) + 1$ and go to 2.

The Outcomes of the Basic R_0 Module.

We claim that 2(b) cannot occur infinitely often and hence $\lim_s k(0,s) = k(0)$ exists. Note that we have only reset $k(0,s)$ if 2(b) pertains in step 3. So suppose $k(0,s) \to \infty$ and hence $\phi_e(k(0,s)) \to \infty$. Then for each q and almost all y, we have $C(\langle q,y \rangle) = 0$.

We write $A =^* B$ to denote that the symmetric difference of A and B is finite. So $C_q =^* \emptyset$ for all q. Furthermore, for all q, we can compute a stage $h(q)$ where

$$[\forall t > h(q)](C_q(\langle y,q \rangle) = 0 \quad \text{for all } y \text{ with } |y| > h(q))$$

where $h(q)$ is the stage where R_0 asserts control of row q.

Finally, we know that for all k,

$$\Phi_e((A \oplus C)^{(\phi_e(k))}) = B_k$$

This allows us to get a reduction $\Delta(A) = B$. For each input $\langle y, k \rangle$, Δ simply computes $B(\langle y, k \rangle)$ for all y with $|y| \leq h(k)$, and $C(\langle z, k' \rangle)$ for all k', z with $k' \leq \phi_e(k)$ and $|z| \leq h(k)$. Then Δ simulates $\Phi_e(A^{(\phi_e(k))}; \langle y, k \rangle)$ if $|y| > h(k)$ with the exception that, if Φ_e calls some $\langle r, k' \rangle$ with $|r| \leq h(k)$ (and necessarily $k' \leq \phi_e(k)$), then Δ uses the table of values for C to provide the answer.

Note that the computations of $\Delta(A; \langle x, k \rangle)$ and $\Phi_e(C; \langle x, k \rangle)$ must agree and hence $\Delta(A) = B$, a contradiction. Thus 2(b) can pertain only finitely often. It follows that there are two outcomes.

Outcome $(0, f)$: 2(a) occurs for some t. Then we win R_0 with finite effect. (*Comment:* Once R_0 is met in this way, say at stage t, then we are completely free to do what we like with all y for which $|y| > t$ without injuring R_0.)

Outcome $(0, \infty)$: 2(a) does not occur. Then ϕ_e is not total. Note that the effect of R_0 is in this case infinite and for some $k = \lim_s k(0, s) - 1$, we will have

$$C^{(\phi_e(k))} =^* \emptyset$$

and furthermore, there is a reduction Δ_0 with time bound $\phi_e(k)|x|^n$ for which

$$\Delta_0(A^{(\phi_e(k))}) = B^{(k)}$$

Note that for the basic module, Δ_0 is simply Φ_e.

The Basic Module for R_1.

This is essentially the same as for R_0 except that for R_1 we wish to set $C(\langle x, k \rangle) = B(\langle x, k \rangle)$. Herein is the basic conflict: an even-indexed requirement R_j asks that lots of rows look like $\emptyset$ and an odd-indexed R_j asks for them to look like B.

Combining Strategies.

Fixed Parameter Tractability and Completeness 211

We cannot perform a delayed diagonalization as in the proof of Ladner's theorem, since we cannot know if $\phi_e(k)$ is defined. The combination of strategies needs the priority method. Let us consider a module for R_1 that works in the outcomes of R_0. We cannot know if this outcome is $(0, f)$ or $(0, \infty)$. Instead we have a strategy based on a guess as to R_0's behavior. Basically R_0 always believes that $k(0, s)$ is $k(0)$, that is, that the current value is the final one. Let $e = e(0)$, $n = n(0)$, $f = e(1)$ and $m = n(1)$.

Whilst R_1 believes that $\phi_e(k(0,0)) \uparrow$, R_1 acts as if R_0 is not there. So if $k(0, 0) = k(0)$ and $\phi_e(k(0,0)) \uparrow$ then we win R_1 for the same reasons as we did for R_0. On the other hand, if $\phi_e(0) \downarrow$ for some least stage s, then R_0 will assert control of $C^{(\phi_e(k(0,0)))}$. For the sake of R_1 we have probably been setting $C(0, x) = B(0, x)$ for all x with $|x| < s$. Since R_0 has higher priority than R_1, R_1 must release its control of C_0 (and indeed of C_j for $j \leq \phi_e(k(0))$) until a stage, if any, occurs where 2(a) pertains to R_0 so that R_0 is satisfied and releases control forever (or it becomes inactive because of a time bound being exceeded). Note that if 2(a) pertains at t, then R_1 is free to reassert control of C_0 for all y of the form $\langle y, 0 \rangle$ with $|y| > t$. Also, in this case, as R_1 is the requirement of highest overall priority remaining, its control cannot be violated and hence it will be met.

On the other hand, while R_0 can hope that 2(a) will pertain to R_1, R_0 may have outcome $(0, \infty)$ and R_0 will never release control of C_0. The key idea at this point is that we begin anew with a version of R_1 believing that $k(0, s + 1) = k(0)$. That is, R_0 will *never again act*.

This version of R_1 can only work with C_q for $q > \phi_e(k(0, s)) = \phi_e(k(0, 0))$. Some care is needed since potentially we need all of B to meet R_1.

An elegant solution to this difficulty is to *shift B* into C above $\phi_e(k(0, s))$. Thus R_1 will ask that

$$C(\langle x, q \rangle) = B(\langle x, q - \phi_e(k(0, s)) - 1 \rangle)$$

for $q > \phi_e(k(0, s))$. It does so until either $k(0, t)$ is reset again, or 2(a) pertains, or the time bounds are exceeded. In the latter cases, it reverts to the $(0, f)$-strategy. In the first case it begins anew on $q > \phi_e(k(0, t))$. Since this restart process only occurs finitely often, it follows that we eventually get a *final* version of R_1 whose actions will not be disturbed.

Thus there is a final version of R_1 that is met as follows. As $\lim_s k(0,s) = k(0)$ exists, there is a value r and a stage s_0 so that for $q \geq r$ and $s > s_0$, R_1 is not initialized at stage s and can assert control on C_q if it so desires. If R_0 has outcome $(0, f)$, then $r = 0$, otherwise $r = \phi_e(k(0) - 1) + 1$. So we know that if R_1 fails then for all j there is a stage $h(j)$ (computable from the parameters r and s_0) where for y with $|y| > h(j)$

$$C(\langle y, r+j \rangle) = B(\langle y, j \rangle) \text{ and}$$

$$\Phi_f(A; \langle y, r+j \rangle) = C(\langle y, r+j \rangle).$$

Thus if R_1 fails again we can prove there is a reduction $\Delta(A) = B$ with running time $O(|z|^m)$ and computable constants. This is a contradiction.

The outcomes for R_1 are thus either $(1, \infty)$ and $(1, f)$. In the former case we know that for a finite number of rows j and for almost all y, $C(\langle y, j \rangle) = B(\langle y, j \rangle)$. But we also know that for such rows there is a reduction Δ_f such that

$$\Delta_f(A; \langle y, j \rangle) = C(\langle y, j \rangle) \text{ in time } O(|y|^m) \text{ and computable constants.}$$

We continue in the obvious way with the inductive strategies. Consider eg R_2. It is confronted with at worst a finite number of rows permanently controlled by R_0 and a finite number by R_1. However, in each case we know that there is a reduction from a computable number of rows af A to these rows, and hence a reduction

$$\Psi_2(A; \langle y, j \rangle) = C(\langle y, j \rangle)$$

for all j cofinally under the control of either R_0 or R_1. Therefore to argue that R_2 is met, we get to use Ψ_2 to help construct a reduction from A to B. That is, for R_i, let $e = e(i)$ and $n = n(i)$. Then inductively we have a reduction and constants $p(2), m(2)$ and $r(2)$ with

$$\Psi_2(A^{m(2)}; \langle x, j \rangle) = C(\langle x, j \rangle)$$

for all $j \leq p(q)$ running in time $m(2)|x|^{r(2)}$. Furthermore, we have a stage s_2 such that for all $k < 3$, R_k ceases further activity.

Thereafter R_2 is free to assert control over any row q of C for $q > p(2)$. If we suppose that R_2 fails, then for each such q, R_2 will eventually assert

Fixed Parameter Tractability and Completeness 213

control of C_q at some stage $h(q)$ to make $C(\langle x, q \rangle) = 0$ for all x with $|x| > h_2(q)$ and we have $\Phi_{e(2)}(C) = B$.

Now to get a reduction Δ from A to B we go as for R_0 except that now if $\Phi_{e(2)}$ makes an oracle question of $\langle y, j \rangle$ for $j \leq p(2)$, we use Ψ_2 to answer this question. Thus we get a reduction Δ_2 that runs in time $O(|x|^{r(2)+n(2)})$, with computable constants and correct use. Thus again $B \leq A$, a contradiction.

The remaining details give no further insight and we leave them to the reader. □

What happens to $\leq_T^q$, $\leq_m^q$ for $q \in \{u, n\}$? Answer: The above proof will go through if we only consider reductions $E \leq_T^q F$ (or m reductions) *via a function $\phi(x)$ with an approximation $\phi(x, s)$ for which there is a recursive function g such that*

$$\{s : \phi(x, s) \neq \phi(x, s+1)\} \leq g(x).$$

That is although we do not know the constants we do know in advance the maximum number of times that our approximation can change. In the context of the Robertson-Seymour application, we would not necessarily know the obstruction set but would know in advance a bound on the possible size of the set. Of course this won't in general happen, leaving the following question apparently open.

(4.3) Question Do any of the other reductions generate a dense structure on the recursive sets? Indeed, for any of the other reductions if, for some k, $W[k] \neq W[k+1]$, then is there an infinite collection of classes between $W[k]$ and $W[k+1]$?

We will briefly observe that we can easily get a density result for certain classes of sets with additional hypotheses. We also look at these questions for the degrees *at large* where we find substantial differences from the classical case. Before we do so we remark that (4.1) has many obvious variations : for instance one can put an infinite antichain between A and B, or recursively presentable lattices, etc. Again these results don't really give new insights and we omit them.

We begin with a weak density result. Some parameterized problems exhibit *concentrated* nondeterminism. A well known example of this is the following.

PLANAR k-COLOURABILITY
Input A planar graph G and an integer k.
Question Is G k-colourable?

Now for any $k \neq 3$ we know that this is linear time and hence certainly in FPT. Yet we also know that it is NP complete. Although it is not in the W- hierarchy unless $P = NP$, it shows that many natural problems are *concentrated* in the sense that for some k, B_k itself is not in P-time. With this motivating example in mind we have:

(4.4) Theorem *For $\leq$ any of the reductions, if A and B are recursive sets, with $A < B$ and such that $B_m \not\leq A$ for some m, then there is a C such that $A < C < B$.*

Proof We give the proof for $\leq_T^u$, the others being similar. Suppose that $B_m \not\leq_T^u A$. We define C_k to be empty for all $k \neq m$. Then we meet

$R_{2\langle e,n\rangle}$: Either $\lim_s \phi_e(m,s)$ does not exist, or
$\Phi_e(A^{(\phi_e(m))}) \neq C_m$, or
there exists x such that the running time
of $\Phi_e(A^{(\phi_e(m))}; \langle x, m\rangle)$ exceeds $\phi_e(m)|x|^n$.

$R_{2\langle e,n\rangle+1}$: Either $\phi_e(m,s)$ has no limit, or
$\Phi_e((A \oplus C)^{(\phi_e(m))}) \neq B_m$, or
the running time is wrong.

The proof is fairly similar to some of our earlier ones and we only sketch it. We first break the requirements down into

$R'_{2\langle e,n\rangle,q}$: Either $\phi_e(m,s)$ changes at least q times, or as before. $R'_{2\langle e,n\rangle+1,q}$ is similarly defined. So for simplicity asume that the time bound for $R'_j = R'_{2\langle e,n\rangle,q}$ is not exceeded, nor is the use bound. We give R'_j priority j. We allow it to assert controll of C_m when it has the priority and it has seen $\phi_e(m,s)$ change exactly q times. At such a stage it will set $C(\langle x, m\rangle) = 0$

until we get a disagreement or we see $\phi_e(m,t) \neq \phi_e(m,s)$ for some $t > s$. Such a stage must exist by hypothesis. It is clear that either we win via some R'_j, q with finite effect, or for all q we see that $\phi(m,s)$ changes q times and hence $\phi_e(m,s)$ has no limit. □

We remark also that density *fails* for the nonrecursive sets. This stands in contrast to the situation for $\leq^p_T$, as was realized by Shinoda. Shinoda[unpubl.] observed that the Ladner density argument can be modified to work for *any* sets A and B with $A <^p_T B$. Recall, for R_{2e} that we wish to keep $C(x)$ equal to 0 until a stage and a z_n are found where we *know* that $\Phi_e(A \oplus C; z_n) \neq B(z_n)$. Now if Φ_e has use bounded by $|x|^{n(e)} + e$, then by stage, say, $2^{2^{|x|^{n(e)}+e}}$, B can figure out, in P-time relative to B, if there is some z_n with $|z_n| = s$ and $\Phi_e(A \oplus C; z_n) \neq B(z_n)$. Thus we inductively promise that we won't switch until we see a stage t which occurs for the first z_n such that $\Phi_e(A \oplus C; z_n) \neq B(z_n)$, where $t = 2^{2^{|z_n|^{n(e)}+e}}$, and similarly for the requirements with odd indices.

This means that the whole construction can be made P-time in B giving the following result.

(4.5) Theorem *(Shinoda[unpubl.])* *For any sets A and B, if $A <^p_T B$ then there exists C with $A <^p_T C <^p_T B$.*

Now we show that (4.5) fails for $\leq^s_T$.

(4.6) Theorem *There exists $A \notin FPT(\leq^s_T)$ such that for all $B <^s_T A$, $B \in FPT(\leq^s_T)$.*

Proof We build an r.e. set $A = \bigcup_s A_s$ in stages to satisfy the following requirements.

$N_{e,n}$: Either ϕ_e is not total, or
$(\exists x, k)(\Phi_e(A^{(\phi_e(k))}; \langle x, k \rangle))$ has running time exceeding $\phi_e(k)|x|^n$, or
$\Phi_e(A) \in FPT(\leq^s_T)$, or
$A \leq^s_T \Phi(A)$.

P_e: $\bar{A} \neq W_e$ where W_e denotes the e-th r.e. set.

We will carefully describe the ideas for the modules before we describe the formal construction. As usual a $P_{e,n}$ or an $N_{e,n}$ ceases activity if the running time ever proves wrong. So we can assume only to be considering such good (Φ_e, ϕ_e) pairs.

We meet the $P_{e,n}$ as follows. We *pick* a fixed *follower* $z = z(e,n) = \langle 0, \langle e, n \rangle \rangle$ targeted for A. We wait until we see a stage s with $\langle 0, \langle e, n \rangle \rangle \in W_{e,s}$ and then put $\langle 0, \langle e, n \rangle \rangle$ into A. This is the basic module but needs modification to live with the $N_{e,n}$.

Now $N_{e,n}$ will have two outcomes. These are labeled $(\langle e,n \rangle, f)$ and $(\langle e,n \rangle, \infty)$. It is the former if ϕ_e is not total, and the latter if ϕ_e is total.

The idea is the following. Associated with almost all elements targeted for A (i.e. of the form $\langle 0, j \rangle$) we have a current (e, n)-*state*. This will initially be f. We *raise* the state to ∞ if we see a stage s and a number q such that $\phi_{e,s} \downarrow$ and
$$\Phi_e(A_s^{j+1}) \neq \Phi(A_s^j)$$
where A_s^k is the result of putting $\langle 0, k \rangle, ..., \langle 0, s \rangle$ into A_s and changing nothing else. The idea is that we cancel, by enumeration into A all followers of the form $\langle 0, j' \rangle$ for $j' \leq s$ and $j' > j$. We then reassign $\langle 0, j \rangle$ the $P_{g,m}$ of highest priority not yet satisfied and having no follower of state ∞. We also promise that if $\langle 0, j \rangle$ enters at stage t then we also cancel $\langle 0, j' \rangle$ by enumeration for all $j < j' \leq t$. We call this the *dump*. For a single $N_{e,n}$ the idea succeeds. For consider the outcomes. Either for almost all $\langle 0, j \rangle$, $\langle 0, j \rangle$ is cancelled or has state ∞; or for almost all j, $\langle 0, j \rangle$ is cancelled or has state f. If the first option pertains we define a reduction Δ taking $\Phi_e(A)$ to A as follows. Suppose we have defined it for all $j \leq k$. Find the first r with $r > k$ and the first stage s where $\langle 0, r \rangle$ has state ∞, and is given the value ∞ at stage s. Then for all r' with $k < r' < r$, $\langle 0, r' \rangle \in A$ if $\langle 0, r' \rangle \in A_s$. Also, there is some $q = q(r)$ such that
$$\Phi_e(A_s; q) \neq \Phi_e(A_s^r; q).$$

Now we define Δ. Inductively, we know that for each $\langle 0, k \rangle < \langle 0, r \rangle$ with state ∞ not yet in A will be associated with a number $q(k)$ such that the value of $\Phi_e(A; q(k))$ determines if $q(k)$ enters A. Then Δ says that for $k < m < r$, $\langle 0, m \rangle \in A$ iff $\Phi_e(A; q(d)) \neq \Phi_e(A_s; q(d))$ for the least d with $q(d) \notin A_s$.

Finally Δ says that $\langle 0, r \rangle \in A$ iff $\Phi_e(A; q(r)) \neq \Phi_e(A_s; q(r))$. This generates a very large constant but note that we get, for each row, $\Phi_e(A) \leq_T^s A$ via a constant time reduction.

If the final state is f then we claim that $\Phi_e(A) \in FPT(\leq_T^s)$. To see this, as the final state is f for almost all j, there is a stage t and a k such that for all $j > k$, $\langle 0, j \rangle$ has state f and if $m < k$ then

$$\langle 0, m \rangle \in A \text{ iff } \langle 0, m \rangle \in A_s.$$

Now we know that for all $s > t$ and all y, j if $|y| \leq s$ and $\phi_{e,s}(j) \downarrow$, then

$$\Phi_e(A_s^j; y) = \Phi_e(A_s; y).$$

It follows that no matter whether $\langle 0, j \rangle$ enters or not, we always get the same answer. Thus to compute $\Phi_e(A; \langle 0, j \rangle)$ find the least stage $s > t, |y|$ where $\phi_e(j) \downarrow$ and compute $\Phi_e(A_s; y)$. This is the correct answer.

For more than one strategy we need to nest the states. To do this requires the so-called tree of strategies $0''$ priority method. For more on this method we refer the reader to Soare[So].

Let $T = \{\infty, f\}^*$, with $\infty <_L f$ inducing the lexicographic ordering on T. We refer to members of T as *guesses*. We shall use the phrase 'initialize'. We take this to mean that all followers, etc currently associated with a requirement are no longer associated. We remind the reader that all computations are bounded by s at stage s.

(4.5) Definition We say that P_e requires attention at stage s if P_e is least such that one of the following holds:

(i) for some follower x of P_e, we have $x \in W_{e,s}$.
(ii) P_e has no follower at stage s.

Construction
Stage s find the least number of the form $\langle 0, j \rangle$ not yet in A such that for some k,

(i) $\langle 0,j \rangle$ has j-state $\sigma * (k, f)$

(ii) There is at least k elements of the form $\langle 0, q \rangle$ not yet in A with $q < j$ having state σ and (potentially followers) or followers.

(iii) $\phi_{e(k),s}(j) \downarrow$.

(iv) $\Phi_{e(k)}(A_{s-1}^{j+1}; y) \neq \Phi_e(k)(A_{s-1}^{j}; y)$ for some y with $|y| < s$.

If such $\langle 0, j \rangle$ exists, declare $\langle 0, j \rangle$ to have state $\sigma * (k, \infty)$ and put $\langle 0, j+1 \rangle, ..., \langle 0, s \rangle$ into A_s. Declare $\langle 0, j \rangle$ as unassigned and initialize any P_q not having a follower with state $\leq_L \sigma * (k, \infty)$. Now find the least $\langle 0, j' \rangle$ with $j' < j$ such that $\langle 0, j' \rangle$ has state $\leq_L \sigma * (k, \infty)$. For any $\langle 0, r \rangle$ with $j' < r < j$, declare $\langle 0, r \rangle$ as *no longer a potential follower EVER AGAIN* (and having no state henceforth.)

Now see if P_e requires attention. If (i) holds via $x = \langle 0, q \rangle$, put $\langle 0, q' \rangle$ into A for $q < q', s$. Initialize all P_k for $k > e$. If (ii) holds find the least $\langle 0, j \rangle$ not yet assigned to a requirement and still a potential follower, and assign $\langle 0, j \rangle$ to P_e.

End of Construction

Verification (sketch) Clearly A is r.e.. Let TP denote the true path of the construction, that is TP is in $[T]$ the collection of all paths through T and is the leftmost one visited $\Delta(A) = B$ with running time $O(|z|^m)$ and computable constants.This infinitely often. Specifically $\lambda \subseteq TP$, and whenever $\sigma \subseteq TP$, then $\sigma*(0, \infty) \subseteq TP$ if $(\exists^\infty s)(s$ is a $\sigma*(0, \infty)$-stage) and otherwise $\sigma * (0, f) \subseteq TP$. We claim that TP exists and each P_e requires attention at most finitely often. This is easily proven by induction on $|\sigma|$. Thus if $\sigma \subseteq TP$ we can go to a stage t where, for all e with $e < |\sigma|$, P_e does not require attention after stage t and for no stage $s > t$ is s a τ-stage with $\tau \leq_L \sigma$ and $\tau \not\subseteq \sigma$. Now at the least σ-stage after t, P_e will be assigned a follower x with state σ if it does not already have one. This follower cannot be cancelled or initialized by choice of t and hence will succeed in meeting P_e.

Finally to see that all the $N_{e,n}$ are met, we can argue almost precisely as we did in the basic module. By the fact that there are only 2^{j+1} many f states, we can see that almost all j eventually get in the same f state. If the relevant state is of the form $\tau * (e, \infty)$ then $A \leq \Phi(A)$ and if this is of the form $\sigma = \tau * (e, 0)$ it is the case that $\Phi_e(A)$ is in $FPT(\leq_T^s)$. The only difference is that, in the latter case for no $(0, j)$ of state σ and no stage $s > t$ for some parameter t is it the case that $\phi_e(j) \downarrow$ and for some y with $|y| \leq s$,

$$\Phi_e(A_s^{j+1}) \neq (A_s^j).$$

Since the only things that survive have state σ it follows that $\Phi_e(A) \in FPT(\leq_T^s)$. □

5. Oracle Results

In this section we shall explore oracle results. These results provide some evidence that methods that relativize are not sufficient to resolve questions such as $FPT(\leq_T^u) =?W[P]$. Of course the exact meaning of this imprecise statement is not quite clear in view of such results as Shamir's $IP = PSPACE$ result (Shamir[Sh]) which is known to fail relative to a random oracle. None the less we feel that the results of this section at least indicate that the relevant separation or collapse results will be hard. This is in the same spirit as Baker, Gill and Solovay[BGS]. We also believe that these oracle separation results support our thesis that the weft hierarchy is infinite.

We begin with an oracle result that supports the thesis that $FPT = ?W[P]$ is independent of $P =?NP$. In view of this result we believe that it is unlikely that there is a proof that $P \neq NP$ implies $FPT \neq W[P]$ unless the hierarchies collapse.

(5.1) Theorem *There is a recursive oracle A relative to which $W[P] = FPT$ yet $P \neq NP$.*

Proof We do this for $\leq_T^s$ and observe that the obvious modifications work for the other reducibilities. Let Q_e denote the e-th P-time relation. Define K^B via

$\langle\langle x, e, 0^n \rangle, k\rangle \in K^B$ if for some y with $|y| = |x|$, y has weight k and

$Q_e^B(y)$ holds in n steps.

In view of the direct relationship between circuits and relations, it is clear that K^B is $W[P]$-complete.

Now let f be any recursive function from N to N. Suppose we build a recursive set A such that for each k and all x with $|x| \leq f(k)$ we have $A(\langle x, k \rangle)$ can be computed in $g(k)$ many steps, and for all y with $|y| > f(k)$ we have $A(\langle y, k \rangle) = B(\langle y, k \rangle)$. We claim that $A \equiv_m^s B$, so that $W[P]^B = W[P]^A$. To see this for the k-th row for the reduction from A to B, say, we first compute $g(k)$ and $f(k)$. As A and B are recursive we can write the corresponding initial segments in a table. Otherwise $\langle x, k \rangle \in A$ iff $\langle x, k \rangle \in B$, and hence $A \equiv_m^s B$.

Now take any B with $W[P]^B = B$. That is, define B via

$\langle \langle x, e, 0^n \rangle, k \rangle \in B$ iff for some y with $|y| = |x|$, y has weight k and $Q_e(y)$ holds in n steps.

Now it will suffice to define f, g and A as above and ensure that $P^A \neq NP^A$. We do this as follows. We must meet the requirements

$$R_k : \Gamma_k(A) \neq C,$$

where $C \in NP^A$ and Γ_k denotes the k-th P-time procedure with use q_k, say. We meet R_k via row $k+1$. We define C so that

$$\langle x, k \rangle \in C \text{ iff } (\exists y)[|y| = |x| \text{ and } \langle y, k \rangle \in C], \text{ and hence } C \in NP^A.$$

At stage k, we will have defined $f(i)$, $g(i)$ for $i \leq k$ and a restraint $r(k)$. Choose x so that $2^{|x|}$ exceeds $q_k(\langle x, k \rangle)$, and $|x| > r(k)$. Now compute $\Gamma_k(A_k^*; \langle x, k+1 \rangle)$ where A_k^* is the result of setting A equal to B on all $\langle x, j \rangle$ for $j \leq k$ and $|x| > f(j)$, and setting $A(\langle x, k \rangle)$ equal to what we have decided at stages $\leq k$ for y with $|y| \leq f(j)$. (So, basically we've decided at stage k the precise contents of A on $\langle x, j \rangle$ for $j < k$.) Set $A_k^*(\langle z, r \rangle) = 0$ for $r \geq k+1$.

Now if $\Gamma_k(A_k^*; \langle x, k+1 \rangle) = 1$ define $\langle x, k+1 \rangle = 0$ for all z with $|z| \leq r(k+1)$ with $r(k+1) \geq r(k)$ and also exceeding all uses seen so far. (This means that future actions will not affect these protected computations). If $\Gamma_k(A_k^*; \langle x, k+1 \rangle) = 0$ then for some y with $|y| = |x|$ we have that $\langle y, k+1 \rangle$ is not queried during the computation of $\Gamma_k(A_k^*; \langle x, k+1 \rangle)$ because $2^{|x|}$ exceeds

Fixed Parameter Tractability and Completeness   221

$q_k(|x|)$. Now put $\langle y, k+1 \rangle$ into A and otherwise set $A(\langle z, k+1 \rangle) = 0$ for all z with $|z| \leq r(k+1)$. Set $f(k+1) = r(k+1)$ and define $g(k+1)$ appropriately.

This ensures that $\Gamma_k(A) \neq C$ via the witness $\langle x, k+1 \rangle$ since inductively all previous restraints are maintained, and hence all previous disagreements are also preserved. Thus $NP^A \neq P^A$) and yet $FPT^A(\leq_m^s) = W[P]^A$. □

(5.2) Corollary There exist recursive oracles A and B with $W[P]^A = W[P]^B = FPT^A = FPT^B$, $A \equiv_m^s B$, $NP^B = P^B$, yet $P^A \neq NP^A$.

Proof Take A and B as in the proof above. We claim that $NP^B = P^B$. This will then give the desired result. Define D via:

$$\langle x, e, 0^{p_e(|x|)} \rangle \in D \text{ iff some computation of } \Phi_e(B; x) \text{ accepts in } n \text{ steps.}$$

Then as usual D is NP^B complete. We claim that $D \in P^B$. To see this simply note that

$$\langle x, e, 0^{p_e(|x|)} \rangle \in D \text{ iff } (\exists j)[j \leq |x| \text{ and } \langle\langle x, e, 0^{p_{h(e)}(|x|)} \rangle j \in B],$$

where $Q_{h(e)}$ is the relation representing Φ_e. Hence $P^B = NP^B$. □

Now we turn to results separating the W hierarchy. Ideally we would like an oracle that separates the whole W hierarchy, showing it infinite. Unfortunately at this stage we don't know how to do this. Using a Baker, Gill and Solovay construction it is not difficult to show that $W[P]$ can be different from FPT in relativized worlds (indeed in random worlds). We do a little better. We construct an A such that $W[1]^A \neq FPT^A$. Before we do this we briefly describe how to interpret oracle results in circuits. This is a matter where there is no universal agreement. One natural idea that we use in our construction is to view an assignment of values to the inputs into oracle and gates as determining a word, and considering the gate as outputting one if the word is in A. For our purposes this seems a reasonable model for an oracle circuit.

To separate $W[P]$ from FPT we use the set C defined as follows.

$$\langle z, k \rangle \in C \text{ iff } (\exists y(|y| = |z| \text{ and } y \in A \text{ and } y \text{ has weight } k).$$

Now C is in $W[1]^A$ via the circuit such that, on input $\langle z, k \rangle$ the circuit to accept $\langle z, k \rangle$ consists of a single oracle gate with inputs $z_1, ..., z_{|z|}$ ordered left to right. Then $\langle z, k \rangle \in C$ iff there is a weight k word accepted by the gate (i.e. in A). It is routine then to build A to meet

$R_{e,n}$: Either ϕ_e is not total, or
there is an $\langle x, k \rangle$ such that $\Phi_A; \langle x, k \rangle) \neq C(\langle x, k \rangle)$, or
the running time is incorrect.

To do this we assign $R_{e,n}$ some row $k = k(n)$ with $m^k > O(m^n)$. Then we meet $R_{e,n}$ as follows. Wait till $\phi_e(k) \downarrow$. Then find an x of sufficient length as not to injure other requirements, and so that $|x|^k$ exceeds $\phi_e(k)|x|^n$. We can then diagonalize via $\langle x, k \rangle$ in the standard way, using the string not addressed in the A computation of length $|x|$ (if it is the case that Φ_e is outputting 0), or doing nothing as the case may be but then restraining the result so as not to be disturbed by future actions. The strategies combine sequentially as usual and the result follows. Thus we have for each of the reducibilities (noting that the above can be easily modified for the others):

(5.3) Theorem *There is a recursive oracle A such that $W[1]^A \neq FPT^A$ (for any of the reducibilities).*

Now we address some remarks to the above. On the positive side it shows that in relativized worlds (with the model used) that $W[1]$ can differ from $mon - W[1]$ although they are the same in the real world. (Here $mon - W[t]$ denotes the version of $W[t]$ obtained by only considering circuits with no inverters.) Thus this provides an example of the failure of relativization, or perhaps suggests that other relativization models might be more appropriate (more on this later.) Furthermore standard techniques would seem to show that this oracle failure will hold relative to a random oracle. This gives an amusing example of the failure of the battered random oracle hypothesis.

On the negative side, the above is rather unsatisfying. The separation is achieved not by analyzing the structure of the circuits involved but rather by the manner by which a procedure can address information from an oracle. In a way it clearly demonstrates the shortcomings of the Baker-Gill-Solovay results too. We feel that it would be infinitely more satisfying to get a separation achieved more by the combinatorics and less on the oracle. This might well lead to real insight into the issues involved.

Open Questions, Generalizations, etc

There are several obvious generalizations one could pursue. One could look at parameterizations of the hierarchy, $PSPACE$, $\#P$ etc. This seems a very interesting exercise and we have some results with Karl Abrahamson [ADF] on games and the number of moves to win and $PSPACE$. Here one asks that the parameterization works in space $O(|x|^\alpha)$ for a fixed alpha. We also have some partial results on analogues of D_P. Here Valiant-Vazirani[VV] works for the analogue D_P for $W[P]$ but not at the $W[1]$ level and weight is lost in the procedure they use to take the hashed formula and convert to CNF.

Quite aside from the above we feel that central questions in the area include whether collapse propagates upward in the hierarchy (i.e. $W[t] = W[t+1]$ implies $W[t] = W[u]$ for all $u > t$.) Another question is to understand the exact relationship of the concepts here with classical notions such as NP. We have some contributions in this area also to be found in [ADF]. Also things such as f.p. crypto would seem very worthwhile. After all one really needs is feasibly one way functions.

Finally the degree structure of, in particular $W(\leq_T^u)$ and $W(\leq_T^n)$ remains to be explored.

References

[ADF] K. Abrahamson, R. Downey, and M. Fellows, "Fixed Parameter Tractability and Completeness IV: $W[P]$ and $PSPACE$," to appear.

[ADF2] K. Abrahamson, R. Downey, and M. Fellows, "Fixed Parameter Intractability II," in STACS'93, (1993) 374-385.

[AS] K. Ambos-Spies, "On the Structure of the Polynomial Degrees of Recursive Sets," Habilitationschrift, Universitat Dortmund, 1984.

[AN] K. Ambos-Spies and A. Nies, "The Theory of The Polynomial Time Many-One Degrees is Undecidable," to appear.

[AN2] K. Ambos-Spies, A. Nies, and R.A. Shore, "The Theory of the Recursively Enumerable Weak Truth Table Degrees is Undecidable,"to appear.

[BGS] T. Baker, J. Gill, and R. Solovay, "Relativizations of the $P =?NP$ Question," Siam J. Comput. 4(1975), 431-442.

[BDG] J. Balcazar,J. Diaz, and J. Gabarro," Structural Complexity" Volumes 1 and 2 Springer Verlag (1987,1989).

[BG] J. F. Buss and J. Goldsmith, "Nondeterminism Within P," to appear, SIAM J. Comput.

[Bo] H. L. Bodlaender, "On Disjoint Cycles," Technical Report RUU-CS-90-29, Dept. of Computer Science, Utrecht University, Utrecht, The Netherlands, August 1990.

[CD] P. Cholak and R. Downey,"Undecidability and Definability for Parameterized Polynomial Time Reducibilities,"(to appear) *Logical Methods* (ed. Crossley, Remmel, Shore and Sweedler) Birkhauser, Boston.

[Do] R.Downey, "Nondiamond Theorems for Polynomial Time Reducibility," J.C.S.S. **45** (1992) 385-395.

[DF1] R. Downey and M.Fellows, "Fixed Parameter Tractability and Completeness, " Congressus Numerantium **87** (1992) 161-187.

[DF2] R. Downey and M. Fellows, "Fixed Parameter Tractability and Completeness I: Basic Results," to appear.

[DF3] R. Downey and M. Fellows, "Fixed Parameter Tractability and Completeness II: On Completeness for $W[1]$," to appear.

[DF4] R. Downey and M. Fellows,"Fixed Parameter Tractability and Completeness," Monograph in Preparation.

[DF5] R. Downey and M. Fellows,"Feasible Parameterized Tractability," to appear, *Feasible Mathematics II* (ed. Clote and Remmel) Birkhauser, Boston.

[FL1] M. Fellows and M. Langston,"On Search, Decision, and the Efficiency of Polynomial Time Algorithms," in STOC'89, (1989) 501-512.

[FL2] M. Fellows and M. Langston, "An Analogue of the Myhill-Nerode Theorem and It's Use in Computing Finite Basis Characterizations," in FOCS'89 (1989) 520-525.

[GJ] M. Garey and D. Johnson, *Computers and Intractability*, Freeman, San Francisco, 1979.

[La] R. Ladner, "On the Structure of Polynomial Tine Reducibility," JACM, **22** (1975) 155-171.

[Me] K. Melhorn, "Polynomial and Abstract Subrecursive Classes," JCSS, **12** (1976) 147-178.

[RS1] N. Robertson and P. Seymour, "Graph Minors XIII: The Disjoint Paths Problem," to appear J. Comb. Th. B.

[RS2] N. Robertson and P. Seymour, "Graph Minors XV: Wagner's Conjecture," to appear, J. Comb. Th. B.

[Sh] A. Shamir, "IP=PSPACE," FOCS **31** (1991) 145-152.

[Sho] J. Shinoda, Personal Communication.

[SS] J. Shinoda and T. Slaman, "On the Theory of $PTIME$ Degrees of Recursive Sets, J.C.S.S. **41** (1990) 321-366.

[ShS] R. Shore and T. Slaman, "The P-T-Degrees of Recursive Sets; Lattice Embeddings, Extensions of Embeddings, and the Two Quantifier Theory," Theoretical Computer Science **97** (1992) 263-284.

[St] L. Stockmeyer, "Planar 3-Colourability is NP-Complete," SIGACT News,5 19-25.

[VV] L.G. Valiant and V.V. Vazirani, "NP is as easy as detecting unique solutions," *Theoretical Computer Science* 47 (1986), 85-93.

Degrees of Unsolvability in Abstract Complexity Theory

Martin Kummer*
Universität Karlsruhe

1 Introduction

It is a wellknown phenomenon in abstract complexity theory that the results hold *"for all x up to finitely many exceptions"* (a.e.). Often these exceptions are closely connected with noneffectiveness. The most prominent example appears in the context of the Speed-up Theorem: If a function f has sufficiently large speed-up then there is no algorithm for speeding up the programs computing f—but there is an algorithm that computes speeded up versions of f that agree with f almost everywhere (Blum [3]). We will investigate in three cases—the Rabin-Blum Theorem [2, 9], the Speed-up Theorem of Blum [2], the Union Theorem of McCreight and Meyer [8]— the degrees of unsolvability of the associated sets of exceptions. For the first two cases we obtain generalizations of results of Blum [3], Fulk [4] and Schnorr [10]. As one might suspect, the relevant degree class is the class of all complete degrees. In the third case we characterize the growth rate of complexity bounds from the Union Theorem. It is somewhat surprising that here the relevant degree class turns out to be the upward closure of the promptly simple degrees of Maass [6]. As an application we show that the growth rate of every total recursive time bound $g(n)$ such that $P = DTIME(g(n))$ must be nonconstructive (in a precise sense).

For background on abstract complexity theory we refer the reader to the recent survey of Seiferas [12]. The general reference for background on recursion theory is Soare [14].

---

*Institut für Logik, Komplexität und Deduktionssysteme, Universität Karlsruhe, Postfach 6980, D-7500 Karlsruhe 1, Germany, (e-mail: kummer@ira.uka.de)

Notation and Definitions $\omega = \{0, 1, \ldots\}$ is the set of all natural numbers. A, M denote subsets of ω. χ_A is the characteristic function of A. $A \upharpoonright i := A \cap \{x \in \omega : x < i\}$. $\max(\emptyset) := 0$. $\langle -, - \rangle$ denotes a computable pairing function such that $x, y \leq \langle x, y \rangle$. $\langle x, y, z \rangle := \langle x, \langle y, z \rangle \rangle$, etc.. The decoding function of the i-th component is denoted by $(-)_i$, e.g. $(\langle x, y \rangle)_1 = x$. $P_n(R_n)$ is the set of all partial (total) recursive functions $f : \omega^n \to \omega$ $(n \geq 1)$. We say that g majorizes f, iff $g(x) \geq f(x)$ a.e.; $f =^* g$ iff $f(x) = g(x)$ a.e. φ denotes a Gödelnumbering of P_1. K is the halting problem: $K = \{i : \varphi_i(i) \downarrow\}$. A partial function $\psi : \omega \to \omega$ is said to be partial recursive in M iff ψ is computable using M as an oracle. Let Φ denote an arbitrary complexity measure with respect to φ, i.e. $\text{dom}(\Phi) = \text{dom}(\varphi)$ and the predicate $P(i, x, y) \equiv \Phi_i(x) = y$ is recursive.

2 The Rabin-Blum Theorem

Rabin [9] proved that there exist decidable problems of arbitrary high complexity, i.e., for any lower bound $h \in R_1$ there is a recursive set A such that any Turing machine which decides A needs more than $h(x)$ steps for almost every input x. Blum [2] generalized this result for arbitrary complexity measures:

THEOREM RB ([2, 9]) *For any $h \in R_1$ there is a recursive set A such that:* $(\forall i)(\varphi_i = \chi_A \to \Phi_i(x) > h(x) \text{ a.e.})$. A *is called* a.e. h-complex.

For investigations of the metatheory of Theorem RB we refer the reader to Fulk [4], Gill, Blum [5] and Smith [13]. For $h \in R_1$ and A a.e. h-complex we call ψ an (h, A)-*bounding procedure* (this terminology was introduced by Fulk [4]) if ψ is a partial function such that:

$$(\forall i)[\varphi_i = \chi_A \to \psi(i) \downarrow \wedge (\forall x > \psi(i))(\Phi_i(x) > h(x))].$$

Fulk [4] proved, using the double recursion theorem, that there exists a recursive function h_0 such that for any $h \in R_1, h \geq h_0$ and any recursive a.e. h-complex set A, there is no partial recursive (h, A)-bounding procedure. It is easy to see that there is a bounding procedure which is partial recursive in K (in fact, it can be chosen even total recursive in K). The next theorem gives the converse.

Theorem 2.1 *There exists a recursive function h_0 such that for every recursive function $h \geq h_0$ and every recursive a.e. h-complex set A: If there exists an (h, A)-bounding procedure which is partial recursive in M, then $K \leq_T M$.*

Proof: Let k be a total recursive one-one fucntion that enumerates K. By the s-m-n theorem we define $f, g \in R_2$ as follows:

$$\varphi_{f(e,i)}(x) = \begin{cases} 0 & \text{if } k(x) = i; \\ \varphi_e(x) & \text{otherwise.} \end{cases}$$

$$\varphi_{g(e,i)}(x) = \begin{cases} 1 & \text{if } k(x) = i; \\ \varphi_e(x) & \text{otherwise.} \end{cases}$$

Let $h_0(n) = 1 + \max\{\Phi_{f(e,i)}(x), \Phi_{g(e,i)}(x) : e, i, x \leq n \wedge k(x) = i\}$. Note that $h_0 \in R_1$. Suppose that $h \in R_1, h \geq h_0$, and A is a recursive a.e. h-complex set. Choose an index e such that $\chi_A = \varphi_e$. If $i \notin K$ then $\varphi_{f(e,i)} = \varphi_{g(e,i)} = \chi_A$. If $i \in K$ then there is a unique n such that $k(n) = i$. Then $\varphi_{f(e,i)}(n) = 1 - \varphi_{g(e,i)}(n)$, and if $n \geq e, i$ then $h_0(n) > \max(\Phi_{f(e,i)}(n), \Phi_{g(e,i)}(n))$, furthermore, either $\varphi_{f(e,i)} = \chi_A$ or $\varphi_{g(e,i)} = \chi_A$. Suppose that ψ is any (h, A)-bounding procedure which is partial recursive in M. Then we can decide K using M as an oracle: For given i we simultaneously enumerate K an compute $\psi(f(e,i))$ and $\psi(g(e,i))$ until either i is enumerated into K or $\psi(f(e,i))$ and $\psi(g(e,i))$ both terminate. Note that one of these cases must occur. In the first case $\chi_K(i) = 1$. In the second case we claim that $\chi_K(i) = 1 \Leftrightarrow i \in \{k(x) : x \leq \max(e, i, \psi(f(e,i)), \psi(g(e,i)))\}$. Suppose for a contradiction that $i = k(n)$ and $n > \max(e, i, \psi(f(e,i)), \psi(g(e,i)))$. Either $\varphi_{f(e,i)} = \chi_A$ or $\varphi_{g(e,i)} = \chi_A$, say the former. Then $\Phi_{f(e,i)}(n) > h(n)$ as ψ is an (h, A)-bounding procedure and $n > \psi(f(e,i))$. On the other hand, $h(n) \geq h_0(n) > \Phi_{f(e,i)}(n)$ as $n \geq e, i$, a contradiction. Thus it follows that K is recursive in M. ∎

Note that Theorem 2.1 subsumes Fulk's result. Our proof appears to be simpler because we avoid the double recursion theorem. Of course, the basic idea is the same as in Fulk's proof.

3 The Speed-up Theorem

Informally, the Speed-up Theorem of Blum [2] states that there exist recursive sets having no fastest decision procedure. For a more precise formulation let us recall some wellknown definitions and facts.
A function $f \in R_1$ is said to have *m-speed-up* $(m \in R_2)$ iff

$$(\forall i)[\varphi_i = f \rightarrow (\exists j)(\varphi_j = f \wedge m(x, \Phi_j(x)) < \Phi_i(x) \text{ a.e.})].$$

(if $\varphi_j = \varphi_j \wedge m(x, \Phi_j(x)) < \Phi_i(x)$ a.e., then we say that *j has m-speed-up on i*)

f has *effective m-speed-up* iff

$$(\exists g \in R_1)(\forall i)[\varphi_i = f \rightarrow \varphi_{g(i)} = f \land m(x, \Phi_{g(i)}(x)) < \Phi_i(x) \text{ a.e.}].$$

(in this case we say that f has effective m-speed-up *via g*)
f has *partial m-speed-up* iff

$$(\forall i)[\varphi_i = f \rightarrow (\exists j)(\varphi_j =^* f \land m(x, \Phi_j(x)) < \Phi_i(x) \text{ a.e.})].$$

f has *effective partial m-speed-up* iff

$$(\exists g \in R_1)(\forall i)[\varphi_i = f \rightarrow \varphi_{g(i)} =^* f \land m(x, \Phi_{g(i)}(x)) < \Phi_i(x) \text{ a.e.}].$$

(in this case we say that f has effective partial m-speed-up *via g*)
The basic facts concerning the possibilities of speed-up are contained in the following theorem due to Blum:

THEOREM B ([2, 3])
a.) For any $m \in R_2$ there is a $\{0,1\}$-valued function $f \in R_1$ which has m-speed-up.
b.) If $m \in R_2$ is sufficiently large then there does not exist a function $f \in R_1$ which has effective m-speed-up.
c.) If $m \in R_2$ is sufficiently large then every function $f \in R_1$ which has partial m-speed-up also has effective partial m-speed-up.

Using part b.) of Theorem B, Schnorr proved a related result:

THEOREM S ([10, 11]) *If $m \in R_2$ is sufficiently large then there is no $f \in R_1$ such that there is $\psi \in P_1$ satisfying:*

$$(3.1) \quad (\forall i)[\varphi_i = f \rightarrow [\psi(i) \downarrow \land (\exists j \leq \psi(i))(\varphi_j = f \land |\{x : m(x, \Phi_j(x)) > \Phi_i(x)\}| \leq \psi(i))]].$$

Informally, Schnorr's Theorem states that the size of a program having m-speed-up on i and the number of arguments where speed-up is not achieved, are not simultaneously bounded by a recursive function. The program size of a program having speed-up, may or may not be bounded by a recursive function, as the following results show:
Blum [3, p. 305] (see also Meyer, Fischer [7, §5]) observed that for every $m \in R_2$ there exists $f \in R_1$ having m-speed-up and $\psi \in R_1$ such that:

$$(\forall i)[\varphi_i = f \rightarrow (\exists j \leq \psi(i))(\varphi_j = f \land m(x, \Phi_j(x)) \leq \Phi_i(x) \text{ a.e.})].$$

Meyer and Fischer [7, Theorem 5] proved that there exists $m \in R_2$ and $f \in R_1$ having m-speed-up such that there is no $\psi \in P_1$ with:

$$(\forall i)[\varphi_i = f \rightarrow (\psi(i)\downarrow \wedge (\exists j \leq \psi(i))(\varphi_j = f \wedge \\ m(x, \Phi_j(x)) \leq \Phi_i(x) \text{ a.e.}))].$$

With a K-oracle all of these bounding functions are computable: For any $m \in R_2, m(x,y) \geq y$ and any $f \in R_1$ having m-speed-up there exists a partial function h which is partial recursive in K such that:

$$(\forall i)[\varphi_i = f \rightarrow [h(i)\downarrow \wedge \varphi_{(h(i))_1} = f \wedge \\ (\forall x \geq (h(i))_2)(m(x, \Phi_j(x)) \leq \Phi_i(x)), \text{ where } j = (h(i))_1]].$$

h is computed as follows: On input i we run through all numbers $n = 0, 1, 2, \ldots$ where $n = \langle j, a \rangle$, and check whether

$$(\forall x \geq a)(m(x, \Phi_j(x)) \leq \Phi_i(x)) \wedge (\forall x < a)(\varphi_j(x)\downarrow) \\ \wedge (\forall x)(\varphi_i(x) = \varphi_j(x)).$$

It is easy to see that this can be checked recursive in K, for any $\varphi_i \in R_1$. As soon as the outcome of this test is positive we output $h(i) = \langle j, a \rangle$. If f has m-speed-up then the computation terminates for each input i which is a φ-index of f. $\square$

Now we consider the degrees of unsolvability of various bounding procedures: Suppose that $f \in R_1$ has effective partial m-speed-up via $g \in R_1$ where $m \in R_2, m(x,y) \geq y$. A partial function ψ is called an (m, f, g)-*bounding procedure* iff

$$(\forall i)[\varphi_i = f \rightarrow (\psi(i)\downarrow \wedge (\forall n \geq \psi(i))(\varphi_{g(i)}(n) = f(n)))].$$

By the preceding remark there exists an (m, f, g)-bounding procedure which is partial recursive in K. The converse is given by the next theorem.

Theorem 3.1 *Suppose that $m \in R_2$ is sufficiently large and $f \in R_1$ has effective partial m-speed-up via $g \in R_1$. If M is any set such that there exists an (m, f, g)-bounding procedure which is partial recursive in M, then $K \leq_T M$.*

Proof: This is proved by a modification of the proof that effective m-speed-up is impossible if m is sufficiently large ([3, Theorem 2], see also [11, Satz 9.3.1] and [12, §5] for alternative presentations). We will therefore first recall the idea of that proof and then indicate the modifications.

The remaining formal details are tedious but completely standard. The interested reader may consult Schnorr [11]. Alternatively, the outlined proofs can be easily formalized in the space complexity measure for Turing machines, and can then be transferred to arbitrary measures using the recursive relatedness of any two measures (and the fact that for the Relatedness Theorem we indeed have recursive bounding functions).
Review of the proof for Theorem B, b.): Suppose that $m \in R_2$ is sufficiently large and $f \in R_1$ has effective m-speed-up via $g \in R_1$. Using the double recursion theorem one defines programs e_0, e_1 and a set $A = \bigcup \{A_x : x \geq 0\}$ of "special inputs" as follows:

Construction
Initialize $A_{-1} = \emptyset$. On input x we compute A_{x-1} and use x steps to check whether for all $y \in A_{x-1}$: $\varphi_{g(e_0)}(y) = \varphi_{g(e_1)}(y) = f(y)$. If the verification cannot be completed (because the computations do not terminate in x steps, or the equalities do not hold), then let $\varphi_{e_0}(x) = \varphi_{e_1}(x) = f(x)$ and $A_x = A_{x-1}$. Otherwise, i.e. if the verification can be completed, let $A_x = A_{x-1} \cup \{x\}$ and do the following: Compute in parallel $f(x)$, $\varphi_{g(e_0)}(x)$ and $\varphi_{g(e_1)}(x)$. If the computation of $f(x)$ converges first then let $\varphi_{e_0}(x) = \varphi_{e_1}(x) = f(x)$. Otherwise let j be minimal such that $\varphi_{g(e_j)}(x)$ converges first, and let $\varphi_{e_j}(x) = f(x), \varphi_{e_{1-j}}(x) = \varphi_{g(e_j)}(x)$.
End of construction

Using the hypothesis $(\forall i)(\varphi_i = f \rightarrow \varphi_{g(i)} = f)$, it follows that $\varphi_{e_0} = \varphi_{e_1} = f$. The set A is infinite and for almost every $x \in A$ there is $j \in \{0, 1\}$ such that $\varphi_{e_j}(x)$ converges in at most ("a little more than") $\Phi_{g(e_j)}$ many steps. Thus, if m is sufficiently large then there is $j \in \{0, 1\}$ such that for infinitely many x: $m(x, \Phi_{g(e_j)}(x)) > \Phi_{e_j}(x)$. This contradiction proves that g does not exist. □

Now we return to the proof of Theorem 3.1: Let $k \in R_1$ be a fixed one-one function such that $\mathrm{rg}(k) = K$. Suppose that $m \in R_2$ is sufficiently large and $f \in R_1$ has effective partial m-speed-up via $g \in R_1$. We define, uniformly in i, programs $e_0(i), e_1(i)$ as above, with the slight modification that on input x nothing is enumerated into A unless $k(n) = i$, for some $n < x$. Clearly, if $i \notin K$ then $\varphi_{e_0(i)} = \varphi_{e_1(i)} = f$. If $i \in K$, say $k(n) = i$, then there must exist a least $x_0 > n$ such that for some $j \in \{0, 1\}$, $\varphi_{e_j(i)}(x_0) \neq f(x_0)$, as otherwise we would obtain the same contradiction as above. It follows that $\varphi_{e_j(i)}(x_0) = \varphi_{g(e_{1-j}(i))}(x_0) \neq f(x_0)$ and $\varphi_{e_{1-j}(i)}(x_0) = f(x_0)$. As the verification cannot be completed on any input $x > x_0$ we find that $\varphi_{e_0(i)}(x) = \varphi_{e_1(i)}(x) = f(x)$, for all $x > x_0$. In particular, $\varphi_{e_{1-j}(i)} = f$ and every (m, f, g)- bounding procedure ψ

must satisfy $\psi(e_{1-j}(i)) \downarrow > x_0 > n$. As in the proof of Theorem 2.1 it is now easy to see that if M is any set such that there exists an (m, f, g)-bounding procedure which is partial recursive in M then $K \leq_T M$. ∎

Corollary 3.2 *Suppose that $m \in R_1$ is sufficiently large, M is r.e. and $f \in R_1$ has m-speed-up. If there exists a function h which is partial recursive in M such that*

$$(\forall i)[\varphi_i = f \rightarrow (h(i) \downarrow \wedge \varphi_{h(i)} = f \wedge m(x, \Phi_{h(i)}(x)) < \Phi_i(x) \ a.e.)],$$

then $K \leq_T M$.

Proof: Let M, m, f, h satisfy the hypothesis of Corollary 3.2. By the Modulus Lemma and the Limit Lemma (see [14, III.3]) there are functions $h' \in R_2$ and $\mu : \omega \rightarrow \omega$ partial recursive in M such that for all φ-indices i of f:

$$\mu(i) \downarrow \wedge (\forall s \geq \mu(i))(h'(i, s) = h(i)).$$

Here we are using that M is r.e. For each i we define the program $g(i)$ that works as follows: On input x, spend x steps in computing $h'(i, 0), h'(i, 1), \ldots$ Let s be maximal such that $h'(i, s)$ has been computed, and let $\varphi_{g(i)}(x) = \varphi_{h'(i,s)}(x)$. □

Clearly, $g \in R_1$ and for each φ-index i of f, $g(i)$ has partial m'-speed-up on i for some m' not much smaller than m. Let $\psi(i)$ be the least x such that $h'(i, 0), h'(i, 1), \ldots, h'(i, \mu(i))$ are computed within x steps. Then ψ is an (m', f, g)-bounding procedure which is partial recursive in M. Thus by Theorem 3.1, K is recursive in M if m' is sufficiently large. The proof is completed by the observation that m' is large if m is large. ∎

We do not know whether the hypothesis that M is r.e. can be omitted in Corollary 3.2. Note that Corollary 3.2 can be viewed as a completeness criterion for r.e. sets. For other such criteria see [14, V.5]. Now we will consider Theorem S. We already noticed that a function ψ as in (3.1) can be computed using a K-oracle. The converse is given by the next result.

Theorem 3.3 *Suppose that $m \in R_2$ is sufficiently large and $f \in R_1$ has m-speed-up. If M is any set such that there is a partial function ψ which is partial recursive in M and satisfies condition (3.1), then $K \leq_T M$.*

Proof: Also in this proof we first recall the idea of Schnorr's proof [10, 11] of Theorem S and then indicate the necessary modifications. The remaining details are left to the reader.
Review of the proof of Theorem S: Suppose for a contradiction that $m \in R_2$ is sufficiently large (in particular $m(x,y) \geq y$), $f \in R_1$ has m-speed-up and $\psi \in P_1$ satisfies (3.1). One shows that f has effective m'-speed-up for some m' not much smaller than m, contradicting Theorem B, b.).

Construction
For each e the program $h(e)$ works as follows: On input x we spend x steps in computing $\psi(e)$. If the computation does not converge then let $\varphi_{h(e)}(x) = \varphi_e(x)$. Otherwise compute the finite set $L(x)$ defined as follows:
$$\{j : j \leq \psi(e) \wedge (\forall y < x)[(\Phi_j(y) < x \wedge \Phi_e(y) < x) \rightarrow \varphi_e(y) = \varphi_j(y)]$$
$$\wedge (\exists^{\leq \psi(e)} y < x)[\Phi_e(y) < x \wedge m(y, \Phi_j(y)) > \Phi_e(y)] \text{ and the latter}$$
inequality can be verified within x steps]$\}$.

Compute in parallel $\varphi_e(x)$ and $\varphi_j(x)$, for all $j \in L(x)$. If $\Phi_j(x) \leq \Phi_e(x)$, for all $j \in L(x)$ and $|\{\varphi_j(x) : j \in L(x)\}| = 1$ then let $\varphi_{h(e)}(x) = \varphi_j(x)$, for $j = \min(L(x))$. Otherwise let $\varphi_{h(e)}(x) = \varphi_e(x)$.
End of construction

Suppose that $\varphi_e = f$. By hypothesis there exists $j \leq \psi(e)$ such that $\varphi_j = f$ and $|\{x : m(x, \Phi_j(x)) > \Phi_e(x)\}| \leq \psi(e)$. Note that j cannot be eliminated from $L(x)$, therefore $\varphi_{h(e)} = f$. $L(x)$ converges to a set L, $1 \leq |L| \leq \psi(e) + 1$, which contains only programs which a.e. compute f and have m-speed-up on e. As $L(x)$ can be computed fast it follows that $h(e)$ has m'-speed-up on e, for $m' \in R_2$ not much smaller than m. This contradicts Theorem B, b.). □

Now we return to the proof of Theorem 3.3: Unlike the proof of Theorem S we cannot use Theorem 3.1 directly, rather we need to incorporate the construction from its proof in an explicit way. Let $k \in R_1$ be a fixed one-one function such that $rg(k) = K$. Suppose that $m \in R_2$ is sufficiently large and $f \in R_1$ has m-speed-up. Define a program transformation $h(e, i)$ in the same way as $h(e)$, with the addition that $\psi(e)$ is replaced by the least n such that $k(n) = i$ (if n does not exist then $\varphi_{h(e,i)} = \varphi_e$). For each i we define, uniformly in i, two programs $e_0(i)$ and $e_1(i)$ by the double recursion theorem as in the proof of Theorem 3.1 with the modification that now we are using $h(e, i)$ instead of $g(e)$.

Suppose that ψ is partial recursive in M and satisfies (3.1). If $i \notin K$ then $\varphi_{e_0(i)} = \varphi_{e_1(i)} = f$, thus $\psi(e_0(i)) \downarrow, \psi(e_1(i)) \downarrow$. If $i \in K$, say

Degrees of Unsolvability in Abstract Complexity Theory   235

$k(n) = i$, then we verify that if ψ is defined on $e_0(i)$ and $e_1(i)$ then $\max(\psi(e_0(i)), \psi(e_1(i))) > n$: Note that by Blum's construction, $\varphi_{e_0(i)} = f$ or $\varphi_{e_1(i)} = f$ (without assuming any properties of h). Thus there are two cases:
a.) $\varphi_{e_0(i)} = f$ and $\varphi_{e_1(i)} = f$. If $\max(\psi(e_0(i)), \psi(e_1(i))) \leq n$ then $h(e_0(i), i), h(e_1(i), i)$ would both compute f and would have m'-speed-up on $e_0(i), e_1(i)$, respectively. As in the proof of Theorem 3.1 there is $j \in \{0,1\}$ such that for infinitely many x: $m'(x, \Phi_{h(e_j(i),i)}(x)) > \Phi_{e_j(i)}(x)$, a contradiction.
b.) $\varphi_{e_0(i)} \neq f$ or $\varphi_{e_1(i)} \neq f$. Say $\varphi_{e_0(i)} = f$ and $\varphi_{e_1(i)} \neq f$. If $\psi(e_0(i)) \leq n$ then $\varphi_{h(e_0(i),i)} = f$, and therefore, by Blum's construction, $\varphi_{e_1(i)} = f$, a contradiction.
As in the proof of Theorems 2.1, 3.1 we can compute k using M as an oracle. ∎

4 The Union Theorem

For $g \in R_1$ the complexity class C_g is defined as follows:

$$C_g = \{\varphi_i \in R_1 : \Phi_i(x) \leq g(x) \text{ a.e.}\}.$$

McCreight and Meyer proved a general result concerning countable unions of complexity classes:

UNION THEOREM ([8]) *If $\{f_i\}_{i \in \omega}$ is a uniformly recursive family of total recursive functions such that $(\forall i)(f_i \leq f_{i+1})$ then there exists $g \in R_1$ such that:*

$$C_g = \bigcup\{C_{f_i} : i \in \omega\}.$$

As a corollary, there exists $g \in R_1$ such that P = DTIME($g(n)$). Note that in this case the time bound is measured in the length of the binary representation of the input. It is well known that any such g must be rather pathological, e.g., by the Time Hierarchy Theorem, it cannot be time constructible. We are interested in the growth rate of g. As for each k there exists a set $A \in$ P such that each Turing machine which decides A has time complexity $\Omega(n^k)$, it follows that g majorizes every polynomial, i.e., $(\forall k)(\exists n_0)(\forall n \geq n_0)(g(n) > n^k)$.
The growth rate of g is described by the function $S_g(k) := \max\{n : n^k > g(n)\}$. Note that S_g is recursive if g is any "natural" function. We will characterize the degrees **d** such that there exists $g \in R_1$ with

P = DTIME($g(n)$) and S_g recursive in d. This degree class will turn out to be the upward closure of the promptly simple degrees.

In order to motivate the construction in the proof of Theorem 4.1 below, let us briefly sketch why S_g is nonrecursive: If S_g were recursive then we could diagonalize in time $g(n)$ over all polynomial time computable sets, much similar as in the proof of the Time Hierarchy Theorem. On input x of length n we spend n steps in computing successively $S_g(1), S_g(2), \ldots$ If k_0 is the greatest k such that $S_g(k)$ is computed within this time bound and $S_g(k) < n$, then we know that $g(n) > n^{k_0}$. Therefore we may spend $n^{k_0} - n$ steps on the diagonalization of the $(n)_1$-th Turing machine. Clearly, in this way we are able to diagonalize every polynomial time bounded machine. □

As similar results hold for may other complexity classes we will first prove a very general result in the context of abstract complexity theory and then obtain the above mentioned results as a special case.

For the convenience of the reader we recall some facts and results concerning promptly simple sets, see Soare [14, XIII] for further material.

Let $\{W_e\}_{e \in \omega}$ denote the standard Gödelnumbering of the r.e. sets. We fix a recursive enumeration $\{W_{e,s}\}_{e,s \in \omega}$ (i.e. $W_e = \bigcup \{W_{e,s} : s \geq 0\}$, $W_{e,s} \subseteq W_{e,s+1}$, and $\{W_{e,s}\}_{e,s \in \omega}$ is a strong array). We assume that $(\forall x, e, s)(|W_{e,s+1} - W_{e,s}| \leq 1$, and $(x \in W_{e,s} \to e, x < s))$. $x \in W_{e, \text{at } s}$ denotes that $x \in W_{e,s} - W_{e,s-1}$, if $s > 0$.

The notion of a *promptly simple* set was introduced by Maass [6]:

Definition ([6]) A coinfinite r.e. set A is *promptly simple* iff there is a function $p \in R_1$ and a recursive enumeration $\{A_s\}_{s \in \omega}$ of A such that for every e: (4.1) W_e infinite $\to (\exists s)(\exists x)(x \in W_{e, \text{at } s} \cap A_{p(s)})$.
A Turing degree is called promptly simple iff it contains a promptly simple set.

The definition of prompt simplicity is independent of the particular enumeration, see [14, XIII.1.3]. We will be using the following characterization of sets having promptly simple degrees:

Fact (see [14, XIII.1.6, 1.7]) Let A be an r.e. set and $\{A_s\}_{s \in \omega}$ a recursive enumeration of A. Then A has promptly simple degree iff there is a function $p \in R_1$ such that for all $s, p(s) \geq s$, and for all e:
(4.2) W_e infinite $\to (\exists x)(\exists s)[x \in W_{e, \text{at } s} \wedge A_s \upharpoonright x \neq A_{p(s)} \upharpoonright x]$.
Equivalently, (4.2) can be replaced by (4.3):
(4.3) W_e infinite $\to (\exists^\infty x)(\exists s)[x \in W_{e, \text{at } s} \wedge A_s \upharpoonright x \neq A_{p(s)} \upharpoonright x]$.

Ambos-Spies, Jockusch, Shore and Soare [1] proved that the class PS of all promptly simple degrees coincides with the class NC of all noncappable degrees (an r.e. degree **a** is noncappable iff it is not half of a minimal pair, i.e. there does not exist an r.e. nonrecursive degree **b** such that $\mathbf{a} \cap \mathbf{b} = \mathbf{0}$), and forms a filter in the class of all r.e. degrees. There exists a low promptly simple degree [6] and a high degree which is not promptly simple (as there exists a minimal pair of high r.e. degrees, see [14, XIV.3.1]).

Now we turn to the analysis of the Union Theorem. First we need an appropriate formalization of the notion that the f_i's define a strictly increasing sequence of complexity classes such that $C_{f_{i+1}}$ contains a.e. f_i-complex functions. In order to avoid complexity gaps we require that the f_i's must be taken from a measured set. Note that by the Honesty Theorem, there exists a measured set such that for every complexity bound there exists, in a uniform way, a complexity bound from the measured set defining the same complexity class.

Definition let h be a recursive function and $\{\psi_i\}_{i \in \omega}$ be a measured set (i.e. the prediacte $P(i, x, y) \equiv \psi_i(x) = y$ is recursive). A uniformly recursive sequence $\{f_i\}_{i \in \omega}$ of total recursive functions is called (h, ψ)-*good* iff there exist recursive functions b, c such that:

(1) $(\forall i)(f_i \leq f_{i+1})$
(2) $(\forall i, x)(x > b(i) \rightarrow h(f_i(x), x) \leq f_{i+1}(x))$
(3) $(\forall i)(f_i = \psi_{c(i)})$.

Theorem 4.1 *For every measured set $\{\psi_i\}_{i \in \omega}$ there exists a recursive function h such that for every (h, ψ)-good sequence $\{f_i\}_{i \in \omega}$ and every $g \in R_1$ with $C_g = \bigcup \{C_{f_i} : i \in \omega\}$:*

$$M_g := \{\langle n, i \rangle : (\exists x > n)(f_i(x) > g(x))\} \text{ has promptly simple degree.}$$

(Note that $S_g(i) := max\{n : f_i(n) > g(n)\}$ and M_g are Turing equivalent if g majorizes every f_i.)

Proof: The outline of the proof is very roughly as follows (for the time complexity measure of Turing machines): Suppose that $\{f_i\}_{i \in \omega}$ has all desired properties and $C_g = \bigcup \{C_{f_i} : i \geq 0\}$. Define a recursive approximation $\{A_s\}_{s \in \omega}$ of M_g:

$$A_s := \{\langle n, i \rangle : n, i \leq s \wedge (\exists x)(n < x \leq s \wedge f_i(x) > g(x))\}.$$

We try to diagonalize in time $g(x)$ over all classes C_{f_i}, for $i \geq 0$. This will be successful if M_g does not have promptly simple degree. The e-th attempt to diagonalize works as follows: On input n, spend n steps in computing successively $A_0, A_1, \ldots$ Suppose that there is an $s < n$ such that A_s has been computed for the first time. Then we compute $W_{e,\text{ at }s}$. Suppose that $x \in W_{e,\text{ at }s}$, and x is the m-th element of W_e in the given enumeration. Call i active, if $x > b(i-1)$ and there is n' such that $\langle n', i \rangle < x$ and $g(z) \geq f_i(z)$, for all $n' \leq z \leq s$. If there is an active $i = (m)_1$ then spend $f_{i-1}(n)$ steps to diagonalize the j-th machine on input n, where $j = (m)_2$.

If h is sufficiently large then we are using less than $h(f_{i-1}(n), n) \leq f_i(n)$ steps. Hence if we are using more than $g(n)$ steps then $g(n) < f_i(n)$ and it follows that $\langle n', i \rangle \in M_g - A_s$, in fact, there is $t \leq n, t > s$, such that $A_s \upharpoonright x \neq A_t \upharpoonright x$. So, if we can guarantee that $A_s \upharpoonright x = A_n \upharpoonright x$, then we are computing a function from C_g.

If W_e is infinite and g majorizes every f_i then we construct in the e-th attempt a function d_e that is not in $\bigcup \{C_{f_i} : i \geq 0\}$. Let $p(s)$ denote that n such that on input n the set A_s is computed for the first time (for simplicity we are assuming here that n is independent of e). If M_g does not have promptly simple degree then, by (4.2) there is e such that W_e is infinite and $(\forall x, s)(x \in W_{e,\text{ at }s} \to A_s \upharpoonright x = A_{p(s)} \upharpoonright x)$. By the above remarks it follows that d_e is contained in C_g. This contradicts the hypothesis $C_g = \bigcup C_{f_i}$. Thus M_g has promptly simple degree.

It is straightforward to convert this sketch into a proof that works for the special case P=DTIME($g(n)$) alluded to above. Now we will present the formal implementation of the general case. Here the intuitive ideas from the outline get somewhat obscured by the technical machinery.

Let $\{\psi_i\}_{i \in \omega}$ be given. First we construct in a universal way the diagonalization inputs. We define a partial recursive function $\eta(e, e_1, e_2, e_3, s)$, for all e, e_1, e_2, e_3, s as follows:

Construction
For abbreviation let $g = \varphi_{e_1}, b = \varphi_{e_2}, c = \varphi_{e_3}, f_i = \psi_{c(i)}$.
Compute $g(i), c(i), b(i)$ for all $i \leq s$. If one of these values is undefined then $\eta(e, e_1, e_2, e_3, s)$ is undefined. Now suppose that all of the computations terminate. If $x \in W_{e,\text{ at }s}$ (then $x < s$) and there exists a number $m = \langle i_1, i_2, i_3, i_4 \rangle$ such that the following conditions are satisfied:
(1) $m < x, i_3 > 0, i_4 = c(i_3 - 1), b(i_3 - 1) < x$,
(2) $m \notin \{\eta(e, e_1, e_2, e_3, 0), \ldots, \eta(e, e_1, e_2, e_3, s-1)\}$,
(3) there exists a number n such that
$\langle n, i_3 \rangle < x$, and $g(z) \geq f_{i_3}(z)$, for $n \leq z \leq s$.
Then output the least such number m. Otherwise output 0.
End of construction

Degrees of Unsolvability in Abstract Complexity Theory   239

Choose a total recursive function $t(e, e_1, e_2, e_3, n)$ such that:

$\text{rg}(\lambda n.\ t(e, e_1, e_2, e_3, n)) = \{0\} \cup \text{rg}(\lambda s.\ \eta(e, e_1, e_2, e_3, s))$,
$(\forall s)[\eta(e, e_1, e_2, e_3, s) > 0 \rightarrow$
$(\exists^{=1} n)\ t(e, e_1, e_2, e_3, n) = \eta(e, e_1, e_2, e_3, s)]$, and
$(\forall s, s', n, n')[(s < s' \land \eta(e, e_1, e_2, e_3, s) = t(e, e_1, e_2, e_3, n) > 0 \land$
$\eta(e, e_1, e_2, e_3, s') = t(e, e_1, e_2, e_3, n') > 0 \rightarrow n < n' \land s < n \land s' < n']$.

intuitively, t is an order preserving *total recursive* one-one enumeration of $\text{rg}(\eta)$ and the output 0 plays the role of the undefined output.
By the s-m-n theorem there is a total recursive function $u(e, e_1, e_2, e_3)$ such that $\varphi_{u(e,e_1,e_2,e_3)}(n)$ is computed as follows (Intuitively, $\varphi_{u(e,...)}$ corresponds to the e-th diagonalization d_e):

Construction
If $t(e, e_1, e_2, e_3, n) > 0$, say $t(e, e_1, e_2, e_3, n) = \langle i_1, i_2, i_3, i_4 \rangle$, then we simulate $\psi_{i_4}(n)$ steps in the computation of $\varphi_{i_1}(n)$. If the computation terminates within the step bound then we output $1 - \varphi_{i_1}(n)$. If the computation does not terminate then we output 0. If ψ_{i_4} is undefined then the output is undefined.
If $t(e, e_1, e_2, e_3, n) = 0$ then we output 0.
End of construction

Now we are going to define a sufficiently large h:
By the Compression Theorem there exists a function $h_0 \in R_2$, nondecreasing in both arguments, such that for every i, if ψ_i is total recursive then $C_{h_0(\psi_i(x),x)}$ contains an a.e. ψ_i-complex ($\{0,1\}$-valued) function.
We define a total recursive fucntion h as follows:

$h(y, x) = h_0(y, x) + \max\{\Phi_{u(e,e_1,e_2,e_3)}(x) : e, e_1, e_2, e_3 \leq x$, and
if $t(e, e_1, e_2, e_3, x) = \langle i_1, i_2, i_3, i_4 \rangle > 0$ then $\psi_{i_4}(x) \leq y\}$.

Here we are using that $\{\psi_i\}_{i \in \omega}$ is a measured set. Finally we verify that h has the required properties:
Suppose that the total functions b, c, f_i, g are given as in the statement of the theorem. Choose indices e_1, e_2, e_3 such that $g = \varphi_{e_1}, b = \varphi_{e_2}, c = \varphi_{e_3}, f_i = \psi_{c(i)}$. Note that $\lambda e, s.\ \eta(e, e_1, e_2, e_3, s)$ and $\lambda e, x.\ \varphi_{u(e,e_1,e_2,e_3)}(x)$ are total recursive functions. By the definition of h, $C_{f_{i+1}}$ contains an a.e. f_i-complex set, thus, $g(x) > f_i(x)$ a.e. Let $A_s = \{\langle n, i \rangle : n, i \leq s \land (\exists x)(n < x \leq s \land f_i(x) > g(x))\}$, $\{A_s\}_{s \in \omega}$ is a recursive enumeration of M_g. We define a total recursive function p as follows:

$p(s) = \min\{n : (\forall e \leq s)\{\eta(e, e_1, e_2, e_3, s') : 0 \leq s' \leq s\} \subseteq$
$\{t(e, e_1, e_2, e_3, n') : 0 \leq n' \leq n\}\}$.

Suppose for a contradiction that A does not have promptly simple degree. Then, by (4.2), there is an e such that W_e is infinite and for all x, s:

$$(4.4) \quad x \in W_{e,\,\text{at}\,s} \;\to\; A_s \upharpoonright x = A_{p(s)} \upharpoonright x.$$

Claim 1: $\varphi_{u(e,e_1,e_2,e_3)} \notin \bigcup\{C_{f_i} : i \geq 0\}$.
Proof: Fix i, and choose x_0 large enough such that $g(x) \geq f_i(x)$, for all $x \geq x_0$. Suppose for a contradiction that $\varphi_{u(e,e_1,e_2,e_3)} \in C_{f_i}$, i.e. there are j, x_1 such that $\varphi_{u(e,e_1,e_2,e_3)} = \varphi_j$ and $\Phi_j(x) \leq f_i(x)$, for all $x \geq x_1$. Let $x_2 = \max(x_0, x_1, b(i))$. As W_e is infinite there exist s and $i_2 \geq x_2$ such that $\eta(e, e_1, e_2, e_3, s) = \langle j, i_2, i+1, c(i)\rangle$, in particular $s > x > i_2$. Then there is an $n \geq s$ such that $t(e, e_1, e_2, e_3, n) = \eta(e, e_1, e_2, e_3, s)$ and, as $\Phi_j(n) \leq \psi_{c(i)}(n)$, we are able to complete the simulation of $\varphi_j(n)$. Thus, $\varphi_{u(e,e_1,e_2,e_3)}(n) = 1 - \varphi_j(n)$, a contradiction. □

Claim 2: $\varphi_{u(e,e_1,e_2,e_3)} \in C_g$.
Proof: Let $n_0 = \max(e, e_1, e_2, e_3)$, and $n_1 = \max\{n : g(n) < h(0, n)\}$. Note that n_1 exists as g majorizes f_1 and $f_1(n) \geq h(0, n)$ a.e. Let $n > \max(n_0, n_1)$. We verify that $g(n) \geq \Phi_{u(e,e_1,e_2,e_3)}(n)$: If $t(e, e_1, e_2, e_3, n) = 0$; then $\varphi_{u(e,e_1,e_2,e_3)}(n) \leq h(0, n) \leq g(n)$. If $t(e, e_1, e_2, e_3, n) > 0$, then there is $s \geq e$, $x \in W_{e,\,\text{at}\,s}$, and i_1, i_2, i_3, i_4 such that $t(e, e_1, e_2, e_3, n) = \eta(e, e_1, e_2, e_3, s) = \langle i_1, i_2, i_3, i_4\rangle$. By the definition of p we get $s < n \leq p(s)$. By the properties (1)-(3) from the definition of η we get $b(i_3 - 1) < n$, and there is $n' < n$ with $\langle n', i_3 \rangle < x$ and $\langle n', i_3\rangle \notin A_s$. Therefore, by (4.4), $\langle n', i_3\rangle \notin A_{p(s)}$, i.e., $g(z) \geq f_{i_3}(z)$, for all $n' \leq z \leq p(s)$, in particular $g(n) \geq f_{i_3}(n)$. Putting these results together we get: $g(n) \geq f_{i_3}(n) \geq h(f_{i_3 - 1}(n), n) = h(\psi_{i_4}(n), n) \geq \Phi_{u(e,e_1,e_2,e_3)}(n)$. □

Claims 1,2 contradict the hypothesis. This contradiction shows that M_g has promptly simple degree. ∎

Conversely, without any additional hypotheses, we can sharpen the Union Theorem such that M_g is recursive in any given promptly simple set:

Theorem 4.2 *For every uniformly recursive sequence $\{f_i\}_{i \in \omega}$ of total recursive functions with $(\forall i)(f_i \leq f_{i+1})$, and any promptly simple set A there exists $g \in R_1$ such that $C_g = \bigcup\{C_{f_i} : i \geq 0\}$ and $M_g \leq_T A$. If in addition there exists $b \in R_1$ such that $(\forall i, x)(x > b(i) \to f_i(x) < f_{i+1}(x))$ then we can achieve $M_g \equiv_T A$.*

Degrees of Unsolvability in Abstract Complexity Theory 241

Proof: We will combine the original proof of the Union Theorem [8] with "prompt permitting" (see [14, XIII.2.1]). Let A be r.e. and promptly simple and let $\{A_s\}_{s\in\omega}$ be an enumeration of A such that there exists a total recursive function p satisfying $p(s) \geq s$, and
(4.5) $(\forall e)[W_e \text{ infinite } \rightarrow (\exists^\infty x)(\exists s)(x \in W_{e,\text{ at }s} \wedge A_s \upharpoonright x \neq A_{p(s)} \upharpoonright x)]$.
We are going to construct a total recursive function g, such that $M_g \leq_T A$ via permitting. We are using uniformly enumerable r.e. sets $\{U_e\}_{e\in\omega}$ which are enumerated simultaneously. By the recursion theorem we may assume that we are given a total recursive function h such that $U_e = W_{h(e)}$, for all e. Furthermore, we may assume by the Slowdown Lemma [14, XIII.1.5] that any element enumerated in U_e appears strictly later in $W_{h(e)}$.

Construction
Initialization: Let guess$(x) = x$, for all x, and let $U_e = \emptyset$, for all e.
Stage s: Let $X := \emptyset$. For $e = 0, \ldots, s$ do the following:
If there is an $i \geq \text{guess}(e), i \leq s$ such that $\Phi_e(s) > f_i(s)$ and $i \notin U_e$ then choose the least such i and enumerate i into U_e. Compute the least $t > s$ such that $i \in W_{h(e), \text{ at } t}$. If $A_{p(t)} \upharpoonright i \neq A_s \upharpoonright i$ then let guess$(e) = s$ and $X := X \cup \{i\}$ ("e is active at s").
If $X = \emptyset$ then let $g(s) = f_s(s)$. If $X \neq \emptyset$ then let $m = \min(X)$ and let $g(s) = f_m(s)$.
End of construction

Claim 1: g majorizes every f_i.
Proof: Suppose the claim is false. Then there exists a least e such that $g(s) < f_e(s)$, for infinitely many s. Thus there are infinitely many stages $s > e$ in which X contains an index i such that guess$(i) < e$. By initialization $i \leq e$, and by the action in stage s, guess(i) is greater or equal s at all larger stages. Therefore there are only finitely many such s, a contradiction. □

Claim 2: For all e: $\phi_e \leq g$ a.e. $\rightarrow (\exists i)\Phi_e \leq f_i$ a.e.
Proof: We show the contrapositive. Suppose that for each i there are infinitely many x such that $\Phi_e(x) > f_i(x)$. Then infinitely many numbers will be enumerated into U_e. By (4.5) we get infinitely many stages s such that e is active at s. In each such stage, $g(s) < \Phi_e(s)$. Thus $g < \Phi_e$ infinitely often. This proves the claim. □

Claim 3: $M_g \leq_T A$
Proof: (Permitting argument) the claim follows from the observation that:
$$n > i \wedge \langle n, i \rangle \in M_g \rightarrow A_n \upharpoonright i \neq A \upharpoonright i,$$

and the fact that M_g is r.e. □

Assume that in addition there exists $b \in R_1$ such that

$$(\forall i, x)[x > b(i) \rightarrow f_i(x) < f_{i+1}(x)].$$

Let A be a promptly simple set. We have shown above that there exists $g \in R_1$ such that $M_g \leq_T A$ and g majorizes every f_i. Let a be a total recursive one-one function such that $\text{rg}(a) = A$. We define the total recursive function $g'(n) = \min(g(n), f_{a(n)}(n))$. As $g'(n) \leq g(n)$ and $g'(n)$ majorizes every f_i it follows that $C_{g'} = C_g$. Also it is easy to see that $M_{g'} \leq_T A$. $A \leq_T M_{g'}$ follows by permitting as

$$(\forall n, i)[i = a(n) \rightarrow (\forall x)(b(i) < x < n \rightarrow \langle x, i+1 \rangle \in M_{g'})],$$

and for each i there are at most finitley ,many n such that $\langle n, i \rangle \in M_{g'}$. This completes the proof of Theorem 4.2. ∎

Finally we want to resume the discussion of the time bounds $g \in R_1$ such that $P = \text{DTIME}(g(n))$. Let $M_g = \{\langle n, i \rangle : (\exists x > n)(g(n) < n^i)\}$, $f_i(n) = n^i$, for $i \geq 1$. Let us fix multitape Turing machines as our model. The proof of Theorem 4.2 can be adapted directly to show that any promptly simple degree **a** contains an M_g. The proof of Theorem 4.1 is also easily adaptable: We may assume that t can be computed in linear time and that $\varphi_{u(e,...)}$ is implemented via a universal simulator of all one-tape machines. The remaining details are left to the reader.

Corollary 4.3 Let $G = \{g \in R_1 : P = \text{DTIME}(g(n))\}$.
a.) $PS = \{dg(M_g) : g \in G\}$ (here $dg(M)$ is the Turing degree of M).
b.) There exists a high r.e. degree **b** such that for no $g \in G$, $dg(M_g) \leq_T$ **b**.
c.) There exists a low degree **c** such that $M_g \in$ **c**, for some $g \in G$.

By Corollary 4.3, b.), the growth rate of any $g \in G$ is "highly" nonconstructive.

References

[1] K. Ambos-Spies, C. G. Jockusch Jr., R. A. Shore, R. I. Soare. An algebraic decomposition of the recursively enumerable degrees and the coincidence of several degree classes with the promptly simple degrees. *Trans. Amer. Math. Soc.*, 281:109–128, 1984.

[2] M. Blum. A machine-independent theory of the complexity of recursive functions. *J. of the ACM*, 14:322–336, 1967.

[3] M. Blum. On effective procedures for speeding up algorithms. *J. of the ACM*, 18:290–305, 1971.

[4] M. A. Fulk. A note on a.e. h-complex functions. *J. Comp. Syst. Sci.*, 40:444–449, 1991.

[5] J. Gill, M. Blum. On almost everywhere complex recursive functions. *J. of the ACM*, 21:425–435, 1974.

[6] W. Maass. Recursively enumerable generic sets. *J. Symb. Logic*, 47:809–823, 1982.

[7] A. R. Meyer, P. C. Fischer. Computational speed-up by effective operators. *J. Symb. Logic*, 37:55–68, 1972.

[8] E. M. McCreight, A. R. Meyer. Classes of computable functions defined by bounds on computation. *In:* Proceedings 1st Ann. ACM Symp. on Theory of Computing, pp. 79–88, ACM, New York, 1969.

[9] M. O. Rabin. Degrees of difficulty of computing a function and a partial ordering of recursive sets. Technical Report 2, Hebrew University, Jerusalem, 1960.

[10] C. P. Schnorr. Does computational speed-up concern programming? *In:* Automata, Languages and Programming (edited by M. Nivat), pp. 585-596, North-Holland, Amsterdam, 1973.

[11] C. P. Schnorr. *Rekursive Funktionen und ihre Komplexität*. B. G. Teubner, Stuttgart, 1974.

[12] J. I. Seiferas. Machine-independent complexity theory. *In:* Handbook of Theoretical Computer Science (edited by J. van Leeuwen), Vol. A, pp. 163–186, North-Holland, Amsterdam, 1990.

[13] C. H. Smith. A note on arbitrarily complex recursive functions. *Notre Dame J. Formal Logic*, 29:198–207, 1988.

[14] R. I. Soare. *Recursively Enumerable Sets and Degrees*. Springer-Verlag, Berlin, 1987.

On the Nonuniform Complexity of the Graph Isomorphism Problem[1]

ANTONI LOZANO and JACOBO TORÁN
Universitat Politècnica de Catalunya

Abstract

We study the nonuniform complexity of the Graph Isomorphism (GI) and Graph Automorphism (GA) problems considering the implications of different types of polynomial-time reducibilities from these problems to sparse sets. We show that if GI (or GA) is bounded truth-table or conjunctively reducible to a sparse set, then it is in P; while if we suppose that it is truth-table reducible without restrictions to a sparse set (or, equivalently, that it belongs to P/poly) then the problem is low for MA, the class of sets with publishable proofs. With respect to nondeterministic reductions, contrasting with the fact that GI and GA belong to the class NP∩(co-NP/poly) [Schö 88a], we show that if the considered problems are bounded truth-table strong nondeterministically reducible to a sparse set, then they are in co-NP. Some of these results are proved using graph constructions that show new properties of the GI and GA problems.

1 INTRODUCTION

The problem of deciding whether two given graphs are isomorphic (Graph Isomorphism, GI) and the related question of deciding whether a graph has a nontrivial automorphism (Graph Automorphism, GA) are examples of the few problems in NP that are not known to be neither in P or NP-complete. It is known that GA is polynomial-time reducible to GI [Ma 79] and no such reduction in the other direction is known. In an attempt to classify

---

[1]This research was partially supported by ESPRIT-II Basic Research Actions Program of the European Community under contract No. 3075 (project ALCOM) and by the DAAD and Spanish Government (Acción Integrada 1992 131-B, 313-AI-e-es/zk).

these problems in a better way, both of them have been extensively studied under different points of view. On one side, deterministic algorithms have been produced that decide GI in polynomial time for graphs with certain restrictions like planarity or bounded valence [Hof 82, HoTa 73, Luk 82], or solve the problem in moderate exponential time in the general case (see [Ba 81]). Another approach has been the study of structural properties of these problems which allow a finer classification of GI and GA in new complexity classes, stressing the differences between these two problems and other sets which are known to be NP-complete. This approach is closely related to the development of the theory of interactive proof systems [GoMiRa 85] and the Arthur-Merlin games [Ba 85, BaMo 88] ($\overline{GI}$ was the first nontrivial example of a set having interactive proofs and of a set in the class AM). We can also cite the results showing that GI is low for the second level of the polynomial-time hierarchy [Schö 88] and proving that GI is low for PP [KöScTo 92] (the same results are true for GA [BoHåZa 87, KöScTo 92]). These facts show specific properties of the GI and GA problems and give evidence that they are not NP-complete since otherwise the polynomial-time hierarchy would collapse to its second level and it would be low for the probabilistic class PP.

In this paper we follow this second approach studying the relationships between the sets GI and GA and the nonuniform complexity classes $\mathcal{C}$/poly of the type defined by Karp and Lipton in [KaLi 80]. Since both problems GI and GA belong to NP∩co-AM and AM has polynomial size generators (AM $\subset$ NP/poly), [Schö 88a], it follows immediately that GI and GA belong to the class NP∩(co-NP/poly). A natural question to ask is whether the inclusion could be strengthened to P/poly, the class of sets with polynomial size circuits, or to other related nonuniform complexity classes. We study the implications of such inclusions.

It is well known that P/poly can be characterized as the class of sets which are polynomial-time Turing or truth-table reducible to sparse sets, P/poly = P(SPARSE) = P_{tt}(SPARSE) [Pi 79],[BooKo 88]. The results about the nonuniform complexity of GI and GA are therefore related to reductions from these problems to sparse sets; we study first in Section 3 the question of whether GI and GA have bounded truth-table (btt) reductions to sparse sets and in Section 4 we study the case of truth-table reducibilities to sparse sets or, in other words, whether these problems have polynomial size circuits.

It has been recently shown that if an NP-complete set A is polynomial-time bounded truth-table reducible to a sparse set, then it is in P [OgWa 90]. A crucial point in the proof is the equivalence under many-one reductions of set A and a related set used for the search of witnesses of words in A, a *left set* for A, and for this, the fact that A is NP-complete is used. Using internal properties of the Graph Isomorphism and Automorphism problems we construct in a natural way a left set for GA which is many-one equivalent to GI. Together with some results about word-decreasing self-reducible sets from [OgWa 90],[OgLo 91] and [Ar et al 92] we obtain that if GI is btt or conjunctive reducible to a sparse set, then GI $\in$ P and if GI is btt strong nondeterministically reducible to a sparse set, then GI $\in$ NP$\cap$co-NP. For the case of the Graph Automorphism problem we show that GA is many-one equivalent to its right set and therefore the above results also hold for GA. The reductions between the sets we are dealing with and their left or right sets bring insight in the relationship between both problems. GI is $\leq_m^P$-equivalent to *Left*-GA and GA is $\leq_m^P$-equivalent to *Right*-GA. From this follows that the problem of obtaining the lexicographically first automorphism in a graph is Turing equivalent to GA while, curiously, obtaining the lexicographically last automorphism is Turing equivalent to GI and therefore it seems to be a harder problem. From our constructions we obtain also for the first time a polynomial-time many-one reduction from GA to GI. This improves the existing reductions since until now it was only known that GA is disjunctive truth-table reducible to GI [Ma 79]. Also, our results solve an open problem about reducibilities from GI to sparse sets proposed by Mahaney [Mah 82, JoYo 88].

In [BaFoNiWi 91] it is proved that if EXPTIME has polynomial size circuits, then EXPTIME has publishable proofs, (EXPTIME $\subseteq$ MA). It can be immediately observed that the argument used in this result works in fact for every set A with an interactive proof protocol in which the prover is only as powerful as the set A being decided; we show that if a set A and its complement $\overline{A}$ have both an interactive proof system in which the prover is only as strong as A, and $A \in$ P/poly, then A is low for MA. The result can be used together with the fact that there are interactive protocols for GI and $\overline{\text{GI}}$ [GoMiWi 85] in which the prover only needs to be able to solve GI, and it follows that if GI has polynomial size circuits then GI is low for MA. This contrasts with the best known unconditional upper bound for GI, which is NP ∩ co-AM [BaMo 88, Schö 88], and with the best analogous result for an NP-complete problem like SAT: if SAT has polynomial size circuits, then it is low for the second level of the polynomial-time hierarchy, Σ_2^p, [KaLi 80]. We present also interactive proof protocols for the Graph Automorphism problem and its complement in which it suffices that the prover can decide GA. From this follows that GA ∈ P/poly implies that GA is low for MA.

Finally in Section 5 we study one last structural property of GI and GA considering whether the functions that can be computed in polynomial time with parallel queries to GI or GA could also be computed in polynomial time by making logarithmically many queries to an arbitrary oracle, i.e., whether $FP^{\|GI} \subseteq FP^{A[\log]}$ for an arbitrary set A. We show that if this is the case then GI belongs to the probabilistic class R. It is known that this result holds also for an NP-complete set like SAT, [Be 88, Tod 91, Se 91], although the technique used in our case is quite different since in our proof we use once more the interactive protocol for $\overline{GI}$. For the case of GA we show that if $FP^{\|GA} \subseteq FP^{A[\log]}$ for some set A, then GA ∈ P. The proof is based on the observation that an automorphism can be constructed by asking parallel queries to GA. This shows again a difference between the Graph Isomorphism and Graph Automorphism problems.

2 PRELIMINARIES

We assume that the reader is familiar with complexity classes like P, NP, R, BPP and with the nonuniform classes P/poly and NP/poly (see [BaDiGa 87]).

Interactive proof systems were introduced simultaneously in [GoMiRa 85] and [Ba 85] (called Arthur-Merlin games in this last reference). Roughly speaking, in an interactive proof system a prover communicates with a probabilistic verifier that works in polynomial probabilistic time, trying to convince him that an input string belongs to a certain language. Usually it is considered that the prover has unlimited computational power. It is well known that this is not strictly necessary. We will consider the prover to be a function that receives as input the history of the communication with the verifier and produces the next message. In this way we can study probabilistic protocols in which the prover belongs to a certain complexity class (of functions). MA and AM are the classes of languages that have an interactive proof system with only one communication round between prover and verifier. In AM the verifier sends first a message to the prover and in MA is the other way around. These classes can also be defined using quantifiers in the following way: a language L is in the class MA if and only if there is a deterministic Turing machine M and a polynomial p bounding its running time such that for every string $x \in \Sigma^*$, if x is in L, then:

$$(\exists y : |y| \leq p(|x|)) [\text{Prob}\{z : \langle x, y, z \rangle \in L(M)\} \geq \frac{3}{4}],$$

and if x is not in L, then:

$$(\forall y : |y| \leq p(|x|)) [\text{Prob}\{z : \langle x, y, z \rangle \in L(M)\} \leq \frac{1}{4}].$$

where $z \in \Sigma^{p(|x|)}$ is picked uniformly at random. L belongs to AM if and only if there is a deterministic Turing machine M with running time bounded by a polynomial p such that for every string $x \in \Sigma^*$:

$$\text{Prob}\{y \mid \exists z, |z| \leq p(|x|), [\langle x, y, z \rangle \in L(M) \Leftrightarrow x \in L]\} \geq \frac{3}{4}$$

where $y \in \Sigma^{p(|x|)}$ is picked uniformly at random. The error probability ($\frac{1}{4}$) can be made exponentially small. The following containments between probabilistic complexity classes are known: R $\subseteq$ BPP $\subseteq$ MA $\subseteq$ AM.

We will use different types of polynomial-time reducibilities like bounded truth-table (btt), truth-table (tt) or Turing reducibility. For a formal definition we refer the reader to [BaDiGa 87]. Given an oracle Turing machine M and a set A, $L(M, A)$ denotes the language accepted by M using the oracle A. $FP^{\|A}$ is the class of functions computed in polynomial time by a deterministic machine that can make parallel queries to set A, and $FP^{A[\log]}$ the class of functions computed in polynomial time by asking $O(\log n)$ many times to the oracle A on inputs of length n. For a relativizable complexity class $\mathcal{C}$, $\mathcal{C}^A$ is the class of languages that can be computed by an oracle machine of type $\mathcal{C}$ using oracle A. Lowness is a complexity measure that was first used in the polynomial-time setting in [Schö 83]. A set A is low for a complexity class $\mathcal{C}$ if $\mathcal{C}^A = \mathcal{C}$, i.e., if A does not give $\mathcal{C}$ any additional power when used as oracle.

We denote by S_n the set of permutations on $\{1,\ldots,n\}$. The tuple $(a_1,\ldots,a_n)$ represents the permutation $\varphi \in S_n$ such that $\varphi(i) = a_i$ for each i. The identity permutation is represented by id. We define a natural ordering in S_n and write $(a_1,\ldots,a_n) < (b_1,\ldots,b_n)$ if $a_1 a_2 \ldots a_n$ (considered now as a string over the alphabet $\{1,\ldots,n\}$) is lexicographically smaller than $b_1 b_2 \ldots b_n$. For a permutation $\varphi \in S_n$, we denote by $S_n^{<\varphi}$ ($S_n^{>\varphi}$) the set of permutations in S_n which are smaller (larger) than φ under the above ordering. The notation $S_n[a \to b]$ represents the set of permutations which map a to b. This representation can be extended to sets of permutations in which more than one value is fixed, and $S_n[a_1,\ldots,a_k \to b_1,\ldots,b_k]$ is the set of permutations which map a_i to b_i for $i = 1 \ldots k$. It is not hard to observe that for any permutation $\varphi \in S_n$ the set of permutations which are smaller than φ, $S_n^{<\varphi}$, can be expressed as the union of at most a quadratic number of subsets of permutations in which some values are fixed ($S_n^{>\varphi}$ can be expressed in a similar way). So, we define $S_n^{<\varphi}$ as

$$\bigcup_{i=1,\ldots,n-1} \bigcup_{\substack{j=1,\ldots,\varphi(i)-1 \\ j \neq \varphi(1),\ldots,\varphi(i-1)}} S_n[1,\ldots,i \to \varphi(1),\ldots,\varphi(i-1),j].$$

We only consider simple undirected graphs $G = (V,E)$ without self-loops. We denote vertices in V by natural numbers, i.e., if G has n vertices then $V = \{1,\ldots,n\}$. With this convention, mappings from the set of vertices of G onto the set of vertices of another graph G' with the same number of vertices can be interpreted as a permutation in S_n. Let $G = (V,E)$ be a graph with $|V| = n$. An automorphism of G is a permutation $\varphi \in S_n$ that preserves adjacency in G, i.e., for every pair of vertices i,j, $(i,j) \in E$ if and only if $(\varphi(i),\varphi(j)) \in E$. The set of automorphisms of G, $Aut(G)$, is a subgroup of S_n.

Two graphs $G_1 = (V_1, E_1)$, $G_2 = (V_2, E_2)$ are isomorphic ($G_1 \cong G_2$) if there is a bijection φ between V_1 and V_2 such that for every pair of vertices $i,j \in V_1$, $(i,j) \in E_1$ if and only if $(\varphi(i),\varphi(j)) \in E_2$. The sets GI and GA are defined as

$$\text{GI} = \{(G_1, G_2) : G_1 \text{ and } G_2 \text{ are isomorphic}\}$$

and

$$\text{GA} = \{G : G \text{ has a nontrivial automorphism}\}.$$

It will be convenient to distinguish certain graph vertices by labels. The notation $G_{[v_1,\ldots,v_k]}$ stands for a copy of G with unique distinct labels attached to the vertices $v_1, \ldots, v_k$. This can be done for example by adding paths of appropriate length to the vertices v_i. If we have two graphs of n vertices with k of them labeled, $G_{[u_1,\ldots,u_k]}$ and $G'_{[v_1,\ldots,v_k]}$, we assume that the labels in vertex u_i from G and in vertex v_i from G' are the same.

We define logical combining functions AND_2 and OR_2 for arbitrary sets, as it is done by Kadin [Kad 88].

Definition 2.1 *A set A has AND_2 (OR_2) functions if there exists a polynomial-time computable function $f(\cdot, \cdot)$ such that for all $x, y \in \Sigma^*$, $[x \in A \text{ and } (or) \; y \in A] \Longleftrightarrow f_A(x, y) \in A$.*

We define the union of graphs in the standard way. Let $G_1 = (V_1, E_1)$ and $G_2 = (V_2, E_2)$ be two graphs whose vertex sets V_1 and V_2 are disjoint. Then, the union of G_1 and G_2 is the graph $G_1 \cup G_2 = (V_1 \cup V_2, E_1 \cup E_2)$.

The following fact is easy to prove.

Observation 2.2 *Given the connected graphs G_1, G_2, H_1, H_2, it holds that $G_1 \cup G_2 \cong H_1 \cup H_2$ if and only if*

$$(G_1 \cong H_2 \text{ and } G_2 \cong H_1) \text{ or } (G_1 \cong H_1 \text{ and } G_2 \cong H_2).$$

We define now a function which encodes two given graphs into a single one. An intuitive representation of this function can be seen in the figure.

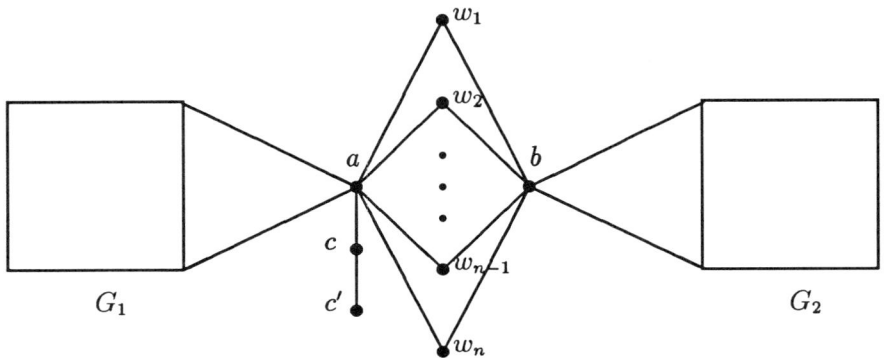

Encoding of graphs G_1 and G_2 in $\mathbf{cod}(G_1, G_2)$.

Definition 2.3 *Let $G_1 = (V_1, E_1)$ and $G_2 = (V_2, E_2)$ be two graphs, and set n to $\max\{|V_1|, |V_2|\}$. We define $\mathbf{cod}$ as the function $\mathbf{cod}(G_1, G_2) = (V, E)$, where*

$$V = V_1 \cup V_2 \cup \{a, b, c, c'\} \cup \{w_1, \ldots, w_n\}$$

and

$$E = E_1 \cup E_2 \cup \{(a, x) : x \in V_1\} \cup \{(b, x) : x \in V_2\}$$
$$\cup \{(a, w_i), (b, w_i) : 1 \leq i \leq n\} \cup \{(a, c), (c, c')\}.$$

From the encoding $\mathbf{cod}(G_1, G_2)$, the original graphs G_1 and G_2 can be recovered in polynomial time. Note that in $\mathbf{cod}(G_1, G_2)$ vertices a and b are uniquely determined since they are the two vertices with highest degree. In order to distinguish a from b, we can look for the unique vertex that is connected neither to a nor to b; this gives us vertex c' that is connected to a through vertex c. Once vertices a and b have been found, the graphs G_1 and G_2 can be easily reconstructed. Function $\mathbf{cod}$ has the following property.

Observation 2.4 *Given the graphs G_1, G_2, H_1, H_2, it holds that*

$$\mathbf{cod}(G_1, H_1) \cong \mathbf{cod}(G_2, H_2) \iff G_1 \cong G_2 \text{ and } H_1 \cong H_2.$$

It is easy to see that Graph Isomorphism has AND_2 functions. Define function $AND_2^{GI}((G_1, G_2), (H_1, H_2)) = (cod(G_1, H_1), cod(G_2, H_2))$. Then, by Observation 2.4, we know that AND_2^{GI} is an AND_2 function for GI.

Graph Isomorphism has also OR_2 functions. This fact was previously obtained by Chang [Ch 89].

Lemma 2.5 GI *has OR_2 functions.*

Proof Define $OR_2^{GI}((G_1, G_2), (H_1, H_2)) = (cod(G_1, H_1) \cup cod(G_2, H_2), cod(G_1, H_2) \cup cod(G_2, H_1))$. This is an OR_2 function for GI since

$OR_2^{GI}((G_1, G_2), (H_1, H_2)) \in GI$

$\iff cod(G_1, H_1) \cup cod(G_2, H_2) \cong cod(G_1, H_2) \cup cod(G_2, H_1)$

$\iff (cod(G_1, H_1) \cong cod(G_2, H_1)$ and $cod(G_2, H_2) \cong cod(G_1, H_2))$, or

$(cod(G_1, H_1) \cong cod(G_1, H_2)$ and $cod(G_2, H_2) \cong cod(G_2, H_1))$

$\iff G_1 \cong G_2$ or $H_1 \cong H_2$.

where we have applied, in first place, the definition of OR_2^{GI} and then, Observations 2.2 and 2.4. □

These functions can be generalized to the case of an arbitrary number of arguments. The following definition has been adapted from [ChKaRo 91].

Definition 2.6 *A set A has OR_ω functions if there exists a polynomial-time computable function f such that $[\exists i \leq n \ x_i \in A] \iff f(x_1, \ldots, x_n) \in A$.*

Definition 3.1 Left-GA $= \{(G, \varphi) : (\exists \sigma : id < \varphi \leq \sigma)[\sigma \in Aut(G)]\}$.

Definition 3.2 Right-GA $= \{(G, \varphi) : (\exists \sigma : id < \sigma \leq \varphi)[\sigma \in Aut(G)]\}$.

In both definitions it is considered that the permutations belong to S_n, where n is the number of vertices of the input graph. We prove that Graph Isomorphism is $\leq_m^P$-equivalent to Left-GA and that Graph Automorphism is $\leq_m^P$-equivalent to Right-GA.

Theorem 3.3 GI is $\leq_m^P$-equivalent to Left-GA.

Proof First we show that GI $\leq_m^P$ Left-GA. Given two graphs with n vertices, G_1 and G_2, consider the graph $G_3 = G_1 \cup G_2$. (W.l.o.g. we can assume that G_1 and G_2 are connected graphs, since if this is not the case we can consider their complementary graphs instead). We can suppose that in G_3 the vertices from G_1 have numbers from 1 to n and the vertices from G_2 range from $n+1$ to $2n$. It is clear that G_1 and G_2 are isomorphic iff there is an automorphism in G_3 mapping the vertices from G_1 to the vertices from G_2, that is, if and only if $(G_3, (n+1, n+2, \ldots, 2n, 1, 2, \ldots, n)) \in$ Left-GA.

Now we prove that Left-GA $\leq_m^P$ GI. Suppose we are given a pair (G, φ) composed of a graph G and a permutation φ. It holds that $(G, \varphi) \in$ Left-GA if there exists a permutation $\sigma \geq \varphi$ such that $\sigma \in Aut(G)$, that is to say, that $(G, \varphi) \in$ Left-GA $\iff (\exists \sigma \geq \varphi)[\sigma(G) = G]$. The reduction function can first test whether $\varphi(G) = G$, and output an image belonging to GI if this is true. If this is false, then

$(G, \varphi) \in$ Left-GA

$\iff (\exists \sigma \in S_n^{>\varphi})[\sigma(G) \neq id \wedge \sigma(G) = G]$

$\iff (\exists \sigma \in \bigcup_{\substack{i=1,\ldots,n-1 \\ j=\varphi(i)+1,\ldots,n \\ j \neq \varphi(1),\ldots,\varphi(i-1)}} S_n[1, \ldots, i \to \varphi(1), \ldots, \varphi(i-1), j]) \, [\sigma(G) = G]$

Lemma 2.7 *If a set A has linear-size OR_2 functions, then it has OR_ω functions.*

Proof Assume that OR_2^A is a linear OR_2 function for A. Call OR_ω^A to the following recursive function defined with a variable number of arguments. For one argument x_1, let $OR_\omega^A(x_1) = x_1$. For two arguments x_1 and x_2, let $OR_\omega^A(x_1, x_2) = OR_2^A(x_1, x_2)$. On n arguments $x_1, \ldots, x_n$, we define $OR_\omega^A(x_1, \ldots, x_n)$ as

$$OR_2^A(OR_\omega^A(x_1, \ldots, x_{\lfloor \frac{n}{2} \rfloor}), OR_\omega^A(x_{\lfloor \frac{n}{2} \rfloor + 1}, \ldots, x_n)).$$

It is easily seen that OR_ω^A, with inputs $x_1, \ldots, x_n$, gives as output a string in A if and only if one of the inputs is in A. Furthermore, in order to check that OR_ω^A can be computed in polynomial time, let $s(m, n)$ be the maximum size of $OR_\omega^A(x_1, \ldots, x_n)$, where $|x_i| \leq m$ for all $i \leq n$. We can observe that the following relations are true for some constant k.

$s(m, 1) \leq m$, and
$s(m, n) \leq k \cdot s(m, \lfloor \frac{n}{2} \rfloor)$, for any $n > 1$.

Solving the recurrence, we get $s(m, n) \leq k^{\log n} m$, which is bounded by $n^{k'} m$ for some constant k'. Then, OR_ω^A is a polynomial-time function and therefore it is an OR_ω function for A. □

Note that a similar lemma can be stated about AND_2 functions. Also, as the OR_2 function for GI used in Lemma 2.5 is linear, we conclude from the previous lemma that GI has OR_ω functions. This fact, together with the known $\leq_d^P$-reduction from GA to GI [Ma 79], leads to a $\leq_m^P$-reduction.

Theorem 2.8 GA *is $\leq_m^P$-reducible to* GI.

3 RESTRICTED tt-REDUCTIONS TO SPARSE SETS

Consider the following definitions of *left* and *right* sets for Graph Automorphism which are defined in the sense of [OgWa 90]. The symbol $<$ refers to the ordering between permutations defined in Section 2.

$$\iff \bigvee_{\substack{i=1,\ldots,n-1 \\ j=\varphi(i)+1,\ldots,n \\ j \neq \varphi(1),\ldots,\varphi(i-1)}} G_{[1,\ldots,i]} \cong G_{[\varphi(1),\ldots,\varphi(i-1),j]}$$

$$\iff G_1 \cong G_2$$

where G_1 and G_2 are the image of the function OR_ω^{GI} applied to the pairs $(G_{[1,\ldots,i]}, G_{[\varphi(1),\ldots,\varphi(i-1),j]})$ for all i and j that satisfy the above conditions. For such i's and j's, the labeling ensures that the graphs $G_{[1,\ldots,i]}$ and $G_{[\varphi(1),\ldots,\varphi(i-1),j]}$ are isomorphic iff G has an automorphism in $S_n[1,\ldots,i \to \varphi(1),\ldots,\varphi(i-1),j]$. □

Theorem 3.4 *GA is $\leq_m^P$-equivalent to Right-GA.*

Proof GA is trivially reduced to *Right*-GA, since for any graph G of n vertices, $G \in$ GA if and only if $(G, \sigma_{\text{last}}) \in$ *Right*-GA, where σ_{last} is the lexicographically largest permutation in S_n.

For the other direction, suppose we are given a pair (G, φ). It holds that $(G, \varphi) \in$ *Right*-GA if there exists a permutation $\sigma \leq \varphi$ in $Aut(G)$, that is to say, that $(G, \varphi) \in$ *Right*-GA $\iff (\exists \sigma \leq \varphi)[\sigma(G) = G]$. The reduction function can first test whether $\varphi(G) = G$, and output an image belonging to GA if this is true. If this is false, then

$$(G, \varphi) \in \textit{Right-GA}$$

$$\iff (\exists \sigma \in S_n^{\leq \varphi})[\sigma(G) = G]$$

$$\iff (\exists \sigma \in \bigcup_{\substack{i=1,\ldots,n-1 \\ j=1,\ldots,\varphi(i)-1 \\ j \neq \varphi(1),\ldots,\varphi(i-1)}} S_n[1,\ldots,i \to$$

$$\varphi(1),\ldots,\varphi(i-1),j]) \; [\sigma(G) = G] \qquad (1)$$

To prove our result we consider some special types of permutations. We say that a set of permutations $S_n[1,\ldots,i \to a_1,\ldots,a_i]$ is *prefix-fixed* if for all $l \leq i$, it holds that $a_l = l$; we say that the set $S_n[1,\ldots,i,j \to a_1,\ldots,a_i,a_j]$, with $i < j$, is *almost-fixed* if it is not prefix-fixed and the smallest element $t \in \{1,\ldots,i,j\}$ such that $t \neq a_t$ is the image of another element (that is to say, $(\exists l \in \{1,\ldots,i,j\})[a_l = \min\{t : t \neq a_t\}]$). For example, the set $S_5[1,2,3,4 \to 1,2,4,3]$ is almost-fixed since element 3 appears as an image.

On input (G, φ), we consider for every set X of permutations from the union in expression 1 three cases depending on whether X is prefix-fixed, almost-fixed, or neither of these types. In each case we associate to X a graph G' in such a way that (a) if there exists a nontrivial automorphism $\sigma \in X$ for G, then $G' \in$ GA, and (b) G' is only in GA if there exists a nontrivial automorphism $\sigma < \varphi$ for G. The theorem follows considering that GA has OR$_\omega$ functions.

Case 1. Assume that X is prefix-fixed. In this case we set G' to $G_{[1,\ldots,i]}$. Clearly there is a nontrivial automorphism for G in X if and only if G' has a nontrivial automorphism.

Case 2. Assume that $X = S_n[1,\ldots,i,j \to a_1,\ldots,a_i,a_j]$, with $i < j$, is almost-fixed. We now set G' to $G_{[1,\ldots,i,j]} \cup G_{[a_1,\ldots,a_i,a_j]}$ and show that (i) if G has an automorphism in X, then G' has a nontrivial automorphism and that (ii) if G' is in GA, then G has a nontrivial automorphism smaller than φ.

First, suppose that G has an automorphism in X. Then, there must be an isomorphism from $G_{[1,\ldots,i,j]}$ to $G_{[a_1,\ldots,a_i,a_j]}$, and this means that there is an automorphism in the union of the two graphs.

For the other direction, assume that $G_{[1,\ldots,i,j]} \cup G_{[a_1,\ldots,a_i,a_j]}$ has an automorphism. We will prove that G has an automorphism smaller than φ. Our assumption implies one of the following three subcases: either there is an automorphism in $G_{[1,\ldots,i,j]}$, or there is an automorphism in $G_{[a_1,\ldots,a_i,a_j]}$, or there exists an isomorphism between these two graphs. In the first subcase, since vertices $1,\ldots,i$ have different unique labels, this automorphism does not move these vertices and then it must be smaller than φ.

In the second subcase, as set X is almost-fixed, the element $m = \min\{t : t \neq a_t\}$ is equal to some a_l, which means that vertex m in graph $G_{[a_1,\ldots,a_i]}$ has a label. Then, an automorphism in $G_{[a_1,\ldots,a_i]}$ cannot move m and this implies that there is an automorphism smaller than φ

in G (note that $m < a_m$, by the definition of m and the fact that we are considering permutations).

In the third subcase (when there is an isomorphism from $G_{[1,...,i]}$ to $G_{[a_1,...,a_i]}$) it is easy to see that there must be an automorphism of G in X since the isomorphism must associate equal labels.

Case 3. Now, assume that $X = S_n[1,\ldots,i \to a_1,\ldots,a_i]$ is neither prefix-fixed nor almost-fixed. Then, the element $m = \min\{t \;:\; t \neq a_t\}$ is not the image of any other element, i.e., there is no $l \leq i$ such that $a_l = m$. In this case, by considering all the possibilities of fixing m, we get a union of almost-fixed sets

$$S_n[1,\ldots,i \to a_1,\ldots,a_i] = \bigcup_{l=i+1,\ldots,n} S_n[1,\ldots,i,l \to a_1,\ldots,a_i,m]$$

This gives us a way of expressing $S_n^{\leq \varphi}$ as a union of almost-fixed sets. Now, we can compute the arguments of the OR_ω function as shown in case 2. □

Observe that *Left*-GA and *Right*-GA can be easily shown to be $\leq_m^P$-equivalent to strict 1-wd self-reducible sets ([OgLo 91]) and to nearly near testable sets ([HeHo 87]), by means of simple bijections between permutations and strings. Also, note that they are $\leq_m^P$-equivalent to *left* and *right* sets, respectively, defined by Ogiwara and Watanabe [OgWa 90]. Therefore, by combining the theorems of this section with Theorem 4.8 of [OgLo 91] (or with the main result in [OgWa 90]) we can state the following corollary.

Corollary 3.5 *If GI (or GA) is $\leq_{btt}^P$-reducible to a sparse set, then it is in P.*

Using the fact that left sets that are conjunctively reducible to sparse sets are in P ([Ar et al 92], [RaRo 92], [Yo 92]) we conclude:

Corollary 3.6 *If GI (or GA) is $\leq_c^P$-reducible to a sparse set, then it is in P.*

Since we tend to think that neither GI nor GA are in P, the above corollaries make us to think that they are probably not reducible to sparse sets under the $\leq_{btt}^{P}$ and $\leq_{c}^{P}$-reducibilities. However, we can derive from Schöning [Schö 88a] that GI and GA are in NP∩(co-NP/poly), and therefore they are $\leq_{T}^{SN}$-reducible to sparse sets. If we consider a more restrictive reducibility like bounded truth-table reductions to sparse sets, then we obtain an interesting consequence from [OgLo 91].

Corollary 3.7 *If GI (or GA) is $\leq_{btt}^{SN}$-reducible to a sparse set, then it is in co-NP.*

We have shown that *Left*-GA is many-one equivalent to GI while *Right*-GA is many-one equivalent to GA. Therefore, the largest automorphism of a graph (in the considered order) can be obtained in FP^{GI} while, curiously, to compute the smallest automorphism it suffices the class FP^{GA}. Another example of a set having a left-set with a different complexity than itself was shown in [KhVa 91] where it is proved that a left set of Planar Graph 4-Coloring is NP-complete, while the problem itself is in P.

Although we did not mention the left and right sets of GI, it is not hard to see using the techniques from this section that both sets are many-one equivalent to Graph Isomorphism.

4 POLYNOMIAL SIZE CIRCUITS

In this section we study the implications of GI and GA having polynomial size circuits. We show that if this is the case then both sets are low for the class MA (the best known bound for GI and GA is NP∩co-AM). This follows from a general theorem that relates polynomial size circuits and sets having interactive proofs.

Theorem 4.1 *Let A be a set such that A and $\overline{A}$ both have an interactive proof protocol in which the prover is in FP^A. If $A \in P/poly$ then A is low for MA.*

Proof Let L be a set in MA^A, we will show that L belongs to MA. By the definition of MA, there is a deterministic oracle Turing machine M and a polynomial p bounding the running time of M such that for every string $x \in \Sigma^\star$, if x is in L, then:

$$(\exists y \; : \; |y| = p(|x|)) \; [\text{Prob}\{\langle x, y, z \rangle \in L(M, A)\} \geq 1 - 2^{-3}],$$

and if x is not in L, then:

$$(\forall y \; : \; |y| = p(|x|)) \; [\text{Prob}\{\langle x, y, z \rangle \in L(M, A)\} \leq 2^{-3}],$$

where $z \in \Sigma^{p(|x|)}$ is randomly chosen under uniform distribution.

Define function l as $l(n) = \lceil \log(p(n)) \rceil + 4$. A and $\overline{A}$ have interactive proof protocols and $A \in \text{P/poly}$. Using the same argument as in [BaFoNiWi 91], the prover can give to the verifier the polynomial size circuit w for instances of A of the desired length, and the verifier simulates the protocol querying the circuit w instead of asking the prover. We can amplify the probability of acceptance of the protocol and it can be concluded that there are two deterministic Turing machines M_1 and M_2 with running times bounded by a polynomial q such that for each length n there exists a string w, $|w| = q(n)$, (a circuit deciding correctly strings of length $\leq p(n)$ such that for any x, $|x| \leq p(n)$,

$(x \in A \Longrightarrow \text{Prob}\{\langle x, w, r \rangle \in L(M_1)\} \geq 1 - 2^{-l(n)})$, and
$(x \in \overline{A} \Longrightarrow \text{Prob}\{\langle x, w, r \rangle \in L(M_2)\} \geq 1 - 2^{-l(n)})$

and also, for each length n, for any string w, $|w| = q(n)$, and any x, $|x| \leq p(n)$,

$(x \notin A \Longrightarrow \text{Prob}\{\langle x, w, r \rangle \in L(M_1)\} \leq 2^{-l(n)})$, and
$(x \notin \overline{A} \Longrightarrow \text{Prob}\{\langle x, w, r \rangle \in L(M_2)\} \leq 2^{-l(n)})$

where $r \in \Sigma^{q(n)}$ is randomly chosen under uniform distribution.

Based on M_1 and M_2 we can define now a probabilistic Turing machine M' that receives as input a string x and a circuit w. If w is a correct circuit for A up to strings of length $\leq p(|x|)$, with high probability M' will simulate correctly machine M with oracle A. On the other hand, if w does not decide A correctly for some queried string, M' will detect it with high probability. M' is described by the next algorithm:

input $\langle x, y, z, w \rangle$ $\{|x| = n, |y| = |z| = p(n), |w| = q(n)\}$

Non-Uniform Complexity of the Graph Isomorphism Problem

```
        simulate M(x, y, z)
        if M queries a to A then {|a| ≤ p(n)}
            guess r ∈ Σ^q(n)
            simulate M₁(a, w, r) and M₂(a, w, r)
            if M₁ accepts and M₂ rejects then
                continue with answer "yes"
            else if M₁ rejects and M₂ accepts then
                continue with answer "no"
            else
                REJECT
            end if  end if
        end if
```

If w is the correct circuit for A and $\langle x, y, z \rangle \in L(M, A)$ the probability that M' accepts $\langle x, y, z, w \rangle$ is greater than or equal to the probability that all the queries (at most p) from M to A are answered correctly in the simulation, and therefore for each length n there exists a string w, $|w| = q(n)$, such that for any x, $|x| = n$,

$$\langle x, y, z \rangle \in L(M, A) \implies \text{Prob}\{M'(x, y, z, w) \text{ accepts}\}$$
$$\geq (1 - 2^{-l(n)+1})^{p(n)} \geq (1 - 1/8p(n))^{p(n)} \geq 1 - 2^{-3}.$$

If $\langle x, y, z \rangle \notin L(M, A)$ then for every circuit w, $M'(x, y, z, w)$ accepts only if w can lead M_1 and M_2 to decide incorrectly in some oracle query, and this can only happen with small probability. For each length n for any string w, $|w| = q(n)$, and any x, $|x| = n$,

$$\langle x, y, z \rangle \notin L(M, A) \implies \text{Prob}\{M'(x, y, z, w) \text{ accepts}\} \leq 2^{-3}.$$

In both cases the probability is taken over the random strings $r \in \Sigma^{p(n)q(n)}$ needed by machine M', under a uniform distribution.

Going back to the set L we have that for a suitable polynomial s, for every string x in Σ^*,

$$x \in L \implies (\exists w : |w| = s(|x|))(\exists y : |y| = s(|x|))$$
$$[\text{Prob}\{M' \text{ following random choices } r \text{ accepts } \langle x, y, z, w \rangle\}$$

$$\geq (1-2^{-3})^2 > \tfrac{3}{4}], \text{ and}$$

$$x \notin L \Longrightarrow (\forall w \,:\, |w| = s(|x|))(\forall y \,:\, |y| = s(|x|))$$
$$[\text{Prob}\{M' \text{ following random choices } r \text{ accepts } \langle x, y, z, w \rangle\}$$
$$\leq 2^{-3} + (1-2^{-3})2^{-3} < \tfrac{1}{4}].$$

where $z, r \in \Sigma^{s(|x|)}$ are randomly chosen under uniform distribution. This is because the probability that M' works correctly for a string x is bounded by the probability of choosing "good" sequences of random choices z and r. This proves that L is in MA and therefore A is low for MA. □

The above theorem works independently of whether the interactive proof protocol for A has one or more than one prover as long as all the provers are only as powerful as A.

The next result follows by the previous one using the two prover protocol for EXPTIME from [BaFoLu 91] and the fact that EXPTIME is closed under complements. This corollary (stated in a different way) was obtained in [BaFoNiWi 91].

Corollary 4.2 *If* EXPTIME $\subseteq$ P/poly, *then* EXPTIME *is low for* MA.

We obtain a second corollary considering the protocol for PSPACE given in [Sh 90], since this class is also closed under complements. This improves a result from [KaLi 80].

Corollary 4.3 *If* PSPACE $\subseteq$ P/poly, *then* PSPACE *is low for* MA.

We apply now the theorem to the Graph Isomorphism and Automorphism problems. In the interactive protocol for $\overline{\text{GI}}$ from [GoMiWi 85] the prover only needs to decide GI. Also using the self-reducibility properties of GI [Schn 76] it is straightforward to design an interactive protocol for GI in which the verifier sends two graphs to the prover and the prover (in FP^{GI}) sends back an isomorphism.

Corollary 4.4 *If* GI $\in$ P/poly, *then* GI *is low for* MA.

This result contrasts with the best analogous known result for an NP-complete set: if SAT $\subseteq$ P/poly, then SAT is low for Σ_2^p.

For the Graph Automorphism problem we obtain the same results as for GI. For this we first show that GA and $\overline{\text{GA}}$ have both interactive proof protocols.

Lemma 4.5 *There is a two round protocol for $\overline{\text{GA}}$ in which the prover is in* FP^{GA}.

Proof On input a graph $G = (V, E)$, the verifier chooses randomly a permutation $\varphi \in S_{|V|}$ and sends the pair $(G, \varphi(G))$ to the prover. The prover sends back a permutation and the verifier accepts iff this permutation is equal to φ.

If $G \in \overline{\text{GA}}$, the prover can find the unique nontrivial automorphism in the graph $G \cup \varphi(G)$ and from it obtain permutation φ. In this case the verifier will always accept. On the other hand, if G has at least one nontrivial automorphism π, then the prover has at least two possible answers since $\pi(G) = \varphi(\pi(G))$. The probability of sending the verifier the correct one is therefore smaller than or equal to $\frac{1}{2}$. For this protocol suffices that the prover is in FP^{GA} since as we will show in Lemmas 5.2 and 5.3, an automorphism can be found querying the set GA. □

GA has also a straightforward protocol in which the verifier sends a graph to the prover and accepts only if he receives back a nontrivial automorphism in the graph. Again by Lemmas 5.2 and 5.3 the prover only needs to be in FP^{GA} in order to find an automorphism. ¿From this and from Theorem 4.1, it follows,

Corollary 4.6 *If* GA $\in$ P/poly, *then* GA *is low for* MA.

5 FUNCTIONS QUERYING GI AND GA

It is known that the class of sets polynomial-time truth-table reducible to NP coincide with the sets that can be computed in polynomial time with at most a logarithmic number of queries to an NP-complete oracle ($\text{P}^{\|\text{SAT}} = \text{P}^{\text{SAT}[\log]}$), [He 87], [BuHa 88], [Wa 87]. However for the case of function classes this fact does not seem to be true. If $FP^{\|SAT} = FP^{SAT[\log]}$, then NP=R and UP=P [Tod 91, Se 91] (also implicit in [Be 88]). These results hold also with the weaker hypothesis that $FP^{\|SAT} = FP^{A[\log]}$ for some set A. The topic is closely related to the concept of polynomial-time enumerators [CaHe 89]; a function f has a polynomial-time enumerator if there is a polynomial-time machine M (the enumerator) which on input x produces a list of values one of which is $f(x)$. The class of functions with polynomial-time enumerators coincides with the functions that belong to $FP^{A[\log]}$ for some set A.

In this section we obtain analogous results as the ones mentioned above for SAT, for the case of GI and GA. We first show that if $FP^{\|GI} = FP^{A[\log]}$ for some set A, then GI $\in$ R. In spite of the parallelism of this result and the one for SAT the technique used in the proof is quite different since the results in [Be 88, Tod 91, Se 91] make use of the randomized reduction introduced in [VaVa 86], and we exploit once more the interactive proof protocol for $\overline{GI}$ from [GoMiWi 85].

Theorem 5.1 *If $FP^{\|GI} \subseteq FP^{A[\log]}$ for an arbitrary set A, then GI $\in$ R.*

Proof By the hypothesis of the theorem we can suppose that for every function $f \in FP^{\|GI}$ there is a polynomial p and an enumerator that on input x produces a list of at most $p(|x|)$ values and $f(x)$ is one of the values in this list. Let the function l be defined by $l(n) = \lceil \log(p(n)) \rceil + 2$. We give a randomized algorithm for $\overline{GI}$. It is based on the same argument as the interactive proof protocol from [GoMiWi 85], substituting the prover by the enumerator.

input graphs (G_0, G_1), with n vertices
guess $w = w_1 w_2 \ldots w_{l(n)} \in \{0,1\}^{l(n)}$
guess $\Pi = (\pi_1, \pi_2 \ldots \pi_{l(n)})$, with $\pi_i \in S_n$
let $L(w, \Pi) = (\langle G_0, \pi_1(G_{w_1}) \rangle, \ldots, \langle G_0, \pi_{l(n)}(G_{w_{l(n)}}) \rangle)$
let $f(L(w, \Pi)) =$
    $\chi_{GI}(\langle G_0, \pi_1(G_{w_1}) \rangle) \ldots \chi_{GI}(\langle G_0, \pi_{l(n)}(G_{w_{l(n)}}) \rangle)$
run the enumerator for f on $L(w, \Pi)$ and obtain the list
    $u_1, \ldots u_{p(n)}$ of values for $f(L(w, \Pi))$
if $w = u_i$ for some i **then**

```
        ACCEPT
else
        REJECT
end if
```

If G_1 and G_2 are nonisomorphic the enumerator will produce the correct string w and the algorithm will always accept.

We study now what happens if the graphs are isomorphic. Given two strings $w, w' \in \{0,1\}^{l(n)}$ and a sequence of $l(n)$ permutations Π, we will say that w' can generate the list $L(w, \Pi)$ if for some sequence of permutations Π' it holds $L(w, \Pi) = L(w', \Pi')$. Observe that if G_0 and G_1 are isomorphic, then any list $L(w, \Pi)$ can be generated by all $2^{l(n)}$ words w' with $|Aut(G_0)|^{l(n)}$ different sequences of permutations each.

The number of possible computation paths the algorithm can choose is $2^{l(n)}(n!)^{l(n)}$ (the possible lists of strings times the possible lists of permutations), and if the graphs are isomorphic, at most $p(n)(n!)^{l(n)}$ of these paths accept. This is because there are $(\frac{n!}{|Aut(G_0)|})^{l(n)}$ different lists $L(w, \Pi)$ (each one is produced $|Aut(G_0)|^{l(n)}$ times by the algorithm) and the enumerator on $L(w, \Pi)$ produces at most $p(n)$ strings w which can generate the list. (As we have explained, all $2^{l(n)}$ words w can generate $L(w, \Pi)$ with $|Aut(G_0)|^{l(n)}$ different sequences of permutations).

The ratio between accepting and possible computations in the algorithm is $\frac{p(n)}{2^{l(n)}} \leq 2^{-2}$. ¿From this acceptance mechanism follows that $\overline{GI} \in$ co-R and therefore GI $\in$ R. □

Using the interactive protocol from Section 3 we could prove this result also for the Graph Automorphism problem. However we can obtain stronger consequences in this case using particular properties of GA. We will show that it is possible to obtain a nontrivial automorphism making only parallel queries to GA. This is proved in two steps, first we show that if a graph has a unique nontrivial automorphism then it can be obtained with parallel queries to GA (it is known that this is also true for any NP-complete problem).

In the second step we show that every graph G can be decomposed into a polynomial number of graphs and if G has a nontrivial automorphism, then at least one of the new graphs has a unique nontrivial automorphism from which it is possible to reconstruct an automorphism for G. This is done based on a construction from [Ma 79] that was also used in [KöScTo 92].

Lemma 5.2 *There is a function $f \in \mathrm{FP}^{\|\mathrm{GA}}$ that receiving as input a graph G, if G has a unique nontrivial automorphism φ, then $f(G) = \varphi$.*

Proof If G has a unique nontrivial automorphism φ, then φ is its own inverse in the automorphism group. Given G, the algorithm for f constructs for each vertex i the graph $G_{[i]}$, and for every pair of vertices i,j $(i \neq j)$ the graph $G_{[i]} \cup G_{[j]}$, and asks the oracle GA (all queries in parallel) which of these graphs have a nontrivial automorphism. The unique non-trivial automorphism maps a vertex i to a different one j if and only if $G_{[i]} \notin \mathrm{GA}$, $G_{[j]} \notin \mathrm{GA}$, and $G_{[i]} \cup G_{[j]} \in \mathrm{GA}$. ¿From the information about all the pairs i,j the unique automorphism can be obtained. □

Lemma 5.3 *Let G be a graph with n vertices. There is a polynomial-time algorithm that on input G produces a list of graphs $G_1,\ldots,G_{p(n)}$ such that if G has at least one nontrivial automorphism, then at least one of the graphs G_i in the list has a unique nontrivial automorphism from which an automorphism for G can be obtained.*

Proof Given $G = (V,E)$ with $|V| = n$, using an argument from [Ma 79], (see also [KöScTo 92]) we can obtain in polynomial time a list of n graphs $G^{(0)},\ldots,G^{(n-1)}$ (a tower of pointwise stabilizers in the automorphism group) such that if G has a nontrivial automorphism, then there is a graph $G^{(i-1)}$ from the list satisfying:

i) $G^{(i-1)}$ has an automorphism φ which maps vertex i to some vertex j, $j > i$ and φ is also an automorphism in G.

ii) For every $j > i$ there is at most one automorphism in $G^{(i-1)}$ mapping i to j.

iii) In $G^{(i-1)}$ there is no nontrivial automorphism mapping i to itself.

Considering these properties, for each graph $G^{(i-1)}$ in the list and for each j, $j > i$ we construct the graph $G^{(i-1)}_{[i]} \cup G^{(i-1)}_{[j]}$, i.e. a graph formed by the union of two copies of $G^{(i-1)}$ one of which has vertex i with a special label and the other copy has vertex j labeled in the same way. Suppose that G has a nontrivial automorphism and $G^{(i-1)}$ is the graph satisfying i), ii) and iii) and therefore for some vertex j there is a unique automorphism φ in $G^{(i-1)}$ mapping i to j, which is also an automorphism in G. It follows that the pair of graphs $G^{(i-1)}_{[i]}$ and $G^{(i-1)}_{[j]}$ are isomorphic and permutation φ is the unique isomorphism between them. Therefore neither $G^{(i-1)}_{[i]}$ nor $G^{(i-1)}_{[j]}$ can have a nontrivial automorphism and the union of both graphs $G^{(i-1)}_{[i]} \cup G^{(i-1)}_{[j]}$, has a unique nontrivial automorphism from which φ can be easily obtained. □

The desired result follows immediately from the above two lemmas.

Corollary 5.4 *If* $\mathrm{FP}^{\|\mathrm{GA}} \subseteq \mathrm{FP}^{A[\log]}$ *for some set A, then* $\mathrm{GA} \in \mathrm{P}$.

Proof ¿From Lemmas 5.2 and 5.3 it follows that there is a function $f \in \mathrm{FP}^{\|\mathrm{GA}}$ that receiving as input a graph G produces a nontrivial automorphism for G if one exists. By the hypothesis, there is a polynomial-time enumerator for f. A deterministic polynomial-time algorithm for GA works as follows: on input G it runs the enumerator for f on G and accepts iff at least one of the values produced by the enumerator is an automorphism for G. □

Acknowledgments We would like to thank V. Arvind, J. Balcázar, R. Beigel, L. Fortnow, B. Gasarch, L. Hemachandra, and J. Köbler for pointing out some errors and suggesting improvements in a previous version of this paper.

References

[Ar et al 92] V. ARVIND, Y. HAN, L. HEMACHANDRA, J. KÖBLER, A. LOZANO, M.MUNDHENK, M. OGIWARA, U SCHÖNING, R. SILVESTRI, AND T. THIERAUF, Reductions to sets of low information content. To appear in *Proceedings of the 19th ICALP* (1992).

[Ba 81] L. BABAI, Moderately exponential bound for Graph Isomorphism. In *Proceedings Conf. on Fundations of Comp. Theory* Springer-Verlag *Lecture Notes in Computer Science 117*, (1981) 34–50.

[Ba 85] L. BABAI, Trading group theory for randomness. In *Proc. 17th ACM Symp. on Theory of Computing*, (1985) 421–429.

[BaMo 88] L. BABAI AND S. MORAN, Arthur-Merlin games: A randomized proof system, and a hierarchy of complexity classes. In *Journal of Computer and System Sciences 36*, (1988) 254–276.

[BaFoLu 91] L. BABAI, L. FORTNOW AND L. LUND, Nondeterministic exponential time has two prover interactive protocols. In *Computational Complexity 1*, (1991) 3–40.

[BaFoNiWi 91] L. BABAI, L. FORTNOW, N. NISSAN, AND A. WIGDERSON, BPP has subexponential time simulations unless EXPTIME has publishable proofs. In *Proceedings of the 6th Structure in Complexity Theory Conference*, (1991) 213–220.

[Be 88] R. BEIGEL, NP-hard sets are p-superterse unless R=NP. Tech. report 88-04, John Hopkins University (1988).

[BaDiGa 87] J.L. BALCÁZAR, J. DIAZ, J. GABARRÓ, *Structural Complexity I*. Springer Verlag, (1987).

[BooKo 88] R. V. BOOK AND K. KO, On sets truth-table reducible to sparse sets. In *SIAM Journal of Computing 17*, (1988) 903–919.

[BoHåZa 87] R. BOPPANA, J. HÅSTAD, AND S. ZACHOS, Does co-NP have short interactive proofs? In *Information Processing Letters 25*, (1987) 127–132.

[BuHa 88] S. BUSS AND L. HAY, On truth-table reducibility to SAT and the difference hierarchy over NP. In *Proceedings of the 3rd Structure in Complexity Theory Conference*, (1988) 224–233.

[CaHe 89] J. Y. CAI AND L. HEMACHANDRA, Enumerative counting is hard. In *Information and Computation 82*, (1989) 34–44.

[Ch 89] R. CHANG, On the structure of bounded queries to arbitrary NP sets. In *Proceedings of the 4th Structure in Complexity Theory Conference*, (1989) 250–258.

[ChKaRo 91] R. CHANG, J. KADIN, AND P. ROHATGI, Connections between the complexity of unique satisfiability and the threshold behavior of randomized reductions. In *Proceedings of the 6th Annual Conference on Structure in Complexity Theory* (IEEE, 1991) 255–269.

[GoMiWi 85] O. GOLDREICH, S. MICALI, AND A. WIGDERSON, Proofs that yield nothing but their validity and a methodology of cryptographic protocol design. In *Proceedings of the 27th Symposium on Foundations of Computer Science*, (1986) 174–187.

[GoMiRa 85] S. GOLDWASSER, S. MICALI, AND C. RACKOFF, The knowledge complexity of interactive proofs. In *Proceedings of the 17th ACM Symposium on the Theory of Computing*, (1985) 291–304.

[He 87] L. HEMACHANDRA, Counting in structural complexity theory, PhD. Thesis, Cornell University (1987).

[HeHo 87] L. HEMACHANDRA AND A. HOENE, On sets with efficient implicit membership tests. In *SIAM Journal of Computing 20(6)*, (1991) 1148–1156.

[Hof 82] C. HOFFMANN, *Group-Theoretic Algorithms and Graph Isomorphism*, Springer-Verlag Lecture Notes in Computer Science 136, (1982).

[HoTa 73] J. HOPCROFT AND R. TARJAN, A $v\log(v)$ algorithm for isomorphism of triconnected planar graphs. In *Journal of Computer and Systems Sciences 7*, (1973) 323–331.

[JoYo 88] D. JOSEPH AND P. YOUNG, Self-reducibility: the effects of structure on complexity. In *Bulletin of the EATCS 36*, (1988) 66–84.

[Kad 88] J. KADIN, Restricted Turing reducibilities and the structure of the Polynomial-Time Hierarchy. PhD thesis, Cornell University, February 1988.

[KaLi 80] R. M. KARP AND R. J. LIPTON, Some connections between nonuniform and uniform complexity classes. *Proceedings of the 12th Annual Symposium on Theory of Computing*, (1980) 302–309.

[KhVa 91] S. KHULLER AND V. VAZIRANI, Planar graph coloring in not self-reducible, assuming P$\neq$NP. In Theoretical Computer Science 88, (1991) 183–189.

[KöScTo 92] J. KÖBLER, U. SCHÖNING, AND J. TORÁN, GI is low for PP. *Proceedings 9th STACS*, (1992) 401–415.

[Luk 82] E. LUKS, Isomorphism of graphs of bounded valence can be tested in polynomial time. In *Journal of Computer and System Sciences 25*, (1982) 42–65.

[Mah 82] S. MAHANEY, Sparse sets and reducibilities. In *Studies in Complexity Theory*, edited by R. Book, (1982) 63–118.

[Ma 79] R. MATHON, A note on the graph isomorphism counting problem. In *Inform. Process. Lett. 8*, (1979) 131–132.

[OgLo 91] M. OGIWARA AND A. LOZANO, On one query self-reducible sets. In *Proceedings of 6th Annual Conference on Structure in Complexity Theory*, (1991) 139–151.

[OgWa 90] M. OGIWARA AND O. WATANABE, On polynomial-time bounded truth-table reducibility of NP sets to sparse sets. In *SIAM Journal of Computing 20*, (1991) 471–483.

[Pi 79] N. PIPPENGER, On simultaneous resource bounds. In *Proceedings of the 20th Symposium on Foundations of Computer Science*, (1979) 307–311.

[RaRo 92] D. RANJAN AND P. ROHATGI, On randomized reductions to sparse sets. In *Proceedings of the 7th Structure in Complexity Theory Conference*, (1992) 239–242.

[Schn 76] C. P. SCHNORR, Optimal algorithms for self-reducible problems. In *3rd Int. Colloq. Automata, Languages and Programming, Edinburgh, University Press*, (1976) 322–337.

[Schö 83] U. SCHÖNING, A low and a high hierarchy within NP. In *Journal of Computer and System Sciences 27*, (1983) 14–28.

[Schö 88] U. SCHÖNING, Graph isomorphism is in the low hierarchy. In *Journal of Computer and System Sciences 37*, (1988) 312–323.

[Schö 88a] U. SCHÖNING, Probabilistic complexity classes and lowness. In *Journal of Comp. and System Sciences 39*, (1988) 84–100.

[Se 91] A. SELMAN, Complexity classes for partial functions. In *Bulletin of the EATCS 45*, (1991) 114–130.

[Sh 90] A. SHAMIR, IP=PSPACE. In *Proceedings of the 30th Symposium on Foundations of Computer Science*, (1990) 2–10.

[Tod 91] S. TODA, On polynomial-time truth-table reducibility of intractable sets to P-selective sets. In *Math. Systems Theory 24*, (1991) 69–82.

[VaVa 86] L.G. VALIANT AND V. VAZIRANI, NP is as easy as detecting unique solutions In *Theoretical Computer Science 47*, (1986) 85–93.

[Wa 87] K. WAGNER, Log query classes, Technical report TR-145, Institut für Mathematik, Universität Augsburg (1987).

[Yo 92] P. YOUNG, How reductions to sparse sets collapse the polynomial-time hierarchy: a primer, Technical report 92-03-07, Dep. of Comp. Sci. and Engineering, University of Washington (1992).

Upper and Lower Bounds for Certain Graph-Accessibility-Problems on Bounded Alternating ω-Branching Programs

Christoph Meinel
FB IV - Informatik
Universität Trier
D-5500 Trier, PF 3825

Stephan Waack
Institut für Mathematik
D-1086 Berlin
Mohrenstr. 39

Abstract

Due to a theory of communication within branching programs that has been developed in [MW91] exponential lower bounds on the size of ordinary, nondeterministic, parity and other sorts of branching programs that are bounded alternating can be derived. In the following we investigate the computational complexity of various *graph–accessibility–problems* in terms of such bounded alternating branching programs.

With respect to the algebraic nature of the lower bound technique and the investigated path problems we consider ω–branching programs and ω–graph–accessibility–problems for certain semiring homomorphism $\omega : I\!N \longrightarrow\!\!\!\!\!\rightarrow R$. (The approach unifies investigations of deterministic, nondeterministic, parity and other counting computation modes.) By means of rank computations of communication matrices we prove for various semiring homomorphisms ω that ω-GAP can not be computed by bounded alternating ω-branching programs within polynomial size. In contrast to these lower bounds we show, that the restrictions ω-GAP_{mon} of ω-GAP to graphs for which a monotone enumeration of the vertices is known in advance can be computed within polynomial size.

1 Introduction

The goal of complexity theory is to determine the amount of computational resources needed to perform certain computational tasks. The major shortcoming being the inability to prove lower bounds for general models of computation. By studying certain restricted models of computation, however, sometimes it is possible to prove such lower bounds. Considering the computational model of brancing programs in [KM91] a theory of communication has been developed that allows to derive exponential lower bounds on the size of ordinary, nondeterministic, parity, or other sorts of branching programs that are *bounded alternating*. The concept of bounded alternating branching programs generalizes various restricted types of branching programs investigated in the past (e. g. OBDDs, real–time, or bounded depth oblivious branching programs).

Bounded alternating branching programs are characterized by the property that there are two disjoint subsets $Y, Z \subseteq X$ in the set X of input variables such that, on each computation path, the number of alternations between reading variables from Y and reading variables from Z is bounded by a constant (or more exactly by $(\#X)^{o(1)}$). Due to the algebraical nature of the lower bound techique the unifying concept of ω–*branching programs*, $\omega : I\!N \longrightarrow\!\!\!\!\rightarrow R$ semiring homomorphism, allows to simultaneously prove results about deterministic, nondeterministic, parity, and other sorts of branching programs. The investigation of bounded alternating ω–branching programs generalizes in at least two important aspects the settings of classical communication complexity theory. On the one hand it ignores a considerable portion of the input variables (the set $X - (Y \cup Z)$ of size $O(\#X)$) on which no sort of assumption is posed. On the other hand, it does not preassume the sets Y and Z. Hence, the complexity of bounded alternating branching programs reflects in a certain sense the amount of communication with respect to any disjoint subsets of the input.

In the following we investigate the complexity of various graph–accessibility–problems *GAPs* in terms of bounded alternating ω–branching programs. This seems to be useful and interesting since

- the graph–accessibility–problems provide paradigmatic complete problems for various logarithmic space–bounded complexity classes, and
- ω–branching programs are a favorite computational model for characterizing and investigating logarithmic space–bounded complexity classes

(see e. g. [Mei89]). Relating the different solutions of the underlying path problem to the algebraic structure of certain semirings we are able to unify investigations substantially by considering ω–*graph–accessibility–problems* ω–*GAPs* for some semiring homomorphisms $\omega : I\!N \longrightarrow R$.

In detail, we show (Theorem 7) that ω-*GAP* can *not* be computed by polynomial size bounded alternating ω-branching programs. This result includes the following statements

- The usual *GAP* is not computable by polynomial size bounded alternating nondeterministic branching programs,
- *NO-PATH-GAP* is not computable by polynomial size bounded alternating co-nondeterministic branching programs,
- *ODD-PATH-GAP* is not computable by polynomial size bounded alternating parity branching programs,
- *MOD-p-GAP* is not computable by polynomial size bounded alternating MOD_p-branching programs.

These lower bounds are, due to our theory of bounded alternating ω-branching programs, derived from some exponential lower bounds on the R-ranks of the communication matrices of ω-*GAP* contributing to the investigation of the communication complexity of graphs problems (e. g. [HMT88, Lov89]), too. By the way, the mentioned lower bounds are the first exponential lower bounds on the size of bounded alternating ω-branching programs for natural problems. So far lower bounds could be proved merely for some artificial problems [MW91] or for more restricted types of branching programs [AM86, KMW89, DKMW90].

The lower–bound–results are nicely contrasted by the fact, that ω-*GAP*$_{mon}$, the restriction of ω-*GAP* to graphs where a monotone enumeration of the vertices (with respect to the edge–relation) is known in advance, is computable by bounded alternating ω-branching programs within polynomial size (Proposition 3). Since the ω-GAP_{mon} as well as the ω-$GAPs$ are complete in the corresponding logarithmic space–bounded complexity classes, this fact demonstrates the computational power of bounded alternating ω-branching programs beside of showing differences in the complexity of the two problems.

The paper is structured as follows. In Section 2, we cite the concept of branching programs and define the unifying concept of ω-branching programs. In Section 3, we consider various ω-$GAPs$ and mention some previous results concerning their completeness for certain logarithmic space–bounded complexity classes. In Section 4, we define bounded alternating ω-branching programs and prove that ω-GAP_{mon} can be computed by polynomial size bounded alternating ω-branching programs. In Section 5 we survey the lower bound technique for bounded alternating ω-branching programs develop in [MW91]. Then, in Section 6 we derive some exponential lower bounds on the R-rank of communication matrices of the ω-GAP which, finally, provide exponential lower bounds on the size of bounded alternating ω-branching programs computing ω-GAP.

2 ω-Branching programs

In order to characterize and investigate nonuniform logarithmic space-bounded complexity classes the combinatorial computation model of branching programs has proved to be of great importance (see e. g. [Mei89]).

Let ω, $\omega : \mathbb{N} \twoheadrightarrow R$ be a semiring homomorphism. An ω-*branching program* P, is a directed acyclic graph where each node has outdegree $\leq \#R$. There is a distinguished node v_s, the *source*, which has indegree 0. Exactly two nodes, v_0 and v_1, are of outdegree 0. They are called 0-*sink* and 1-*sink* of P, respectively. Nodes of outdegree $\#R$ are labelled by variables x over R, $x \in X = \{x_1, \ldots, x_n\}$, or remain unlabelled as the remaining nodes of P. Unlabelled nodes are called *nondeterministic nodes*. Each edge starting in a labelled node is labelled by an element r, $r \in R$, such that no two edges have the same label.

Each input $a = (a_1 \ldots a_n) \in R^X$ defines some *computation paths* in P : Starting at the source v_s, a nondeterministic node v of P is connected with one of its successor nodes. If v is labelled by the variable x_i then v is connected with the successor node of v that is reached from v via the edge labelled by a_i. A computation path is said to be *accepting* if it ends up in the 1-sink v_1. The number of accepting computation paths is denoted by $\text{acc}_P(a)$, $\text{acc}_P(a) \in \mathbb{N}$. Now, P is said to *accept* $a \in R^X$ iff $\omega(\text{acc}_P(a)) = 1_R$.

Obviously, ω–branching programs generalize ordinary branching programs which are programs without nondeterministic nodes where each non-sink node is labelled by a Boolean variable. Hence, for each input, there is exactly one computation path which is accepting if it ends up in the 1-sink.

Moreover, ω-branching programs generalize Ω–branching programs introduced in [Mei88], $\Omega \subseteq \mathbb{B}_2$, and MOD_p-branching programs introduced in [DKMW90], $p \in \mathbb{N}$: If one considers the following semiring homomorphisms $\omega_\vee, \omega_\oplus, \omega_p$,

$$\omega_\vee : \mathbb{N} \twoheadrightarrow \mathbb{B} = [\{0,1\}; \vee, 0, \wedge, 1],$$
$$n \mapsto \begin{cases} 1 & \text{iff } n > 0 \\ 0 & \text{otherwise,} \end{cases}$$

$$\omega_\oplus : \mathbb{N} \twoheadrightarrow \mathbb{F}_2 = [\{0,1\}; \oplus, 0, \wedge, 1],$$
$$n \mapsto \begin{cases} 1 & \text{iff } n \text{ is odd} \\ 0 & \text{otherwise,} \end{cases}$$

$$\omega_p : \mathbb{N} \twoheadrightarrow \mathbb{F}_p = [\{0, \ldots, p-1\}; +(\text{mod } p), 0, \cdot(\text{mod } p), 1],$$
$$n \mapsto n \text{ mod } p,$$

for each prime p, $p \in \mathbb{N}$, then it can easily be seen that

- disjunctive $\{\vee\}$-branching programs are $\omega_\vee$-branching programs (they accept if at least one computation path is accepting),

- parity $\{\oplus\}$-branching programs are $\omega_\oplus$-branching programs (they accept if the number of accepting computation paths is odd), and

- MOD_p-branching programs, p prime, are ω_p-branching programs (they accept if the number of accepting computation paths equals 1 modulo p).

Due to this correspondence and the results of [PŽ83, Mei88, DKMW90] we immediately obtain the following characterization of the nonuniform logarithmic space-bounded complexity classes $\mathcal{L} = L/poly$, $\mathcal{NL} = NL/poly$, co-$\mathcal{NL}$ = co-$NL/poly$, $\oplus\mathcal{L} = \oplus L/poly$, MOD_p-$\mathcal{L} = (MOD_p$-$L)/poly$ by means of (sequence of) polynomial size ω-branching programs. Recall, the size of an ω-branching program P, Size(P), is the number of non-sink nodes of P. By $\mathcal{P}_{\omega\text{-}BP}$ we denote the complexity class of all languages A, $A \subseteq R^*$ that can be accepted by sequence of polynomial size ω-branching programs.

Fact 1.

1. $\mathcal{P}_{BP} = \mathcal{L}$ [PŽ83],
2. $\mathcal{P}_{\omega_\vee\text{-}BP} = \mathcal{NL}$ [Mei88],
3. $\neg(\mathcal{P}_{\omega_\vee\text{-}BP}) = $ co-$\mathcal{NL}$ [Mei88],
4. $\mathcal{P}_{\omega_\oplus\text{-}BP} = \mathcal{P}_{\omega_2\text{-}BP} = \oplus\mathcal{L}$ [Mei88], and
5. $\mathcal{P}_{\omega_p\text{-}BP} = MOD_p\text{-}\mathcal{L}$, p prime, [DKMW90]. □

3 ω-Graph accessibility problems

The classical *graph–accessibility–problem* $GAP = \{GAP_N\}_{N \in \mathbb{N}}$ consists of the decision whether there is a path in a given directed (acyclic) n-node graph $G = (V, E)$, $V = \{1, \ldots, n\}$ and $E \subset V \times V$, that leads from node 1 to node n. As usual, let G be given by its adjacency matrix $G = (a_{ij})_{1 \le i,j \le n, i \ne j}$ with

$$a_{ij} = a(i,j) = \begin{cases} 1 & \text{if } (i,j) \in E, \\ 0 & \text{otherwise.} \end{cases}$$

Let $X = \{x_{ij}, 1 \leq i,j \leq n, i \neq j\}$ and $N = n^2 - n$. $GAP_N : \{0,1\}^X \longrightarrow \{0,1\}$ is defined by

$$(x_{ij}) \longrightarrow \begin{cases} 1 & \text{if there is a path in the graph described by } (x_{ij}) \text{ from 1 to } n, \\ 0 & \text{otherwise.} \end{cases}$$

GAP has been extensively studied in the past. For example, it has been shown to be complete for the complexity class NL of nondeterministic logarithmic space-bounded computations (complete via logspace reductions [Sav70], projection translations [Imm87], and via p-projection reductions for nonuniform NL [Mei86]). Soon it could be realized that certain modified GAPs such as

- *NO-PATH-GAP* (a graph is accepted if there is no path from 1 to n),
- *ODD-PATH-GAP* (a graph is accepted if there is an odd number of paths from 1 to n), or
- *MOD-p-GAP*, p prime, (a graph is accepted if the number of paths from 1 to n equals 1 modulo p)

have similar properties with respect to complexity classes like co-NL, $\oplus L$, MOD_p-L defined by logarithmic space-bounded computations under modified acceptance modes. Interestingly, the structural background of all such results is of algebraic nature. Relating the different solutions of the path problem in the modified GAPs under consideration to the algebraic structure of certain semirings we are able to unify considerations substantially.

Let $R = [R; +_R, O_R, \cdot_R, 1_R]$ be a semiring with a *null*, O_R, and a *one*, 1_R, and let ω,

$$\omega : I\!N \longrightarrow\!\!\!\!\!\rightarrow R,$$

be a semiring homomorphism from $I\!N = [I\!N; +, 0, \cdot, 1]$ onto R. We define the ω-*graph-accessibility-problem* ω-$GAP = \{\omega$-$GAP_N\}_{N \in I\!N}, N = n^2 - n$, by

$$\omega\text{-}GAP_N(G) = 1 \quad \text{iff} \quad \omega(\#[1 \stackrel{G}{\to} n]) = 1_R,$$

where $\#[i \stackrel{G}{\to} j]$ denotes the number of (directed) paths leading, in G, from node i to node j.

Now, if ω_V, $\omega_\oplus$, and ω_p, (p prime) denote the semiring homomorphisms defined in Section 2 then we get

$$GAP = \omega_V\text{-}GAP,$$
$$ODD\text{-}PATH\text{-}GAP = \omega_\oplus\text{-}GAP,$$
$$MOD\text{-}p\text{-}GAP = \omega_p\text{-}GAP.$$

Hence, ω-$GAPs$ provide a comfortable frame for investigating modified $GAPs$. Beside of ω-$GAP = \{\omega\text{-}GAP_N\}_{N \in \mathbf{N}}$ the complementary problems

$$\neg(\omega\text{-}GAP) := \{\neg\omega\text{-}GAP_N\}_{N \in \mathbf{N}}$$

are of interest. While $\neg(\omega_\oplus\text{-}GAP)$ and $\omega_\oplus\text{-}GAP$ as well as $\neg(\omega_p\text{-}GAP)$ and $\omega_p\text{-}GAP$ are computationally equivalent with respect to a variety of computation devices [Mei89, DKMW90] this is not true for $\omega_V\text{-}GAP$ and $\neg(\omega_V\text{-}GAP)$, well-known as $NO\text{-}PATH\text{-}GAP$,

$$NO\text{-}PATH\text{-}GAP = \neg(\omega_V\text{-}GAP).$$

In the following w.l.o.g. we generally assume the directed acyclic graphs to be of outdegree ≤ 2 [Mei89]. Recall, an enumeration of the nodes of a directed acyclic graph G is said to be *monotone* if, for each edge (i,j) of G, it holds $i < j$. If a graph G is given monotone enumerated then it can be determined by the upper triangular $(a_{ij})_{1 \leq i < j \leq n}$ of the adjacency matrix. If we restrict the graph–accessibility–problem to such monoton e enumerated graphs G we write ω-GAP_{mon}. In particular, we have

$$\omega_V\text{-}GAP_{\text{mon}} = GAP_{\text{mon}}2,$$
$$\neg(\omega_V\text{-}GAP\text{mon}) = NO\text{-}PATH\text{-}GAP_{\text{mon}}2,$$
$$\omega_\oplus\text{-}GAP_{\text{mon}} = ODD\text{-}PATH\text{-}GAP_{\text{mon}}2,$$
$$\omega_p\text{-}GAP_{\text{mon}} = MOD\text{-}p\text{-}GAP_{\text{mon}}2, \; p \text{ prime, and}$$
$$id_{\mathbf{N}}\text{-}GAP_{\text{mon}} = GAP_{\text{mon}}1$$

if, in the case $\omega = id_{\mathbf{N}}$, we restrict ourself to the consideration of directed graphs of outdegree 1.

Due to the branching program descriptions of the nonuniform logarithmic space-bounded complexity classes cited in the last section it has been proved that each of these classes can be characterized by the completeness of a certain *graph–accessibility–problem* (for details we refer to [Mei89]).

Fact 2.
Let $\omega : \mathbb{N} \twoheadrightarrow R$ be a semiring homomorphism of $\mathbb{N}$ onto R. Then ω-GAP and ω-GAP_{mon} are (p-projection) complete in $\mathcal{P}_{\omega-BP}$. In detail,

1. $GAP1$ and $GAP_{mon}1$ are (p-projection) complete in $\mathcal{L} = \mathcal{P}_{BP}$,
2. GAP and $GAP_{mon}2$ are (p-projection) complete in $\mathcal{NL} = \mathcal{P}_{\omega_\vee-BP}$,
3. NO-$PATH$-GAP and NO-$PATH$-$GAP_{mon}2$ are (p-projection) complete in co-$\mathcal{NL} = \neg(\mathcal{P}_{\omega_\vee-BP})$,
4. ODD-$PATH$-GAP and ODD-$PATH$-$GAP_{mon}2$ are (p-projection) complete in $\oplus \mathcal{L} = \mathcal{P}_{\omega_\oplus-BP}$,
5. MOD-p-GAP and MOD-p-$GAP_{mon}2$ are (p-projection) complete in MOD_p-$\mathcal{L}$ for each p, p prime. □

4 Bounded alternating ω-branching programs

In order to establish differences in the computational power of polynomial size ω-branching programs we consider bounded alternating ω-branching programs.

Let X be a set of variables over a finite semiring R, $\#X = n$, and let $Y, Z \subseteq X$, $\#Y = \#Z = \varepsilon n$, $\varepsilon > 0$, be disjoint subsets of X. An ω-branching program P, $\omega : \mathbb{N} \twoheadrightarrow R$ semiring homomorphism onto R, that tests variables of X is said to be of *alternation length α with respect to Y and Z* if each path of P can be divided into α segments in which, alternating, variables of Y or variables of Z are not tested. A sequence of ω-branching programs P_n is called *bounded alternating* if there exist two disjoint subsets $Y, Z \subseteq X$,

$\#Y = \#Z = \varepsilon n$, $\varepsilon > 0$, such that P_n is of alternation length $n^{o(1)}$ with respect to Y and Z.

By $\mathcal{P}_{ba\,\omega-BP}$ we denote the set of languages $A \subseteq R^*$ that can be accepted by sequences of polynomial size, bounded alternating ω-branching programs. For the particular semiring homomorphisms $\omega = \omega_\vee$, $\omega_\oplus$, ω_p (p prime), we write, due to Fact 1, $\mathcal{NL}_{ba}$, $\oplus\mathcal{L}_{ba}$, and $MOD_p\text{-}\mathcal{L}_{ba}$ instead of $\mathcal{P}_{ba\,\omega_\vee-BP}$, $\mathcal{P}_{ba\,\omega_\oplus-BP}$, and $\mathcal{P}_{ba\,\omega-BP}$, respectively.

Although polynomial size, bounded alternating ω-branching are definitively less powerful than unbounded alternating ones [MW91] they are quite powerful computation devices. This can be demonstrated considering the monotone versions $\omega\text{-}GAP_{mon}$ of the $\omega\text{-}GAPs$, that, due to Fact 2, are known to be complete in $\mathcal{L}, \mathcal{NL}, co\text{-}\mathcal{NL}, \oplus\mathcal{L}$, and $MOD_p\text{-}\mathcal{L}$, respectively.

Proposition 3.
Let $\omega : \mathbb{N} \longrightarrow R$ be a semiring homomorphism. Then $\omega\text{-}GAP_{mon} \in \mathcal{P}_{ba\,\omega-BP}$. In detail,

1. $GAP_{mon}1 \in \mathcal{P}_{ba\,BP} = \mathcal{L}_{ba}$.
2. $GAP_{mon}2 \in \mathcal{P}_{ba\,\omega_\vee-BP} = \mathcal{NL}_{ba}$.
3. $NO\text{-}PATH\text{-}GAP_{mon}2 \in \neg(\mathcal{P}_{ba\,\omega_\vee-BP}) = co\text{-}\mathcal{NL}_{ba}$.
4. $ODD\text{-}PATH\text{-}GAP_{mon}2 \in (\mathcal{P}_{ba\,\omega_\oplus-BP}) = \oplus\mathcal{L}_{ba}$.
5. $MOD\text{-}p\text{-}GAP_{mon}2 \in (\mathcal{P}_{ba\,\omega_p-BP}) = MOD_p\text{-}\mathcal{L}_{ba}$ for each prime p.

(Observe, that the stated containment of the monotone versions of the ω-GAPs in the ba–BP complexity classes does not imply their completeness since these classes are not closed against projection reductions.)

Proof.
It can be checked easily that the ω-branching program described by its stages in Figure 1 computes $(\omega\text{-}GAP_{mon}2)_N$, $N = n^2 - n$. Since this ω-branching program is bounded alternating with respect to

$$Y = \{x_{ij} : 1 \le i \le \frac{n}{2}\} \text{ and } Z = \{x_{ij} : \frac{n}{2} < i \le n\},$$

(indeed it alternates merely one time) we are done. □

$S_i, 1 \leq i < n$:

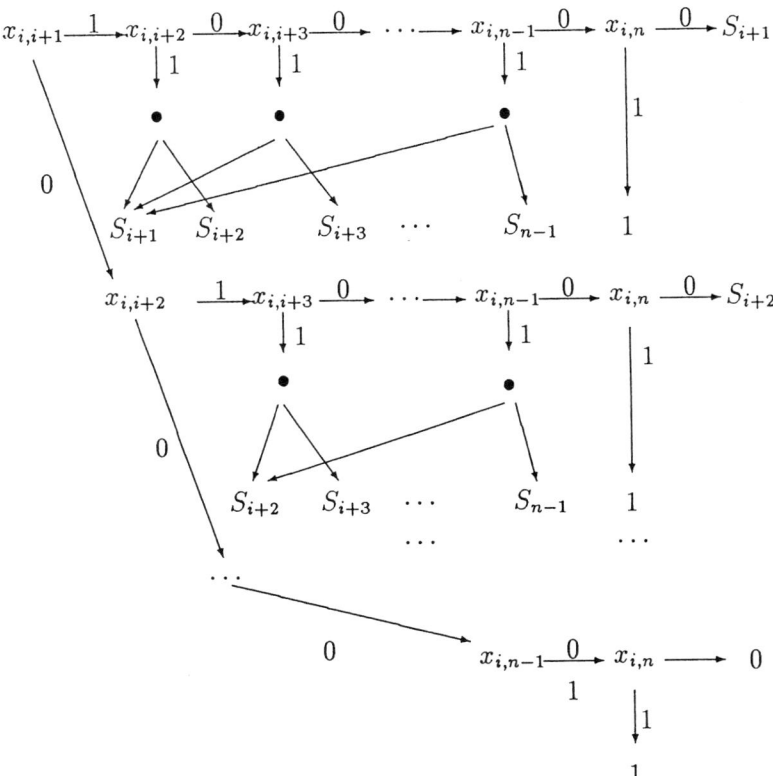

Figur 1.

5 The lower bound technique

In [MW91] a technique for proving exponential lower bounds on the size of bounded alternating ω-branching programs has been developed. This lower bound technique is mainly based on considerations of invariants of the communications matrix $M_{Y,Z}(f)$ of a function f. If

$$f : R^Y \times R^Z \longrightarrow \{0,1\}$$

then $M_{Y,Z}(f)$ is a $(\#R)^{\#Y} \times (\#R)^{\#Z}$ matrix which is defined, for $a_Y \in \{0,1\}^Y, a_Z \in \{0,1\}^Z$, by

$$(M_{Y,Z}(f))_{a_Y,a_Z} = f(a_Y, a_Z).$$

The matrix-invariants used are certain matrix-ranks. Recall, if R is a semiring and if $M \in I\!M_R(n,m)$ is a $n \times m$ matrix over R then the rank of M, $\text{rank}_R(M)$, is the minimum k such that M can be written as $M = A \cdot B$ with $A \in I\!M_R(n,k)$ and $B \in I\!M_R(k,m)$.

In order to derive exponential lower bounds on the size of bounded alternating ω-branching programs, $\omega : I\!N \longrightarrow\!\!\!\!\!\rightarrow R$, that compute a function f, due to the following lemma, it suffices to derive exponential lower bounds on the R-ranks of the communication matrices of certain subfunctions of f. If $Y, Z \subseteq X$ are disjoint subsets of X, and if $c \in \{0,1\}^{X-(Y \cup Z)}$ then we denote by

$$f^c : R^{Y \cup Z} \longrightarrow \{0,1\}$$

the subfunction

$$f^c(y,z) := f(c,y,z) \text{ for all } y \in R^Y, z \in R^Z.$$

Lemma 4 (The lower bound criterion) [MW91].
Let $f : R^X \longrightarrow \{0,1\}$ be a discrete function, let $Y, Z \subseteq X$ be two disjoint subsets of X, and let P be an ω-branching program of alternation length α with respect to Y and Z that computes f. Then

$$\text{Size}(P) \geq (\text{rank}_R M_{Y,Z}(f^c))^{1/3\alpha}$$

for each $c \in R^{X-(Y \cup Z)}$. □

6 Some lower bounds for the rank of communication matrices

In the following we derive exponential lower bounds on the R-rank of certain communication matrices of ω-GAP, $\omega : I\!N \longrightarrow\!\!\!\!\!\rightarrow R$. In order to do this we start with an easy observation concerning the *SEQUENCE EQUALITY TEST* $SEQ = \{SEQ_{2n}\}_{n \in \mathbb{N}}$ for sets Y, Z with $\#Y = \#Z = n$,

$$SEQ_{2n}: R^Y \times R^Z \longrightarrow \{0, 1\},$$

defined by

$$SEQ_{2n}(y, z) = 1 \text{ iff } y = z.$$

Proposition 5.

Let R be a finite semiring. Then it holds

$$\operatorname{rank}_R M_{Y,Z}(SEQ_{2n}) = (\#R)^n. \quad \square$$

Now, by means of suitable *SEQ*s, for each pair Y, Z,

$$Y, Z \subset X = \{x_{ij} : 1 \leq i, j \leq n, i \neq j\},$$

of disjoint subsets, $\#Y = \#Z = \Omega(\#X)$, we derive exponential lower bounds on the rank of communication matrices $M_{Y,Z}(\omega\text{-}GAP)$. The conceptual idea is to identify, by means of p-projections respecting Y and Z, submatrices M' of $M_{Y,Z}(\omega\text{-}GAP)$ corresponding to certain SEQ_{2m} for some $m = \Omega(n)$. Since, due to Proposition 5, the rank of M' is exponential in n the same must be true for the rank of $M_{Y,Z}$.

Recall, a mapping π_m

$$\pi_m : \{x_1, \ldots, x_n\} \longrightarrow \{u_1, \bar{u}_1, \ldots, u_m, \bar{u}_m, 0, 1\}$$

is called a *p-projection* from $M \subseteq \{0,1\}^m$ to $N \subseteq \{0,1\}^n$ if

$$M(u_1, \ldots, u_m) = N(\pi_m(x_1), \ldots, \pi_m(x_n))$$

and if m and n are polynomially related. π_m is said to *respect* Y, Z if

$$(\pi_m(Y) \cup \overline{\pi_m(Y)}) \cap (\pi_m(Z) \cup \overline{\pi_m(Z)}) = \emptyset.$$

Proposition 6.

Let R be a semiring and $\omega : \mathbb{N} \twoheadrightarrow R$ be a semiring homomorphism and let $X = \{x_{ij} : 1 \leq i, j \leq n, i \neq j\}$, $N := \#X = n^2 - n$.

For each pair $Y, Z \subseteq X$ of disjoint subsets with $\#Y = \#Z = \theta(N)$ there are subsets $Y' \subseteq Y$ and $Z' \subseteq Z$, $\#Y' = \#Z' = \Omega(n)$, such that for each partial assignment $c \in \{0,1\}^{X-(Y' \cup Z')}$

$$\mathrm{rank}_R\ M_{Y',Z'}(\omega\text{-}GAP_N^c) = \exp(n).$$

Poof.
Due to Proposition 5 it suffices to give a p-projection reduction from SEQ_{2m} to $\omega\text{-}GAP_N$ for some $m = \Omega(n)$.

Let $Y, Z \subseteq X$ be two disjoint subsets of size $\#Y = \#Z = \xi N$, $0 < \xi \le 1$. Easy graph-theoretical considerations show that there are subsets $Y' \subseteq Y$, $Y' = \{(a_i, b_i), (a_i, c_i) : 1 \le i \le m\}$, and $Z' \subseteq Z$, $Z' = \{(d_i, e_i), (f_i, g_i) : 1 \le i \le m\}$, of size $2m = \Omega(n) = \Omega(\sqrt{N})$, consisting of pairwise vertex-disjoint edges with $1, n \not\in \{a_i, b_i, c_i, d_i, e_i, f_i, g_i : 1 \le i \le m\}$.

Using these subsets Y' and Z' we can define a projection reduction $\pi = \pi_{2m}$

$$\pi : \{y_\iota : \iota \in \{(i,j) : 1 \le i, j \le n, i \ne j\} \longrightarrow \{x_1, \overline{x_1}, \ldots, x_{2m}, \overline{x_{2m}}, 0, 1\}$$

from SEQ_{2m} to $\omega\text{-}GAP_N$ by

$$\pi(y_\iota) := \begin{cases} 1 & \text{if } \iota \in \{(1, a_1), (b_i, d_i), (c_i, f_i), (e_m, n), (g_m, n)\}, \\ x_i & \text{if } \iota = (a_i, b_i), \\ \overline{x_i} & \text{if } \iota = (a_i, c_i), \\ x_{m+i} & \text{if } \iota = (d_i, e_i), \\ \overline{x_{m+i}} & \text{if } \iota = (f_i, g_i), \\ 0 & \text{otherwise.} \end{cases}$$

Obviously, π respects Y' and Z' since Y' and Z' are subsets of disjoint sets, and since

$$\pi^{-1}(\{x_1, \overline{x_1}, \ldots, x_m, \overline{x_m}\}) = \{y_\iota : \iota \in Y'\},$$
$$\pi^{-1}(\{x_{m+1}, \overline{x_{m+1}}, \ldots, x_{2m}, \overline{x_{2m}}\}) = \{y_\iota : \iota \in Z'\}.$$

Due to the definition, the $(\pi(y_\iota))$ are graphs of outdegree 1. They have exactly one path from 1 to n if the sequences $(x_1, \ldots, x_m)$ and $(x_{m+1}, \ldots, x_{2m})$ are

equal and none if the are not. Hence, for each semiring homomorphism ω onto R, we have
$$\omega - GAP_N(\pi(y_\iota)) = 1 \quad \text{iff} \quad \#[1 \xrightarrow{(\pi(y_\iota))} n] = 1.$$
Since
$$\omega - GAP_N(\pi(y_\iota)) = SEQ_{2m}(x_1, \ldots, x_{2m})$$
for all $(x_1, \ldots, x_{2m}) \in \{0,1\}^{2m}$, $\pi = \pi_{2m}$ indeed definies a projection reduction from SEQ_{2m} to ω-GAP_N.

Figure 2 illustrates the case $m = 3$. The dotted arrows depend on the literals they are labelled with. For example, the edge (a_i, c_i) exists iff $\overline{x_i} = 1$. All other edges are fixed in the described manner. □

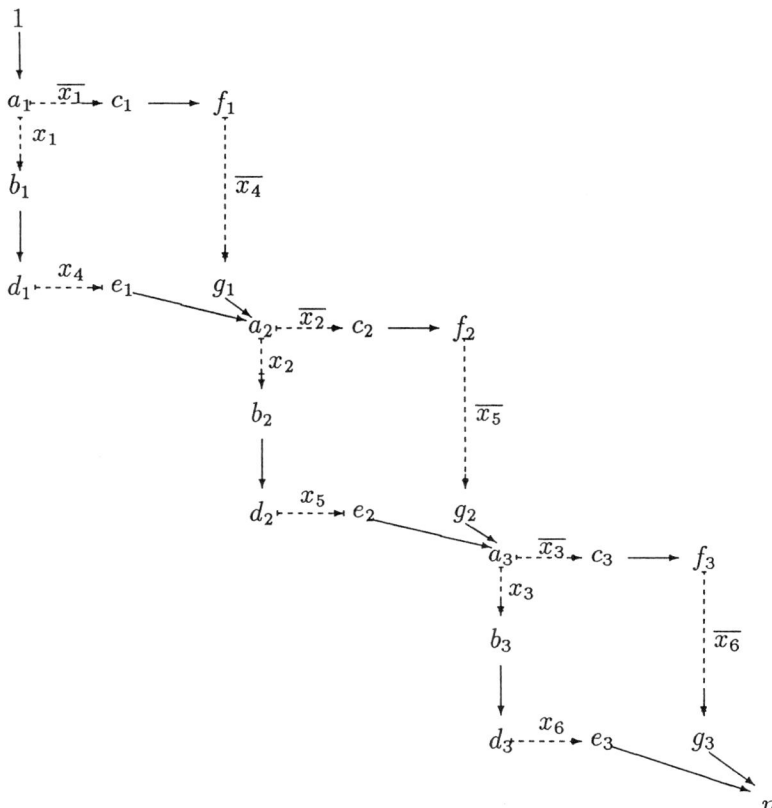

Figure 2.

7 Conclusions

The results of the last section imply, due to the lower bound criterion of Lemma 4, exponential lower bounds on the size of bounded alternating ω-branching programs that solve ω-GAP.

Theorem 7.
Let $\omega : \mathbb{N} \twoheadrightarrow R$ be a semiring homomorphism onto a finite semiring R, and let P be a bounded alternating ω-branching program, that computes ω-GAP_N. Then
$$\text{Size}(P) = \exp(\Omega(\sqrt{N})). \quad \square$$

Theorem 7 implies the following corollary, that contrasts Proposition 3.

Corollary 8.

1. $GAP1 \notin \mathcal{L}_{ba}$,
2. $GAP2 \notin \mathcal{NL}_{ba}$,
3. $NO\text{-}PATH\text{-}GAP2 \notin co\text{-}\mathcal{NL}_{ba}$,
4. $ODD\text{-}PATH\text{-}GAP2 \notin \oplus \mathcal{L}_{ba}$, and
5. $MOD_p\text{-}GAP2 \notin MOD_p\text{-}\mathcal{L}_{ba}$. $\square$

Due to Fact 2, the logarithmic space-bounded complexity classes related to bounded alternating ω-branching programs and those related to unbounded ones can be separated from each other by means of the ω-$GAPs$.

Corollary 9.
Let $\omega = \omega_\vee, \omega_\oplus$, or ω_p (p prime). Then
$$\mathcal{P}_{ba\ \omega-BP} \subsetneqq \mathcal{P}_{\omega-BP}.$$

In detail,
$$\mathcal{L}_{ba} \subsetneqq \mathcal{L}, \quad \mathcal{NL}_{ba} \subsetneqq \mathcal{NL}, \quad co\text{-}\mathcal{NL}_{ba} \subsetneqq co\text{-}\mathcal{NL},$$
$$\oplus \mathcal{L}_{ba} \subsetneqq \oplus \mathcal{L}, \text{ and } MOD_p\text{-}\mathcal{L}_{ba} \subsetneqq MOD_p\text{-}\mathcal{L}, p \text{ prime.} \quad \square$$

Finally, we remark only that the complexity classes $\mathcal{L}_{ba}, \mathcal{NL}_{ba}, co\text{-}\mathcal{NL}_{ba}, \oplus\mathcal{L}_{ba}$, $MOD_p\text{-}\mathcal{L}_{ba}$ (p prime) related to bounded alternating ω-branching programs are separated from each other. The proof of the separation results is based on considerations of the ω-ORTHOGONALITY TEST that are similar to Prop. 3 (upper bounds) and Prop. 6 (lower bounds). For details we refer to [MW91].

References

[AM86] N. Alon, W. Maass: Meanders, Ramsey Theory and Lower Bounds, Proc. 27th ACM STOC, 1986, 30–39.

[DKMW90] C. Damm, M. Krause, Ch. Meinel, S. Waack: Separating Restricted MOD_p-Branching Program Classes, Informatik-Preprint 3; Humboldt-Universität Berlin, 1990.

[DKMW92] C. Damm, M. Krause, Ch. Meinel, S. Waack: Separating Counting Communication Complexity Classes, Proc. STACS'92, LNCS 577, 281–292, Paris, 1992.

[Imm87] N. Immerman: Languages that Capture Complexity Classes, SIAM J. Comput.; Vol. 16, No. 4, 1987, 760–778.

[HMT88] A. Hajnal, W. Maass, G. Turan: On the Communication Complexity of Graph Problems. Proc. 20th STOC (1988) 186–191.

[KMW89] M. Krause, Ch. Meinel, S. Waack: Separating Complexity Classes Related to Certain Input Oblivious Logarithmic Space Bounded Turing Machines, Proc. 4th IEEE Structure in Complexity Theory, 1989, 240–259.

[KW89] M. Krause, S. Waack: On Oblivious Branching Programs of Linear Length, Proc. FCT'89, LNCS 380, 287–296. (To appear in Information and Computation.)

[Lov89] L. Lovasz: Communication Complexity: A Survey. Technical Report CS-TR-204-89, Princeton University.

[Mei86] Ch. Meinel: P-Projection Reducibility and the Complexity Classes L(nonuniform) and NL(nonuniform), Proc. MFCS'86, LNCS 233, 527–535.

[Mei88] Ch. Meinel: Polynomial Size Ω-Branching Programs and their Computational Power, Proc. STACS'88, LNCS 294, 81–90.

[Mei89] Ch. Meinel: Modified Branching Programs and Their Computational Power, LNCS 370, Springer Verlag, 1989.

[MW91] Ch. Meinel, S. Waack: Separating Complexity Classes Related to Bounded Alternating ω-Branching Programs. Preprint No. 276/1991, TU Berlin, FB Mathematik, 1991. (To appear in Mathematical Systems Theory.)

[PŽ83] P. Pudlak, S. Žak: Space Complexity of Computations, Techn. Report Univ. of Prague, 1983.

[Sav70] W. Savitch: Relationships Between Nonderterministic and Deterministic Tape Complexities, J. Comp. Sys. Sci., Vol. 4, No. 2, 1970, 177–192.

Associative Storage Modification Machines

John Tromp
Centrum voor Wiskunde en Informatica
P.O. Box 4079, 1009 AB Amsterdam
The Netherlands
Email: tromp@cwi.nl

Peter van Emde Boas
ILLC, Departments of Mathematics and Computer Science
University of Amsterdam
Plantage Muidergracht 24, 1018 TV Amsterdam
The Netherlands
Email: peter@fwi.uva.nl

Abstract

We present a parallel version of the storage modification machine. This model, called the Associative Storage Modification Machine (*ASMM*), has the property that it can recognize in polynomial time exactly what Turing machines can recognize in polynomial space. The model therefore belongs to the Second Machine Class, consisting of those parallel machine models that satisfy the parallel computation thesis. The Associative Storage Modification Machine obtains its computational power from following pointers in the reverse direction.

1985 AMS(MOS) Subject Classification: 68Q05, 68Q10, 68Q15.
CR Categories: B.3.2, B.4.3, D.4.1, D.4.4.
Keywords and Phrases: machine model, storage modification machine, pointer machine, parallel computation thesis, complexity theory, pspace, second machine class, simulation.

1 Introduction

The Storage Modification Machine (*SMM*) is a machine model introduced by Schönhage in 1977 [16]. The model has its predecessor in the Kolmogorov-Uspenskii machine (*KUM*) [10]. Schönhage advocates his model as *a model of extreme flexibility*.

The model resembles the Random Access Machine (*RAM*) [1] as far as it has a stored program and a potentially infinite memory structure where it stores its data. Whereas the *RAM* uses an infinite sequence of storage registers, each capable of storing an arbitrarily large integer, the *SMM* operates on a directed graph by creating nodes and (re)directing pointers. The main difference between the *SMM* and the *KUM* is that the *KUM* operates on undirected instead of directed graphs.

We can approximately model an *SMM* by a Pascal program that uses records of pointers to records to describe the directed graph[1]:

type pointer = ^node;
   node   = record a,b: pointer end;
var head   : pointer;

In contrast with Pascal, pointers are not allowed to be *nil* or undefined; they must always point to some node. The (finite) set of pointer names, in the example $\{a, b\}$, is called the *alphabet* of *directions*, denoted Δ. The pointers in the graph are labeled with the elements of Δ such that each node in the digraph has, for each direction $\delta \in \Delta$, exactly one outgoing δ-pointer. The graph thus has regular outdegree $|\Delta|$. To complete the analogy between an *SMM* and a Pascal program, the latter must be restricted to the use of only one variable; the pointer head. By repeated application of the Pascal new statement, the program can create an arbitrarily large data structure. This is addressed with expressions like head^.b^.a^.b^.b^.a. Similarly, the *SMM* addresses its storage with words (strings) over Δ, like *babba*. In the *SMM* model, there is only a conceptual head pointer—at any time, one node, call the *center*, is distinguished as the one from where addressing starts. Thus the centre, whose identity can change dynamically, is addressed by the empty

---
[1]In Pascal ^T denotes the type 'pointer to T'; a value of this type is the address of an object of type T. Indirection through a pointer is written as p^, which refers to the object at which p points.

word ϵ, and other nodes are addressed by following pointers starting from the center.

It has been established that from the perspective of computational complexity theory the *SMM* (if equipped with the correct space measure [12, 21]) is computationally equivalent to the other standard sequential machine models like the Turing machine and the *RAM*. This equivalence amounts to the fact that these models simulate each other with polynomially bounded overhead in time and constant factor overhead in space, thus satisfying the so-called *invariance thesis* [17, 22].

For most sequential models there have been proposed parallel machine models based on the classical sequential version. For the Turing machine Savitch [15] has proposed a parallel version based on parallel recursive branching; a model based on nondeterministic forking on a shared set of tapes was described by Wiedermann [24, 25], but this model turns out to be polynomially equivalent in time and space with the standard sequential devices. The richness of parallel models based on the *RAM* is even much greater, which makes it hard, if not impossible to refer to a small set of representative models. There are models based on shared memory and alternative models based on local storage and message passing. Hybrid combinations occur as well. Within each class there exist more refined distinctions like the resolution strategy for resolving write conflicts in shared memory models, the available arithmetic instructions and the mechanism for restricting the number of processors activated during a computation. Moreover, there exist sequential models which become computationally equivalent to parallel models due to their power to create and manipulate exponentially large values in a linear number of steps in the uniform time measure. Also, by exploiting the alternating mode of computation [5], some standard sequential devices become computationally equivalent to the parallel machines.

For a more detailed survey of parallel models we refer to [20, 22]. For the purpose of the present paper it suffices to give some impression of the overall landscape of parallel machine models.

It turns out that most parallel models proposed in the literature belong to the so-called *Second Machine Class* consisting of machine models which obey the *Parallel Computation Thesis*. This thesis expresses that the class of languages recognized in nondeterministic polynomial time on the parallel device is equal to the class *PSPACE* of languages recognized in polynomial space on a sequential device. Conversely all languages in *PSPACE* are recognized in deterministic polynomial time on the parallel machine. In our reading the Parallel Computation Thesis entails the equivalence of deterministic and nondeterministic polynomial time on the parallel model. The models for which the thesis was originally formulated obey this more restricted thesis as well. And indeed those models for which nondeterministic polynomial time seems to exceed *PSPACE* nowadays are held to be more powerful.

Not all parallel models obey the above parallel computation thesis. Some weak models turn out to be polynomial time equivalent to the sequential models (the parallel Turing machine proposed by Wiedermann, and its equivalents [24, 25] being a typical example). Other models, like the *P–RAM* presented by Fortune and Wyllie [7] deviate from the thesis by recognizing exponentially time bounded languages in polynomial nondeterministic time on the parallel device; some parallel devices even recognize arbitrary languages in constant time [13]. The second machine class therefore represents a frequently occurring version of the power of uniform unrestricted parallelism rather than the union of all possible parallel machine models. Second machine class members can be characterized as providing the right mixture of exponential growth potential together with the proper degree of uniformity. The exponential growth potential is required for the implementation of the transitive closure algorithm on a directed graph of exponential size (which models the computation graph of some *PSPACE*-bounded machine), or the direct solution of the *PSPACE*-complete problem *QBF* in polynomial time. The uniformity is required for performing the simulation of a polynomial-time computation of the nondeterministic version of the parallel machine in polynomial space. See [22] for more details on the standard strategies for proving membership in the second machine class.

In this paper we propose (as far as we know for the first time) a parallel version of the storage modification machine which belongs to the second machine class. To our knowledge few parallel versions of pointer machines have been investigated in the complexity theory literature. The earliest reference known to us concerns a parallel version of the Kolmogorov-Uspenskii machine which was proposed by Barzdin [2, 3]. This machine operates like an irregular cellular array of finite state automata in a graph which is dynamically changed by the individual nodes interacting with their neighbourhood. A single computation step resembles a parallel rewrite step in a graph grammar derivation.

In this model all nodes are active in every computation step; if their neighborhood matches the pattern required by the instruction the node will transform its environment. The Hardware Modification Machine (*HMM*) introduced by Dymond and Cook [6] behaves in a similar way. This model indeed has been investigated for its complexity behavior. From Lam and Ruzzo [11] it follows that the machine is equivalent with constant factor time overheads with a restricted version of the *P—RAM* of Fortune and Wyllie. From this result one can observe that the *HMM* represents another example of the class of devices which are located beyond the second machine class - its nondeterministic version accepts *NEXPTIME* in polynomial time.

The computational power of our *ASMM* model originates from the possibility of traversing pointers in their *reverse* order. By using reverse directions, an *ASMM* can address, from a given node x, all the nodes that are associated with x by pointing to x (hence the name[2]). More than one node can be reached on a path by traversing pointers in the reverse direction. Note that at this point it is crucial that we have based ourselves on the *SMM* rather than the older *KUM* model; in an undirected graph traversing pointers in the reverse direction makes no sense.

As in the standard *SMM* model the finite control accesses the storage structure by means of a single center node. The power of traversing reversed pointers is used only in two types of instructions: the *new* and the *set* instruction. The first argument of the above two instructions is a path which now may contain reverse pointers. This path therefore no longer denotes a single node but a set of nodes (which in fact may be empty). The action described by the instruction now will be performed for all nodes in this set in parallel. The second argument of the *set* instruction is required to be a path consisting of forward pointers only; it therefore always denotes a single node. Therefore the action performed by the two instructions above is deterministic.

Our model may be considered to be a member of the class of sequential machines which operate on large objects in unit time and obtain their power of parallelism thereof. Other models of this character are the vector machines of Pratt and Stockmeyer [14], the *MRAM* proposed by Hartmanis and Simon [9] and simplified by Bertoni et al. [4], and also the *EDITRAM* presented by Stegwee et al. [18, 22].

---

[2]compare with *content-addressable associative memory*

Evidently our model is one among a number of possible alternatives for designing a parallel version of the *SMM* model. In the conclusion of this paper we discuss two more alternatives suggested by an anonymous referee.

Following [22] we denote the class of languages accepted in polynomial time by the *ASMM* model by *ASMM–PTIME*. The class of languages accepted in polynomial time by nondeterministic *ASMM* devices is denoted by *ASMM–NPTIME*. The class *PSPACE* as indicated above, denotes the class of languages recognized in polynomial space on a Turing machine. The fact that the *ASMM* is a true member of the second machine class is now expressed by the equality:

ASMM–PTIME = *ASMM–NPTIME* = *PSPACE*

In the proof of this equality we use the well known *PSPACE*-complete problem:

QUANTIFIED BOOLEAN FORMULAS (QBF) [19] :

QUANTIFIED BOOLEAN FORMULAS:
INSTANCE: A formula of the form $Q_1 x_1 \ldots Q_n x_n [P(x_1, \ldots, x_n)]$, where each Q_i equals $\forall$ or $\exists$, and where $P(x_1, \ldots, x_n)$ is a propositional formula in the boolean variables $x_1, \ldots, x_n$.

QUESTION: does this formula evaluate to *true*?

2 The SMM and the ASMM models

Our *ASMM* model is based on the Storage Modification Machine as introduced by Schönhage in 1970 [16]. The *SMM* model resembles the *RAM* model as far as it has a stored program and a similar flow of control. It has a single storage structure, called a Δ-*structure*. Here Δ denotes a finite alphabet consisting of at least two symbols. We denote the reverse of a direction $a \in \Delta$ as $\bar{a}$. Furthermore, $\bar{\Delta} = \{\bar{\alpha} | \alpha \in \Delta\}$ is the set of reverse directions and we let $\tilde{\Delta} = \Delta \cup \bar{\Delta}$.

A Δ-structure X is a finite directed graph each node of which has $k = |\Delta|$ outgoing edges which are labeled by the k elements of Δ. In Schönhage's formalization, a Δ-structure is a triple (X, c, p), where X denotes the finite set of nodes, $c \in X$ is the *center*, and $p : X \times \Delta \to X$ is the pointer mapping; $p(x, \alpha) = y$ means that the α-pointer from x goes to y.

There exists a map p^* from Δ^* to X defined as follows: For the empty string ϵ one has $p^*(\epsilon) = c$, and otherwise $p^*(wa) = p(p^*(w), a)$ is the end-point of the a-labeled pointer starting in $p^*(w)$.

The map p^* does not have to be surjective. Nodes which can not be reached by tracing a word w in Δ^* starting from the center c will turn out to play no subsequent role during the computations of the *SMM*. In the *ASMM* model pointers can be traversed in the opposite direction, and therefore these nodes no longer can be disregarded as being garbage.

The storage of an *SMM* or an *ASMM* is a dynamically changing Δ-structure, which initially consists of a single node, the center. The *ASMM*'s operation is described by a *program*, which is a finite sequence of *labels* and *instructions*. Labels can be used in control flow statements; they should occur exactly once in case the machine is deterministic. *Nondeterminism* is introduced by allowing multiple occurrences of the labels referred to in jump or conditional jump instructions. Consequently we only consider nondeterminism in the flow of control. An alternative would be to design instructions that manipulate the data in a nondeterministic manner, but such instructions easily lead to a more powerful model.

In the text below we separate labels and instructions by a colon, whereas instructions are ended by semicolons.

The instruction repertoire of the *SMM* and the *ASMM* includes the *common* instructions (the λ's are labels and $\beta \in \{0, 1\}$)

input λ_0, λ_1;
output β;
goto λ;
halt;

The *input* instruction reads an input bit β and transfers control to λ_β. The other instructions are straightforward.

Furthermore there exist three *internal* instructions which operate on memory - in this case a Δ-structure X. For the *SMM* the arguments in these instructions are strings over Δ. For the *ASMM* the single argument of *new* and the first argument of *set to* are strings over $\tilde{\Delta}$; the other arguments (second argument of *set to* and both arguments of the *if* instruction) are strings over Δ. All arguments are finite strings which are written literally in the program. We first describe their meaning for the *SMM*:

1. *new W*: creates a new node which will be located at the end of the path traced by W; if $W = \epsilon$ the new node will become the center; otherwise the last pointer on the path labeled W will be directed towards the new node. All outgoing pointers of the new node will be directed to the former node $p^*(W)$

2. *set W to V*: redirects the last pointer on the path labeled by W to the former node $p^*(V)$; if $W = \epsilon$ this simply means that $p^*(V)$ becomes the new center; otherwise the structure of the graph is modified.

3. *if* $V = W$ (*if* $V \neq W$) *then*⟨*instr*⟩: depending on whether $p^*(V)$ and $p^*(W)$ coincide or not, the conditional instruction ⟨*instr*⟩ (conditional jump suffices) is executed or skipped.

In the *ASMM* model the Δ-structure can be addressed by words (also called *paths*) over the alphabet of normal and reverse directions $\tilde{\Delta}$. Every word $W \in \tilde{\Delta}^*$ addresses the (possibly empty) set of all the nodes reachable from the center by following the consecutive directions and reverse directions in W.

The notion of 'addressing' is formalized by the mapping $P : \tilde{\Delta}^* \to 2^X$, defined by:

$$P(\epsilon) = \{c\}$$
$$P(W\alpha) = \{p(x, \alpha) | x \in P(W)\}$$
$$P(W\bar{\alpha}) = \{x | p(x, \alpha) \in P(W)\}.$$

Note: It will often be convenient to give a name to an address path $V \in \Delta^*$. In the code fragments presented in this paper, we will use paths having such a name v as a prefix, in addition to fully explicit paths. This serves two purposes. First, fixed nodes that have been given descriptive names can be addressed by their name rather than some arbitrary path (we say that a node is fixed iff it has a constant address). Second, if we are using one of the pointers from a fixed node to traverse part of the graph, it can be given a name that more closely resembles its function: that of a variable. We will use variable names without specifying which path they stand for, omitting the details of the creation of spare nodes to provide the required[3] pointers.

A node x is said to be *directly* addressable if it is reachable from the center by normal (non-reversed) directions, i.e. $\exists V \in \Delta^* : P(V) = \{x\}$.

In order to facilitate the descriptions of the internal instructions, we define a mapping $Q : \tilde{\Delta}^* \to 2^X$, from a path to the set of nodes from which the last pointer on this path originates, by:

$$\begin{aligned} Q(\epsilon) &= \emptyset \\ Q(W\alpha) &= P(W) \\ Q(W\bar{\alpha}) &= P(W\bar{\alpha}). \end{aligned}$$

The *new* and *set* change the Δ- structure from (X, c, p) to (X', c', p') as follows:

new W;
   Here, $W \in \tilde{\Delta}^*$ determines where new nodes are inserted. If $W = \epsilon$, then a new center c' is created such that $X' = X \cup \{c'\}$ and $p'(c', \delta) = c$ for all $\delta \in \Delta$. Otherwise, if $W = U\tilde{\alpha}$ ($\tilde{\alpha}$ is either α or $\bar{\alpha}$), then for every node $u \in Q(W)$ a new node x_u is created such that $X' = X \cup \{x_u | u \in Q(W)\}$, $p'(u, \alpha) = x_u$, $\forall \delta \in \Delta \, p'(x_u, \delta) = p(u, \alpha)$, and $c' = c$. All other pointers remain unchanged.

set W *to* V;
   Here, $W \in \tilde{\Delta}^*$ determines which pointers are redirected to the node determined by $V \in \Delta^*$. If $W = \epsilon$, then $c' = P(V)$ becomes the new center. Otherwise, if $W = U\tilde{\alpha}$, then for every node $u \in Q(W)$, $p'(u, \alpha) = P(V)$ and $c' = c$. In both cases X' is the restriction of X to the nodes which are reachable from c'.

The third internal instruction is the *if* statement. Since both paths in this instruction consist of forward pointers only, the meaning of this instruction is equal for the *SMM* and the *ASMM*.

---

[3]In the case of the *ASMM*, when we use an address like $v\bar{x}$ with v a variable name, it is desirable for v not to be an x-pointer, i.e. that the address that v stands for doesn't end with the direction x.

The *time complexity* we use is simply the number of instructions executed. We do not concern ourselves with the *space complexity*; see [12, 21] for a discussion of the space complexity of the *SMM*.

3 An illustration of the power of associativity

We demonstrate the power of the *ASMM* model by showing the capability to manipulate arbitrarily large sets in constant time.

The model allows the following natural representation of sets. If W is a word over Δ, and $\alpha \in \Delta$ a direction, then $P(W\bar\alpha)$ is the set of all nodes having their α-pointer directed to the node $P(W)$. Assume that our alphabet is $\Delta = \{A, B, C, \alpha, \beta, \gamma\}$ and that the A, B, and C-pointers from the center go to three different nodes $P(A)$, $P(B)$ and $P(C)$, none of which is the center. We will now consider the sets $P(A\bar\alpha)$, $P(B\bar\beta)$ and $P(C, \bar\gamma)$ and see how the standard set operators can be applied to them by using appropriate *set to* instructions. We have chosen A, B and C to be directions so that the instructions with which we will implement the set operators cannot affect the addressing of the nodes $P(A)$, $P(B)$ and $P(C)$. As long as no such interference exists, we can generalize to the case where A, B and C are not elements of Δ but words over Δ.

The instruction *set* $A\bar\alpha\beta$ *to* B; has the effect of adding to $P(B\bar\beta)$ the set $P(A\bar\alpha)$, while the instruction *set* $A\bar\alpha\beta$ *to* ϵ; removes from $P(A\bar\alpha)$ the nodes which are also in $P(B\bar\beta)$.

The figure below now shows how the standard set operators, shown as assignment statements in the boxes, can be implemented in terms of *set to* instructions. The center ϵ is used to direct pointers away from A or C.

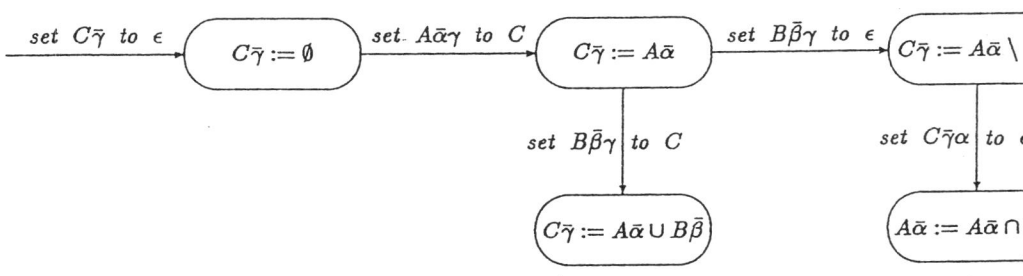

The following program illustrates how in linear time a set $P(\bar{\alpha})$ of exponential size can be constructed (with a singleton alphabet):

$new\ \bar{\alpha};$
$set\ \bar{\alpha}\bar{\alpha}\ to\ \epsilon;$
$\vdots$
$new\ \bar{\alpha};$
$set\ \bar{\alpha}\bar{\alpha}\ to\ \epsilon;$

Initially only the center exists, so all nodes point to the center. If at some point 2^k nodes exist, all of which point to the center, then after the *new* instruction, each of these 2^k nodes now points to one of 2^k newly created nodes, which again point to the center. Next the *set* instruction makes all 2^{k+1} nodes point to the center. Hence after k repetitions of these two instructions the size of the set $P(\bar{\alpha})$ has become 2^k.

In the next section we will see how these and similar constructions are used to process large amounts of data in parallel.

4 $PSPACE = ASMM-PTIME = ASMM-NPTIME$

The proof of membership in the Second Machine Class is usually split into two parts:

Lemma 1 $PSPACE \subseteq ASMM-PTIME$

We prove this by sketching an $ASMM$ which solves the $PSPACE$-complete problem QBF in polynomial time.

Lemma 2 $ASMM-NPTIME \subseteq PSPACE$

We prove this by showing how to simulate t steps of a nondeterministic $ASMM$ on a Turing machine using $O(t^2)$ space.

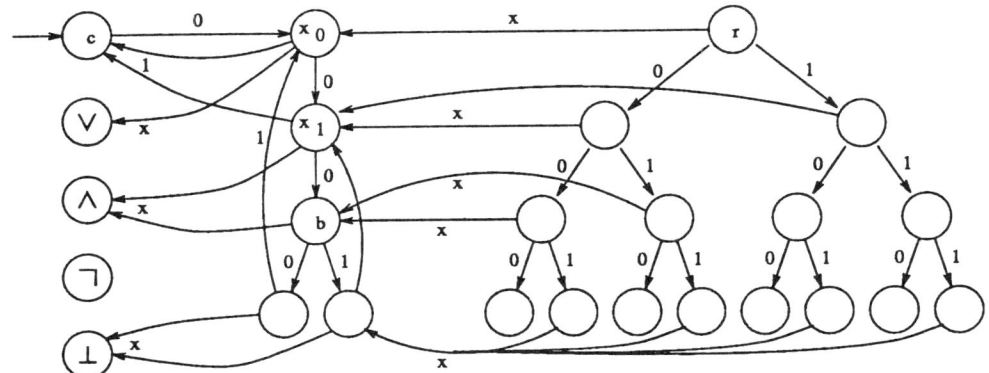

Figure 1: storage structure for $\exists x_0 \forall x_1 : x_0 \wedge x_1$

4.1 $QBF \in ASMM-TIME(n^2)$

The $ASMM$ algorithm we present for solving QBF in polynomial time proceeds in 8 stages. Let $X = \{x_0, \ldots, x_{k-1}\}$ be the set of variables in the formula of length n, let $\Delta = \{0, 1, x\}$ and let c be the center. Basically, the algorithm expands the formula by rewriting the quantifiers, one by one, innermost first, as follows:

$$\forall x_i F(x_i) \Longrightarrow F(0) \wedge F(1),$$
$$\exists x_i F(x_i) \Longrightarrow F(0) \vee F(1).$$

The resulting, fully expanded formula, can be viewed as a tree. It consists of a complete binary tree T of depth k, with an instance of the formula body B rooted at each leaf of T. In each such instance, the variable references can be replaced by the truth values assigned to them along the path from the root to the leaf. The algorithm does little more than to build and evaluate this tree.

Figure 1 depicts the structure built for the example formula $\exists x_0 \forall x_1 : x_0 \wedge x_1$. The part on the right represents the expanded formula (where some of the x-pointers have been omitted for clarity). We now briefly summarize each of the 8 stages:

1. Build a list of nodes $c, x_0, x_1, \ldots, x_{k-1}, b$ linked through the 0-pointer. Using the 0, 1-pointers, build a representation of the formula body as a

binary tree B rooted at b. The non-leaf nodes of B represent the connectives (and, or, not) while the leaves represent instances of variables. Create the nodes $\wedge, \vee, \neg, \bot$ that represent the different types of nodes in B.

2. Build a complete binary tree T of depth k using the 0,1-pointers. For a node at depth i, its 0-subtree represents the case $x_i = 0$ and its 1-subtree the case $x_i = 1$.

3. Build 2^k copies of B rooted at the leaves of T.

4. For every leaf u of B representing an instance of x_i, let the 2^k copies of u direct their x-pointer to either u or c depending on the connective of u's parent and the value assigned to x_i.

5. For every non-leaf u of B, let the 2^k copies of u direct their x-pointer to either u or c depending on the connectives of u and its parent.

6. For every x_i, let the 2^i nodes of T at level i direct their x-pointer to either x_i or c depending on the quantifiers of x_i and x_{i-1}.

7. Evaluate all copies of B in parallel.

8. Evaluate T.

Recalling section 2, we address the nodes $\wedge, \vee, \neg, \bot$ by their name, and use variable-names v, w for the purpose of traversing X and B.

The data structure constructed can be roughly divided in two parts: the linearly sized input representation (on the left in Fig1), and the exponentially sized formula expansion on the right. For a node $v \in X \cup B$, let $C(v)$ its set of copies. For $v = x_i$, these are the 2^i nodes at depth i in T, while nodes in B all have 2^k copies. The two parts, left and right, are connected only by the x-pointers that go from a node in some $C(v)$ to either v or to the centre c. Thus the following invariant holds throughout the execution:

$$\forall v \in X \cup B: \ v\bar{x} \subseteq C(v).$$

Furthermore, the x-pointer from r, the root of T, remains anchored to x_0 until evaluation is completed.

For a clear understanding of the construction, it is important to distinguish between truth-values and their representation. Conceptually, the algorithm works with truth values, 1 (true) and 0 (false). The leaves of the 2^k copies of B are assigned truth values in the obvious way according to which variable they represent. The other nodes are assigned default truth values, which are 0 for a disjunction, and 1 for a conjunction or a negation (recall that the quantifiers have been transformed into disjunction and conjunction). Next, a bottom-up process repeatedly changes defaults, that are in disagreement with their evaluated children, into the correct evaluation. In order to facilitate this process, we use a mixed representation of truth values, as follows.

Call a node $u' \in C(u)$ *active* if its x-pointer is directed to u or *passive* if its x-pointer is directed to c. A truth value is represented by the active state iff that truth value disagrees with the default of its parent, and with the passive state otherwise. To summarize:

parent type	$\wedge$	$\vee$	$\neg$
parent default	1	0	1
active value	0	1	1

As an example, suppose a $\neg$ node $u' \in C(u)$ has a parent $\vee$ node. The parent gets a default value of 0, which is to be changed into a 1 iff either of its children evaluates to 1. Thus, 1 is the active value for u'. The value 0 is passive for u', since it agrees with the default 0 value of its $\vee$ parent. Since a $\neg$ node is 1 by default, u' is active by default, hence its x-pointer is initially directed to u.

The representation of the truth value at a node u therefore depends on the type of the logical connective associated to the parent of u in the tree. This holds also for the nodes in the tree T which are associated to the variables x_i. In this tree the copies of the variables have been treated as logical connectives according to the type of the quantifyer binding this variable. Note that by keeping the x-pointer of r directed to x_0, it is made active by default.

In the algorithm above stages 1, 2 and 3 are used for building the tree; during stage 4 the truth values are assigned to all variable occurrences in the copies of B, and in stages 5 and 6 all intermediate nodes are given their default values. During the final two stages the entire tree is evaluated.

We next describe each of the above stages in some more detail.

In stage 1 the input is examined and used to construct a linearly sized list and tree representing the formula. We represent the type of a node $u \in X \cup B$ by directing its x-pointer to one of the special nodes $\vee, \wedge, \neg, \bot$. As noted before, these four symbols will also be used as paths addressing the nodes. The leaves of B are of type $\bot$ and have their 1-pointer directed to the appropriate x_i. Existentially quantified x_i have $type(x_i) = \vee$ and universally quantified x_i have $type(x_i) = \wedge$.

When traversing the list of x_i, the algorithm needs to be able to detect its end. Since the nodes in B already use the 1-pointer to point to their children (or single child in case of a $\neg$ node), we have the x_i direct their 1-pointer to the centre, and thus by comparing vx with ϵ can tell whether v addresses a node in X or in B.

In stage 2 the parallel power of the machine is used to build an exponentially large tree in linear time. This is achieved by the piece of code below:

```
            new v;                            create r, root of T
            set vx to 0;                      classify it
            set v to 0;                       start X traversal
λ :  new vx̄0;                                 0-children for C(xᵢ)
            set vx̄0x to v0;                   classify in C(xᵢ₊₁)
            new vx̄1;                          1-children for C(xᵢ)
            set vx̄1x to v0;                   classify in C(xᵢ₊₁)
            set v to v0;                      advance to xᵢ₊₁
            if v1 = ε then goto λ;            repeat for all xᵢ
```

The construction of 2^k copies of B in stage 3 proceeds analogously. Note that by now all the leaves of T have their x-pointer directed to b. Traversing B in preorder, we do the following at each node v:

```
            if vx = ⊥ goto λ₂;                do nothing at leaves
            if vx = ¬ goto λ₁;                ¬ node has no 0-child
            new vx̄0;                          create 0-child
            set vx̄0x to v0;                   classify in C(v0)
λ₁:  new vx̄1;                                 create 1-child
            set vx̄1x to v1;                   classify in C(v1)
λ₂:
```

In stage 4, all the x-pointers in the copies of leaves of B are installed. Again omiting the details of how to traverse B, let w be a leaf of B ($wx = \bot$), and $w1 = x_i$ the variable it represents. We show how to install the x-pointers in all copies $C(w)$ of w. We assume that w has the active value 1. The case for 0 is analogous. The code fragment

set v to 0;	start traversal of the x_j
λ : set $v0\bar{x}$ to ϵ;	clear $C(x_{j+1})$
if $v \neq w1$ then set $v\bar{x}0x$ to $v0$;	if $i \neq j$ then take both children
set $v\bar{x}1x$ to $v0$;	if $i = j$ then take only 1-children
set v to $v0$;	advance to x_{j+1}
if $v1 = \epsilon$ then goto λ;	repeat for all x_j

ends with $b\bar{x}$ equal to the set of leaves of T which have 1, the active value, assigned to x_i. In a similar fashion we can traverse the path from b to w, to end up with the active nodes of $C(w)$ pointing to w and the passive ones pointing to c.

The next two stages, 5 and 6, prepare the evaluation by giving default values to copies of non-leaves of B and nodes in T. This is most easily done by first setting $v\bar{x} = C(v)$ for all v in X or internal in B. Then, in a second pass, those copies that should be passive by default are taken care of by executing the following instruction for the appropriate v in X and B:

set $v\bar{x}$ to ϵ;	clear $v\bar{x}$

Now all that's left to be done is the evaluation itself. This is done bottom up—by a post-order traversal of B and then from x_{k-1} back to x_0. With the other cases being analogous, we restrict ourselves to the evaluation of an $\wedge$-node $w \in X \cup B$. Let v be its parent. The default value of w is 1, which is passive if v has type $\wedge$, or active if v has type $\vee, \neg$. The value of w should become 0 if either of its children has value 0, which is active for them. It should now be clear that the code fragment

if $vx = \wedge$ then goto λ_1;	passive or active default?
set $w0\bar{x}\bar{0}x$ to ϵ;	change to passive if 0-child disagrees
set $w1\bar{x}\bar{1}x$ to ϵ;	change to passive if 1-child disagrees
goto λ_2;	

λ_1 : *set* $w0\bar{x}\bar{0}x$ *to* w;   change to active if 0-child disagrees
   set $w1\bar{x}\bar{1}x$ *to* w;   change to active if 1-child disagrees
λ_2 :

evaluates node w. The technique used here is essentially the same as in section 3 for computing a union. Because of our symmetric representation, it works for both $\vee$ and $\wedge$.

When evaluation is complete, the root of T, r, will have its x-pointer directed to either x_0 or c, depending on the truth value of the input formula and on the default truth value for x_0 which in turn depends on its quantifier. In order to test whether the x-pointer from r is directed to x_0, we will use the fact that $x_0\bar{x} \subseteq \{r\}$ and $p(x_0, x) \in \{\wedge, \vee\}$. The instruction

set $0\bar{x}xx$ *to* ϵ;

changes $p(x_0, x)$ to ϵ iff $0\bar{x}$ is nonempty, which is equivalent to $p(r, x) = x_0$. Combining all this information yields the value of the formula.

Regarding the time complexity, the most time-consuming stage is number 4, where for each leaf of B, both X and B are traversed, requiring at most n^2 steps. Hence the complete algorithm runs in quadratic time.

4.2 $ASMM-NTIME(t) \subseteq SPACE(t^2)$

The simulation which proves this inclusion is relatively straightforward and employs previously known methods [14, 9]. We can write down in polynomial space a trace of the computation containing information on the sequence of instructions executed. Since the machine being simulated is nondeterministic this trace is guessed. Next it is verified by means of a system of recursive procedures and some other arrays containing polynomially sized information that this trace indeed represents an accepting computation. The *if*, *new* and *set to* statements pose the main problems, since their impact on the Δ-structure requires repeated recomputations of the current state of the Δ-structure. In polynomial space we cannot explicitly store the possibly exponentially large Δ-structure of the ASMM-machine, so an implicit representation is called for. This will consist of three arrays, and three mutually recursive functions. The arrays are

1. $instr[i]$ holds the instruction executed at step i

2. $nodes[i]$ holds the number of nodes at time i

3. $center[i]$ holds the center at time i

The simulation starts at time 0 and has step i ($i \geq 1$) leading to time i. Each array is of length t, the number of steps to be simulated, and each array element fits in t bits since the number of nodes can at most double after each step. Every node will have a unique number, and the resulting ordering of nodes is used for numbering nodes created by a *new* instruction. More precisely, a *new* W; instruction at step i is simulated as follows:

If $W = \epsilon$, then $center[i] = nodes[i-1]$ and $nodes[i] = nodes[i-1]+1$.

Otherwise, if $W = U\tilde{\alpha}$, then $center[i] = center[i-1]$ and $nodes[i] = nodes[i-1] + |Q(W)|$. Semantically, if $Q(W) = \{x_0 < x_1 < \ldots < x_{k-1}\}$, then at time i, $p(x_j, \alpha) = nodes[i-1] + j$, for $j < k = |Q(W)|$.

For all other instructions, $nodes[i] = nodes[i-1]$ and $center[i] = center[i-1]$, except that the instruction $set\ \epsilon\ to\ V$; sets $center[i]$ to $P(V)$. In order to compute $P(V)$ and to simulate the *if* instruction, we use the following functions:

$p(x, \alpha, i)$ returns the number of the node $p(x, \alpha)$ at time i

$P(x, W, i)$ returns whether $x \in P(W)$ at time i

$Q(x, W, i)$ returns whether $x \in Q(W)$ at time i.

These functions satisfy the equations

$$\begin{aligned}
Q(x, \epsilon, i) &= \mathit{false} \\
Q(x, U\alpha, i) &= P(x, U, i) \\
Q(x, U\bar{\alpha}, i) &= P(x, U\bar{\alpha}, i) \\
P(x, \epsilon, i) &= (x == center[i]) \\
P(x, U\alpha, i) &= (\exists\ 0 \leq y < nodes[i] : P(y, U, i) \wedge p(y, \alpha, i) == x) \\
P(x, U\bar{\alpha}, i) &= P(p(x, \alpha, i), U, i) \\
p(x, \alpha, 0) &= 0
\end{aligned}$$

which shows that they can be easily computed, apart from the case $p(x, \alpha, i)$ for positive values of i. The action of p in this case depends on the value of $instr[i]$, the only interesting values of which are new and set.

Consider first the case $instr[i] = new\ W$. If $x \geq nodes[i-1]$ then (using $Q(y, W, i)$) the difference $x - nodes[i-1]$ can be used to find the y in $Q(W)$ which 'generated' and now points to x (unless $W = \epsilon$, in which case $p(x, \alpha, i) = center[i-1]$). Now $p(x, \alpha, i) = p(y, \alpha, i-1)$. On the other hand, suppose $x < nodes[i-1]$. If $W = U\tilde{\alpha}$ (i.e. α-pointers may have changed) and $Q(x, W, i-1)$, then x has generated $p(x, \alpha, i) = nodes[i-1] + |\{y < x | q(y, W, i-1)\}|$. Otherwise $p(x, \alpha, i) = p(x, \alpha, i-1)$.

Second and last, consider the case $instr[i] = set\ W\ to\ V$. If $W = U\tilde{\alpha}$ and $Q(x, W, i-1)$, then $p(x, \alpha, i)$ is the unique y satisfying $P(y, V, i-1)$. Otherwise $p(x, \alpha, i) = p(x, \alpha, i-1)$.

These functions can easily be coded on a Turing Machine using recursion (stackframes). The recursion depth is bounded by ct, where c is a constant depending only on the maximum path length of the ASMM program. Each stackframe holds a return address and some node numbers and counters each of which fits in t bits. Together with the three arrays, space $O(t^2)$ suffices for the simulation of t steps of the ASMM.

5 Conclusion

Of all the parallel models which have been shown to belong to the Second Machine Class, the *ASMM* is the first to obtain its power from the use of associative addressing, thus making it an interesting addition to the realm of Second Machine Class devices. It provides another example that a small modification of a machine model can enforce a substantial increase in computational power. In [4] it was shown that this increase is provoked by adding multiplicative instructions to the unit-time standard *RAM* model. Similarly the *EDITRAM* model obtains its power from introducing a few edit operators that are available on most real life text editors anyhow. In the *ASMM* model it turns out that traversing pointers in the reverse direction is all we need to obtain full parallel power. At the same time, the fact that the storage structure of the *ASMM* is manipulated by a finite program that interacts with the Δ-structure by means of a single center seems to be the main reason why the machine has not become too powerful. As shown by Lam and Ruzzo [11], a model where the nodes become independently active finite automata becomes equivalent with a restricted version of the *P–RAM* of Fortune and Wyllie. This suffices for making the nondeterministic version more powerful than *PSPACE* (except for the unlikely case that *PSPACE = NEXPTIME*). This situation resembles the relation between the *SIMDAG* described by Goldschlager [8], where a single processor broadcasts its instructions to a collection of peripheral processors and the *P–RAM* model of Fortune and Wyllie [7] where the local processors are independent.

Clearly there are other models which could serve as a parallelized version of the *SMM*. In our model the set-to instruction is rather limited. Since its second argument addresses a single node, it cannot be used for setting different pointers to different destinations. This severely limits the scope of proofs that our machine is indeed so powerful. A more conventional approach, based on the construction of the transition graph of a polynomial space bounded Turing machine, and the computation of its transitive closure by pointer jumping—as suggested by the referee—is rendered infeasible by the limitation of the set-to instruction. Overcoming this limitation would require a different flavour of set-to instruction. A natural possibility is to allow the conventional set-to instruction of the *SMM* to be executed in parallel with respect to many different 'centers', the latter being specified by a third argument which is a string in $\tilde{\Delta}$. This model has some drawbacks, however. One is the possibility of conflicts arising when a pointer must be set to one node when addressed through one center, and to another node when addressed through another center. Resolving this problem would probably detract from the elegance of the model, one of its prime features. Another problem is that it becomes harder to manage all the pointers, since there is no simple way in which to direct a bunch of them to some fixed node where they can be 'out of the way'. Thus it is not a strict generalization of our model, although it should be possible to simulate our set-to instruction with this new one by keeping around an extra direction to always point to the real center.

Another modification suggested by the referee amounts to replacing the flow-of-control nondeterminism by a nondeterministic data manipulation instruction. For example both arguments in the *set W to V* instruction may become strings over $\tilde{\Delta}$; the effect of this instruction is that each node addressed by W redirects its outgoing pointer towards one of the nodes addressed by V.

This instruction makes it possible to guess some truth value for an arbitrarily large set of propositional variables in a single instruction, and suggests a proof that NP is included in $ASMM-NLOGTIME$ (assuming that the model also is upgraded to allow the input to be read in logarithmic time). Consequently this model would fail to be a member of the Second Machine Class; since the purpose of this paper is the design a version of the SMM belonging to the Second Machine Class we abstain from investigating this suggestion in more detail.

We like to use this opportunity to acknowledge for these suggestions of the referee and his other useful remarks which we have used in revising the manuscript.

References

[1] Aho, A.V., Hopcroft, J.E. and Ullman, J.D., *The Design and Analysis of Computer Algorithms*, Addison-Wesley Publ. Comp., Reading, Mass., 1974.

[2] Barzdin', Ya. M., *Universal pulsing elements*, Soviet Physics-Doklady 9 (1965) 523–525.

[3] Barzdin', Ya. M., *Universality problems in the theory of growing automata*, Soviet Physics-Doklady 9 (1965) 535–537.

[4] Bertoni, A., Mauri, G. and Sabadini, N., *Simulations among classes of random access machines and equivalence among numbers succinctly represented*, Ann. Discr. Math. 25 (1985) 65–90.

[5] Chandra, A.K., Kozen, D.C. and Stockmeyer, L.J., *Alternation*, J. Assoc. Comput. Mach. 28 (1981) 114–133.

[6] Dymond, P.W. and Cook, S.A., *Hardware complexity and parallel computation*, Proc. 21st Ann. IEEE Symp. Foundations of Computer Science, 1980, pp. 360–372.

[7] Fortune, S. and Wyllie, J., *Parallelism in random access machines*, Proc. 10th Ann. ACM Symp. Theory of Computing, 1978, pp. 114–118.

[8] Goldschlager, L.M., *A universal interconnection pattern for parallel computers*, J. Assoc. Comput. Mach. 29 (1982) 1073–1086.

[9] Hartmanis, J. and Simon, J., *On the structure of feasible computations*, in Rubinoff, M. and Yovits, M.C. (Eds.), Advances in Computers, Vol. 14 , Acad. Press, New York, 1976, pp. 1–43.

[10] Kolmogorov, A.N. and Uspenskii, V.A., *On the definition of an algorithm*, Uspehi Mat. Nauk 13 (1958) 3–28 ; AMS Transl. 2nd ser. 29 (1963) 217–245.

[11] Lam, T.W. and Ruzzo, W.L., *The power of parallel pointer manipulation*, Proc. 1st Ann. ACM Symp. Parallel Algorithms and Architectures, 1989, pp. 92–102

[12] Luginbuhl, D.R. and Loui, M.C., *Hierarchies and space measures for pointer machines*, Inf. and Comput., 1993, to appear; also: Report UILU-ENG-88-2245, Department of Electr. Engin., University of Illinois at Urbana-Champaign, 1988.

[13] Parberry, I., *Parallel speedup of sequential machines: a defense of the parallel computation thesis*, SIGACT News 18, nr. 1, 1986, pp. 54–67.

[14] Pratt, V.R. and Stockmeyer, L.J., *A characterization of the power of vector machines*, J. Comput. Syst. Sci. 12 (1976) 198–221.

[15] Savitch, W.J., *Recursive Turing machines*, Inter. J. Comput. Math. 6 (1977) 3–31.

[16] Schönhage, A., *Storage modification machines*, SIAM J. Comput. 9 (1980) 490–508.

[17] Slot, C. and van Emde Boas, P., *The problem of space invariance for sequential machines*, Inf. and Comp. 77 (1988) 93–122.

[18] Stegwee, R.A., Torenvliet, L. and van Emde Boas, P., *The power of your editor*, Report RJ 4711 (50179), IBM Research Lab., San Jose, Ca., 1985.

[19] Stockmeyer, L., *The polynomial time hierarchy*, Theor. Comp. Sci. 3 (1977) 1–22.

[20] van Emde Boas, P., *The second machine class 2: an encyclopaedic view on the Parallel Computation Thesis*, in: Rasiowa, H. (Ed.), Mathematical Problems in Computation Theory, Banach Center Publications, Vol. 21, Warsaw, 1987, pp. 235–256.

[21] van Emde Boas, P., *Space measures for storage modification machines*, Inf. Proc. Lett. 30 (1989) 103–110.

[22] van Emde Boas, P., *Machine models and simulations*, in: van Leeuwen, J. (Ed.), Handbook of Theoretical Computer Science, North-Holland Publ. Comp. 1990, pp. 1–66.

[23] Wagner, K. and Wechsung, G., *Computational Complexity*, Mathematische Monographien Vol. 19, VEB Deutscher Verlag der Wissenschaften, Berlin (DDR), 1986, also: Reidel Publ. Comp., Dordrecht, 1986.

[24] Wiedermann, J., *Parallel Turing machines*, Techn. Rep. RUU-CS-84-11, Dept. of Computer Science, University of Utrecht, Utrecht, 1984.

[25] Wiedermann, J., *Weak parallel machines; a new class of physically feasible parallel machine models*, I.M. Havel & V. Koubek (Eds.), proc. Mathematical Foundations of Computer Science 1992, Springer Lecture notes in Computer Science 629 (1992) pp. 95–111.

DATE DUE

150387

APR 2 2 1994

DEMCO, INC. 38-2931